# HISTORY OF TECHNOLOGY

# HISTORY OF TECHNOLOGY

## VOLUME 34, 2019

*Edited by Ian Inkster*

Special Issue:
History of Technology in Latin America

*Edited by David Pretel, Ian Inkster and Helge Wendt*

BLOOMSBURY ACADEMIC
LONDON • NEW YORK • OXFORD • NEW DELHI • SYDNEY

BLOOMSBURY ACADEMIC
Bloomsbury Publishing Plc
50 Bedford Square, London, WC1B 3DP, UK
1385 Broadway, New York, NY 10018, USA

BLOOMSBURY, BLOOMSBURY ACADEMIC and the Diana logo are trademarks of Bloomsbury Publishing Plc

First published 2020



A catalogue record for this book is available from the British Library.

A catalog record for this book is available from the Library of Congress.

ISBN: HB: 978-1-3500-8559-6
ePDF: 978-1-3500-8560-2
eBook: 978-1-3500-8561-9

ISSN: Print: 0307-5451
Online: 2514-7838

Series: History of Technology, Volume 34

Typeset by RefineCatch Limited, Bungay, Suffolk
Printed and bound in Great Britain

# CONTENTS

# Technology in Latin American History: Perspectives, Scales and Comparisons

DAVID PRETEL, IAN INKSTER AND HELGE WENDT

Anyone teaching or researching the Latin American history of technology has probably faced the following criticism: Latin America is not home to an innovative technological culture; it is not a cohesive technological region – so why study this? These commonly held views may explain why (until recently) the history of technology in Latin America has been considered of secondary importance. There are at least two additional reasons that have constrained the development of the field: one, a narrow definition of technology associating it with formal science and disruptive high-tech innovation; two, the invisibility of the scattered Latin American historiography written in Spanish and Portuguese. Certainly, the latter connects with the tacit idea that narratives from the standpoint of the 'Global South' are less theoretically and methodologically sophisticated. That said, the interest and research in the role of technological change in the economic, political and social history of the region are not new nor small in scope. It is a growing field, particularly in regard to the development of English-language scholarship.[1] However, as Juan José Saldaña observes, technology and technological activity still receive relatively modest attention from historians working in Latin America compared to other subdisciplines.[2]

As demonstrated in recent historiographical surveys by Kreimer and Vessuri as well as Medina and Lemon, among others, there has been a scholarly shift in the historical study of Latin American science and technology.[3] Over the past three decades, there has been a new trend in the field, one that privileges the study of technologies in a socio-cultural context, factoring in local communities of expertise, hybrid knowledge and domestic technical capacities in infrastructure, agricultural production, nuclear energy and computers, just to name the foremost sectors. However, such historiographical essays – from the disciplinary position of the history of science and technology and science and technology studies – exclude seminal contributions by economic and business historians. Take, for instance, the work of Edward Beatty on technology transfer and patents in Porfirian Mexico, Alan Dye's analysis of the rise of continuous-process technologies for mass sugar production in Cuba between 1899 and 1929, and Aurora Gómez-Galvarriato's case study of the mechanization of the textile industry during the Mexican Revolution.[4] Similarly, older histories of invention and technical education – such as Ramón Sánchez Flores' monumental work on Mexico – are also ignored in recent historiographical surveys.[5]

Although important in its own right, the full debate over what lies behind the lack of attention to the Latin American history of technology is beyond the scope of this editorial introduction. Whatever the explanation may be, recent historiography is making clear that the history of technology in Latin America is not only an important field of enquiry but one that addresses broader historiographical issues of the region: from national innovation systems to commodity production in mining and agriculture, and from Cold War science to everyday technologies and infrastructures. The contributors to this special issue study these and many other central themes in historical perspective. By covering such varied topics, this volume offers a number of views of Latin America's technological past. It brings together authors approaching the history of technology from varied disciplines, including the history of science, economic history, historical sociology, and science and technology studies. The contributors to this collection do not explicitly share a common research agenda or historiographical perspective, nor do they offer a comprehensive synthesis of the field. If anything, the authors share the appreciation that a historical examination of technology in Latin America might offer a novel entry point into some of the region's most pressing debates. Instead of advancing a common view – no matter how critical or persuasive this might be – this volume has the modest objective of offering essays that explore some of the most intriguing aspects of Latin America's technological history from the nineteenth century through to the present day. It reflects the plurality of research questions, methodologies and perspectives that have characterized recent historiography on Latin American science, technology and industrialization. Taken together the essays presented here shed light on the complex history of technology in Latin America, which is characterized by a tension between autonomy and dependence as well as transformations at different scales, that is, at the intersection of local, national, regional and global histories. This special issue itself is not only proof of the diverse background of the authors but, more important, of the ambiguous and contradictory role of technology in Latin American history. We believe that the eclectic nature of this special issue recognizes the plurality of Latin America's history of technology, which is characterized by, in María Portuondo's words, a mingling of 'triumphs with failures, interdependence with dependence, and progress with decay'.[6]

## BETWEEN DEPENDENCE AND AUTONOMY

In 1979, the influential economic historians Ciro Cardoso and Héctor Pérez Brignoli wrote that technological dependence was the most characteristic feature of Latin

American industrialization since the region's political independence.[7] Similarly, for Andre Gunder Frank technology – together with foreign investment – was a form of imperial control in Latin America. In his influential book *Capitalism and Underdevelopment in Latin America* (1967), Gunder Frank wrote: 'American technology is becoming the new source of monopoly power and the new basis of economic colonialism and political neo-colonialism'.[8] These and other authors portrayed the industrial structure of the region as dependent on the transfer of costly foreign industrial technologies due to the lack of domestic capacities and competitive manufacturing. Much of the early works in the history of technology followed a similar line of thinking until at least the 1980s. Until that decade, historical accounts mostly started from the premise that coloniality, neo-colonial situations and foreign economic hegemony were the primary forces explaining Latin American technological development.[9] Of course, the structural dependency theory did not always remain wedded to the former premise – dependency theory had a life of its own, even if it can be seen as an extension of earlier studies of Latin America's colonial heritage. Such a politicized economic perspective likewise influenced Latin American social and philosophical studies of Latin American science during those decades.[10]

In particular, structuralist and 'dependence' theories, widely influential between the 1950s and the early 1980s, characterized Latin American economies as semi-peripheral and 'victims' of the historical deterioration of the terms of trade for primary products. Such approaches typically viewed the world economy as asymmetrical and in need of endogenous industrial strategies; they favoured technological autonomy. Such scholarship was concerned with what held back growth and development in economic terms, attempting to demonstrate how Latin America's technological underdevelopment resulted from unbalanced commercial relations. The problem with this approach was that it did not provide enough empirical historical data to fully support their structural claims in the long term.[11] Most social scientists, especially economists and sociologists, tended to present abstract interpretative models of regional dependence that gave only passing acknowledgement to historical contingencies, intra-regional disparities and the precise nature of the technologies employed in the region.

Since the late 1980s, however, explicit references to Latin America's technological dependence have faded away from the historical scholarship. Latin America's technological underdevelopment and world asymmetries in knowledge production have been assiduously avoided. Historians have turned their attention to local histories of technology and their diversity.[12] Through a constructivist approach, scholars have moved to the analysis of both socially embedded technologies and alternative epistemologies in historical perspective.[13] Case studies, micro-history and qualitative analysis are the preferred methodologies. This 'new' narrative is primarily concerned with the historicizing of technological encounters, indigenous knowledge systems, co-production of technologies, creole expertise and hybrid practices. From this perspective, technology is no longer considered exogenous to Latin American societies. Representative of this trend is the volume *Beyond Imported Magic: Essays on Science, Technology and Society in Latin America* edited by Eden Medina, Ivan de Costa Marques and Christina Holmes in 2013, which provided a fine corrective to studies viewing technology used in the region as foreign and transplanted from abroad.

To an extent, the entire focus shifted from technology as a possible development engine to technology as an embedded attribute of local and regional cultures. As valuable as it may be to bringing to light the cultural dimension of technology, such recent studies failed to establish the relationship between indigenous knowledge and Latin American development. At the same time, even when international circulation is acknowledged, the global economic dimension is for the most part ignored. Although cultural studies of the Latin American history of technology are becoming more dominant, there are still many economic historians paying attention to the trends of trade and transfer of mechanized technologies in the most visible economic sectors such as railways, mining, agriculture and product manufacturing.[14] Recent economic historiography contrasts with earlier structuralist studies in that it is mostly concerned with a long-term quantitative study of the patterns of technological change, trade and transfer rather than with presenting an abstract macro-model of world economic relations and pervasive underdevelopment.

Following the insights of Christopher Freeman and Bengt-Åke Lundvall, among others, empirical studies of national systems of innovation have also reached maturity – another scholarly development of note.[15] For this approach, national institutions, actors and policies are the most critical forces in understanding technological developments, particularly addressing questions such as national investment in research and development (R&D), industrial policy and the institutionalization of engineering capacities.[16] Such literature – which concentrates on the second half of the twentieth century – does not negate the idea of technological dependence itself but rather ceases to see it as the teleological result of world economic dynamics. The 'national innovation system' perspective grew as a critical reaction to dominant neoliberal policies adopted in the region starting in the 1990s. The tension between Latin America's technological autonomy and dependence is an implied element of these 'late' structuralist narratives but in a much less politicized fashion and using different terminology. Of course, the international perspective is not absent from this approach, but Latin American technologies and projects are not merely represented as deriving from the region's relationships with the United States and Europe.

Among the emerging themes in the field, studies examining the appropriation of foreign technologies in the Latin American context are especially insightful and offer a corrective to previous studies framed by the structuralist logic. Take, for instance, Bernhard Rieger's consideration of the nationalist representation of the Volkswagen Beetle in Mexico and the tension between factory culture and the broader national imaginary.[17] A related strand of research is the study of the contingent process of Americanization throughout the twentieth century. New technologies and products were cornerstones of the spread of the American dream and consumer culture in Latin America after the Second World War.[18] But there are earlier examples as well, like the Americanization of railway technology in countries such as Mexico and Chile.[19] The limits of structuralist accounts are also evident when we look at a number of cases indicating the technological autonomy of the region. The development of computer and nuclear industries in Brazil, Mexico and Argentina make fascinating case studies of this.[20] Admittedly, whether these cases prove the attainment of a relative technological autonomy in Latin America or are just exceptions remains an open question.

In the past few years, the field abounds with studies on technical experts and expertise, such as engineers, scientists, managers and agronomists. The making of creole steam technologies and chemical innovations for sugarcane production in nineteenth-century Cuba has received particularly close attention. Works by, among others, Rood, Curry-Machado, Pretel and Fernández-de Pinedo have illustrated the collaboration of a number of experts in the adaptation of foreign technologies to Cuban environmental conditions and the racial management of slave labor.[21] By doing so, these studies question the idea of technologies and organizational processes passively imported from abroad. Also examining expertise in Latin America, in this case from below, Soto-Laveaga's multi-scale analysis of barbasco root exploitation in rural Mexico between the 1950s and 1970s reveals how peasants' chemical knowledge influenced global science – the chemistry of synthetic steroid hormones – and how global scientific chemistry impacted rural labour and agriculture dynamics.[22] Such an emphasis on the role of creole expertise in the production of export commodities was already advanced in the pioneering work *States of Nature* (2002), where historian of science Stuart McCook illustrated the development of a set of hybrid botanical and agricultural practices in the Caribbean throughout the Long Nineteenth Century.[23]

From a different perspective, recent literature has stressed the participation of technical and scientific experts in the political conflicts of the region. Take, for example, Justin Castro's book on the part played by radio broadcasting technology and wireless engineers in the advent of the Mexican Revolution and subsequent political changes.[24] Also analysing the Mexican Revolution, Guillermo Guajardo shows how uses of railway infrastructures and equipment in wartime reflected a certain ambivalence, which stemmed from the conflicting visions of peasants and industrial workers.[25] In the case of Mexico, such contested views of technology and large engineering projects can be traced back to the construction of the desagüe (drainage system) of Mexico City during the colonial period, as historian Vera Candiani has illustrated in her recent pathbreaking book, *Dreaming of Dry Land: Environmental Transformation in Colonial Mexico City*.[26] The political dimension of technological expertise delineated in these works may well be extended to other Latin American spheres, such as the penetration of indigenous peoples' lands during colonial and post-colonial times.

Although the role of users of technology has caught some scholarly attention, it remains an underexplored topic that deserves closer attention. Ana María Otero-Cleves, for example, shows how Colombian peasants' and artisans' consumption of foreign machetes in the nineteenth century influenced product design abroad.[27] Certainly, recent scholarship has preferred the examination of the consumption of 'Promethean' technologies and large infrastructures connected to the twentieth-century culture of modernity. Joel Wolfe and Héctor Mendoza, for instance, have studied the place of the automobile not only in the material progress of Brazil and Mexico, respectively, but as a means of building national identity and the state, including territorial integration.[28] Similarly, large infrastructures and technological systems – particularly railways – are traditional topics in national-oriented histories of Latin America, although in-depth attention to technological issues is scarce.[29] Some promising new research on dams and telephone networks focusing on users, maintenance and political contestation is lending new insight to this theme.[30] What is still lacking, however, is a more in-depth examination of the social significance, cultural appropriation and use of everyday technologies and consumer goods such as construction materials, product technologies, agricultural tools and household gadgets.

## LATIN AMERICAN TECHNOLOGIES IN GLOBAL HISTORY

The history of Latin American technology is underrepresented in the growing – and somewhat successful among wider audiences – global history written in the past two decades. Instead, colonial and Atlantic perspectives or – for the post-independence period – national and local approaches to Latin American history are preferred to global ones.[31] Constrained by scholars' research bias and institutional limits, global histories in which Latin American technologies are at the centre of the story remain rare. The lack of dialogue between global history and Latin American history is not exclusive to the study of technology and science. As historian Mathew Brown notes, both subdisciplines – global history and Latin American history – have developed in relative isolation, at least in their study of the post-1800 period. After the political independence of Latin American nations, neither the region nor its newly founded nations maintain their centrality in global histories.[32]

It seems clear that a full and proper understanding of the history of technology in Latin America requires not only an understating of the local, national or regional scales but also an appreciation of global dynamics. The global approach can reveal the central place of Latin American histories of technologies in world history as opposed to former nation-centred narratives or histories revolving around European colonialism. A global perspective in the Latin American history of technology does not, however, entail returning to a narrative where foreign actors are the only players behind Latin America's technological transformations. It requires placing Latin American case studies in international contexts, for example by investigating how local technologies and scientific institutions shaped and were shaped by global developments. In such studies, neither globalization, nor globality, nor localization or localism possesses autonomous analytical capacity. It is not an abstract 'globalization' that has profoundly influenced socio-economic spheres, but separately definable processes, which take place in large-scale interlocal environments.[33] This global perspective is not entirely new; already in the 1990s, several authors were making clear that limited rates of industrial innovation in Latin America do not imply a lack of active participation in the adaptation, reproduction and transformation of imported technologies to specific settings.[34]

Historically, global technologies and infrastructures have been adapted and transformed upon their arrival or installation in Latin America. Local technologies, indigenous knowledge, domestic political processes and distinctive geographical and environmental conditions shaped – and still shape – these diverse Latin American technological cultures, systems and institutions. Both the adaptation of foreign technologies and the invention of new technological devices are embedded in social, economic and epistemic dynamics in a given society, while at the same time constrained by the relations between societies.[35] Globalization is an historical process affecting a large number of other spheres as diverse as, in Lyn Carter's words, 'information, capital, labour, markets, communications, technological innovations and ideas'.[36] Therefore, for the study of Latin American history, strict distinctions between technology and science (and other forms of knowledge) are problematic.[37] However, at the same time, Latin America's scientific production and its part in the global network of (big) science are central issues that should be considered in a global history of science.[38] In this sense, the study of the emergence of national engineering and scientific communities throughout the twentieth century – and their relationship with both national politics and foreign expert communities – deserves further research.[39]

The case of silver production in Spanish America is especially telling. Silver mining and trade were crucial developments in the making of a global economy between the sixteenth and nineteenth centuries, as well as in laying the foundations of European–Asian economic relations during this period.[40] As Saul Guerrero demonstrates in his recent book, *Silver by Fire, Silver by Mercury,* the process of refining silver ore using mercury was brought to perfection in the New World at a scale unknown in Europe, with new milling equipment, chemical recipes and recycling equipment created locally.[41] Guerrero makes clear that the production of distinctively local knowledge and techniques was instrumental in the making of a global economy of silver production.

There is a temptation to see the 'global' as a dominant historical force, one that imposes its main characteristics on different localities around the world. However, as Jürgen Renn, among others, has shown, global scientific and technological knowledge are bound to the local conditions of their reproduction.[42] For example, in the case of the British invention of the reverberatory furnace during the seventeenth century, it seems clear that this technology for smelting metallic ores like copper or iron developed in multi-local contexts over several decades. Much later on, during the nineteenth century, massive investments and breakthroughs in melting technologies took place once again in a number of metal-ore smelting sites (in Australia and Chile, for example).[43] The relationship between the development of such furnaces and the expansion of a global copper market, as studied by Chris Evans and Olivia Saunders, show that often no clear distinctions between economic and epistemic processes of global–local interactions can be drawn.

In a similar example, the blast furnace experienced global proliferation. This British invention became a standard technology in some parts of Europe, as in France and Prussia in the eighteenth century. Later it was used to produce iron and steel in Mexico and Brazil.[44] Looking at these examples, it is apparent that refining technologies became – despite many difficulties – globalized practical knowledge used at distinct mining sites throughout the world. In their local environments, these furnaces adapted to the precise geological requirements of production through the mediation of local expert cultures and institutions that governed knowledge. In these and other examples, the geographical spread of technologies transformed ancillary knowledge-governing institutions in Spanish America. The circulation of knowledge and expertise in the Atlantic often weakened the institutional ties within the Spanish empire and facilitated the emergence of other political entities organizing the production of technologies, such as creole corporations.[45]

The notion of economies of knowledge (or the epistemic matrix) embraces the complex interplay of various fields of anthropogenic enquiry and non-human processes at several scales. The history of biophysics in Cuba, for instance, displays such a multi-layered evolution. Scientific progress in Cuba since the 1960s was not merely transplanted from more economically advanced countries, nor was it an imitation. Scientists from Cuba, Eastern European countries, the Soviet Union and several Western European countries worked together in different techno-scientific projects to strengthen Cuban physics.[46] The collaboration of Cuban academia with foreign research institutes from the 1960s onwards lay behind the institutionalization of theoretical physics in Cuba, including the development of high-quality laboratory technology. Foreign participation in Cuban geo-physics was adapted to Cuban geographical and geological conditions. The transnational dimension of Cuban science can also be identified in other sectors, such as biomedical research, with the production of lasers and nuclear laboratory equipment for medical analysis.[47] However, Cuban science is not the only example of such transnational scientific collaboration in Latin America. Take the rise of the Uruguayan meatpacking industry, for instance. The story of Liebig's Extract of Meat Company in the River Plate of Fray Bentos (Uruguay) in the 1860s, as shown by Lucía Lewowicz, reveals the interplay of national, local and international forces in the making of this manufacturing firm. Practical science-based innovations were developed at this factory through transfers of knowledge, commercial partnership and the adaptation of manufacturing processes.[48]

## LATIN AMERICAN HISTORIES OF TECHNOLOGY IN THE ASIAN AGE

The twenty-first century is often seen as the 'Asian century'. The renewed efflorescence in Asian studies in European and North American universities is clear, among other fields, in the history of technology. In the context of an Asian turn, one may wonder whether the expanding history of technology in Latin America can learn from the themes, approaches and scope of recent histories of technology in Asia. Of course, the technologies of 'the rest' have not all been the same, and their histories cannot be equated, but looking at the elements what appears now to have undergirded East Asian economic success might help sharpen focus and nuance future debate about the history of technology in Latin America.[49]

If we are to presently limit ourselves to the original historical queries relating technological dynamics to economic modernization and increased social welfare, the first generalization of real import takes the form of a grand irony for all analysts who have rejected the notion of any benign colonial expansion of technologies from the nineteenth century onwards – whatever the long-term impacts may be, both technological transfers from more technologically advanced systems and local statist interventions appear to have been inseparable from economic modernization in

Japan, the Newly Industrializing Countries (or NICs, including Taiwan, South Korea, Hong Kong and Singapore), and most recently mainland China. This is far from postulating that the observed patterns of change were somehow optimal or culturally superior, or that they were the only changes possible. But they were the ones that effectively prevented the wholesale breakdown of the global economic system during the last decade.

A second irony arises out of the first: there is no unproblematic evidence that relatively speedy economic modernization through technology transfers, adaptations and transformations has irrevocably harmed the outstanding cultural institutions and norms of these Asian societies. Whilst all these nations have gone through authoritarian political regimes at some point, most have liberalized on the heels of technological modernization, whilst China has retained communism as an official ideology though liberalizing a series of economic and social institutions during and following economic modernization from 1978. The UN Development Programme global index for 2015 ranked Singapore fifth and Hong Kong twelfth, both above the UK, and Japan and South Korea seventeenth and eighteenth, above France, Belgium and Italy.[50] Calculated for the period of 2015–2020, Hong Kong, Japan and Singapore were ranked second to fourth in global female life expectancy, South Korea coming in at number seven, and in the same period each of Asia's NICs (plus Japan) were ranked among the top 12 nations in lowest infant mortality, with Singapore holding first place jointly with Luxembourg. Nor has initial technological dependency resulted in present-day technological incapacity – based on 79 global indicators, Singapore ranks above Germany in the international innovation index, and South Korea, Hong Kong and Japan rank above France. In the early twenty-first century, these cannot be called failing, lagging or moribund socio-economic systems when compared broadly with the technological innovators of Europe and the United States, with their assumed early historical advantages.[51]

We can offer a third irony: none of these nations had somehow escaped nineteenth- and early-twentieth-century Western aggression and colonialism. There is no doubt that prior to their technological development, all Asian nations in this group suffered from colonialism or division within Western spheres of influence, or severe military and commercial threats from gunboat diplomacy. Furthermore, through programmes of Westernized, heavy industrial and military expansion (especially in the 1890s and 1930s), one of this group, Japan, colonized Taiwan (as Formosa), Korea (as Chosen), and much of China (as Manchukio), a late colonial enterprise impossible in the absence of a military-industrial complex speedily developed by Meiji and Showa governments determined to take their place amongst the comity of industrialized nations.

May we resolve these ironies to some extent by throwing some light on the contrasting cases of Latin America? Outstanding features of the East Asian pattern revolved around technology transfer and the role of the state, as well as institutional innovation. Situating Latin American history within the wider comparative context may well illuminate some broader issues of the history of technology in this region. Looking again at the UN Development Programme global index, Latin American countries appear far from Hong Kong, South Korea, Japan and Singapore. Chile makes the first appearance, ranking forty-first, followed by Argentina (forty-seventh),

Uruguay (fifty-fifth), Cuba (seventy-third), and Mexico (seventy-fourth). This relatively slow development correlates with the relatively low rates of regional expenditure in R&D and limited innovation capacity in domestic industries. The question of whether long-term technological dependence and national innovation policies explain this overall situation in Latin America remains an open question, one that still must be addressed by historians in conjunction with more culturally oriented studies of technology in Latin America.

We might go so far as to suggest that it was institutional innovation that emerged as the principal factor behind such late-developer success among Asian countries, creating novel historical conjunctures out of the technological innovation, state interventions and technology transfers associated with such nations. Similarly to the history of Asia, the linkages between institutional innovation and Latin American paths of technological development could be further explored. A long time ago Joseph Schumpeter had argued that technological change (changes of production techniques as well as new product innovations) should be seen as bound up with a broader technological modernization composed of conquests of new markets, novel organizational methods, for instance, in distribution, and new legal devices, as well as new uses of new and old resources or wastes, such as gold tailings. If we acknowledge a version of the Schumpetarian model for technological change in history, then much of Chinese history – as below – not only becomes more 'rational' in a Western perspective, but may well be thought of as at least equally productive.[52] These are elements that effectively embrace such varied institutional innovations since 1851 as intellectual property regimes, international exhibitions and associated forms of active agency, educative and training institutions increasingly integrated with scientific research and investigation, and the myriad of research and development organizations that have emerged from private sector enterprise, and many of which are discussed in the present special issue of *History of Technology* devoted to Latin America.[53]

Of course, institutions work in a great manner of ways both generally in economic history and specifically in the history of technological change. Indeed, as summarized so well by Douglass North, history 'is largely a story of institutional evolution in which the historical performance of economies can only be understood as a part of a sequential story'.[54] This is very well illustrated by China both in the time of its depletion and in the very recent years of its economic emergence. At a time when the Western economies outreached and then impinged upon the technological history of China, governance in the empire tended to seek institutional innovations rather than economic innovations as solutions to problems of economic development and social welfare. An early thesis argued that the institutional innovations in the rural economy that allowed China to mobilize its labour resources for economic surplus from the latter half of the nineteenth century were induced by an economic stagnation and a search for more effective control over local and regional resources.[55] More recent work has suggested a great complexity linking governance, institutions and technological change in the way in which Chinese political history has unfolded.[56]

Innovation without disruption was perhaps a far more ingrained worldview of Chinese governance than was any search for technological change. China's very late emergence from around 1978 was similarly characterized by technological change as but one ingredient of a huge programme of modernization in which new institutions of property rights, education and training and information and knowledge transfers figured very largely. Many commentators have viewed the process as limited by the determination to retain communist ideology and governance as a stabilizing regime in the absence of liberal and enlightenment traditions.[57] Yet economic progress associated with very speedy technology advancement in China has been as fast as in any miracle growths elsewhere (for instance Japan or Germany), far faster than that of the early starters of the nineteenth century, and most unlike those cases. The Chinese case has cut through both relative isolation from and the antagonism of the prosperous traders of the developed world, and has done so in the face of the decided crisis of global development since 2008. Indeed, it is possible to see the variations of Chinese political rhetoric and regime since earlier in the twentieth century as political institutional innovations enabling a flow of technology transfers as well as a steady improvement of overall technological capability.[58]

Similarly, in the history of Latin America, institutions have also played a central role in its uncertain technological history, where the study of institutions and institutional change are presented as essential drivers of technological dynamics. This perspective allows researchers to discuss the central question of whether an alternative institutional arrangement would have generated economic growth in Latin America at a faster rate than was the case.[59] The history of patents is among the topics where the relationships between institutional innovation and technological progress is most readily apparent. Historical records for the nineteenth and twentieth centuries reveal that foreign patenting represented a large stake in Latin American innovation systems. Latin American patent systems have been relatively small – and with important cross-national variations – but at the same time prone to foreign interventions, dependencies and heated public controversies.[60] It seems clear that the political elites of the newly independent republics saw property rights as an institutional pillar of the modern state. That is, patent rights were seen as a necessary institutional reform to foster national progress. Despite recent doubts about the ultimate benefits of patent protection for the development of Latin America, several nations of the region were early signers of the 1893 Paris Convention on Industrial Property. This tendency is also apparent when we look at the negotiations concerning intellectual property at the Pan-American Conference starting in the late nineteenth century. The political discussions about intellectual property rights in Latin America undoubtedly intensified in the post-TRIPS era (1994 to the present), particularly for questions relating to bioprospecting, pharmaceutical innovation and the protection of indigenous knowledge and material culture.[61]

Late development approaches to technological modernization and its relations with formal knowledge tend to focus on the role of the modernizing state in formulation and funding of institutions of education, training and applied research. For a further maturity of Latin American history of technology, the extensive historiography on Japan's late development can be specially illuminating. The outstanding industrial success of Meiji Japan, from 1868, was associated with a complex mixture of private and public sector activity, a massive influx of Western technologies and knowledge agents, a great attention to new legal and regulatory regimes and a surprisingly low level of public expenditure.[62] Japan's success in transferring, emulating, settling and adapting Western-originated industrial technologies emerged from co-requisite developments in non-conventional, usually non-government institutions based on fast emerging urban knowledge and technology institutions. The success of late development in the Japanese case seems to have depended upon informal processes of emulation, knowledge competition and sharing, and informal association, much of which was beyond direct state funding and tutelage.[63]

There can be little doubt that in Meiji Japan the growth of urban-based associations and communities of knowledge was to an extent conditioned by government action, for they depended on new civilities within new social locations and on a general social understanding that such elements of foreign interaction and proximity were approved of and entered into by powerful members and networks of the governing elite. So, the moving frontier that over time absorbed and translated Western knowledge and social understandings was neither a mere official artefact nor a solely spontaneous outgrowth of urbanism and cultural challenge from outside. Whereas the latter was the reflection of a public sphere replete with social and economic innovations, the former appears to have acted as a more consciously constructed chain of sites for the transfer and advancement of useful but transcending knowledge.

By contrast with the case of Japan, the history of industrial modernization and late development of Latin America throughout the twentieth century is replete with examples of failed government intervention in both the transfer of technology and the building of domestic technological capability. But as such Latin America's technological failures have additional explanations, from excessive power of the private sector to the lack of enforcement of the law, from the limited absorption of technology through foreign trade, to the extractive nature of foreign investments.

## TECHNOLOGY IN LATIN AMERICAN HISTORIES

Focusing on the nineteenth and twentieth centuries, the contributors to this special issue reflect on a range of central themes from diverse theoretical and methodological perspectives. Many articles go beyond the national scale, either by concentrating on the regional dimension or by developing comparative and international approaches to local case studies. Admittedly, the thematic and geographical coverage of the volume is necessarily limited. While the authors examine case studies for South, Central, and North America, including the Caribbean, some countries are less represented than others. Mexico and Brazil, the largest economies of the continent, are considered extensively. Case studies for Argentina, Cuba, Venezuela and Chile are also examined throughout the volume.

Five topics are extensively treated in this special issue. First, the co-production of industrial and scientific technologies through transnational circulation, including the dynamics of resistance, adaptation and appropriation. Second, the role of technology and expertise in political changes, that is, the study of techno-politics, where engineers and scientists are active historical agents. Third, the interplay of environmental and technological imperatives in the rural histories of the region, such as conflicts at commodity frontiers and shifts in agrarian production. Fourth, the consideration of everyday technologies, mass transport systems and infrastructures, including telephones, ports, electricity and computers. Fifth, the contrasting histories of expert knowledge circulation and knowledge popularization, such as the way new technologies and practical science are communicated.

Commodity production for global markets has been historically associated with technological transformations. Daniel B. Rood considers the development of railways, ports and warehouses in mid-nineteenth-century Cuban cities, particularly in Cárdenas and the surrounding sugar-producing region of Matanzas. The article combines a local history of the expansion of infrastructures with a look at the broader picture of the national sugarcane industry and global capitalist transformations. Its major historiographical innovation is showing that port infrastructures were not uniform throughout Cuba but the result of changing needs and local particularities. Elite merchant-planters established a logistical system adapted to the local conditions of cane production, such as which kinds of products were available. In doing so, planters realigned local geographies of production as well as larger Atlantic circuits of trade.

The relevance of the regional context to technological developments is highlighted in Helge Wendt's article on coal production in late-nineteenth-century Mexico. Wendt reviews the developments of geological knowledge, extraction techniques and industrial production in the states of Coahuila, Nuevo León and Puebla. His contribution concentrates on two related dimensions of coal production: first, on the global interrelations in knowledge production, particularly the central role of foreign geologists' knowledge and US companies' investment in the export of Mexican coal; second, on the disruptions to or constraints on knowledge transfer, exchange and diffusion to Mexico. Finally, Wendt's comparison of the northern coal region with southern coal works outlines the distinct patterns of regional industrial and mining developments.

Christiane Berth argues in her contribution that the expansion of rural telephony in Latin America is not a case of passive diffusion from abroad but of 'entangled knowledge'. This entanglement involved local demand, national planning and global developments in the telephone market. During the early and mid-twentieth century, local experts, national bureaucrats and telephone users were active agents in the building of telephone networks in places as remote as Chiapas. In some rural areas of Latin America, telephone cooperatives became a successful model, one alternative to centralized national models. Telephone networks have social relevance for rural communities, which explains their rapid development despite the lack of expert knowledge, foreign investment and national programmes.

Another aspect of the development of new technological devices is reflected in Diana J. Montano's article on the popular reception of tramways in early-twentieth-century Mexico City. This new transport system increased the number of deadly accidents considerably, spurring the invention of new safety devices, or so-called lifeguards. Montano gives a detailed account of several of these innovations and the criticisms of the public, who felt they were left without any safety measures against this newly introduced, deadly technology. Within an increasingly densified urban space, these inventions represented challenging new technological objects that were widely discussed in Mexican newspapers. The tests of new security devices became public spectacles and were commented upon sarcastically by Mexican newspapers.

Patents of such lifeguards are evidence of rich innovation activity and a changing technological culture in this period.

Technological change has been a significant force behind the shifting dynamics at commodity frontiers in Latin America and the Caribbean. David Pretel examines the rise and fall of three commodities in the Yucatán Peninsula between the mid-nineteenth century and the Second World War: henequen, chicle and logwood. It was during the years of the Caste War (1846–1901), a long-lasting Maya revolt, that commodity frontiers became deeply linked with technological trajectories. The interplay of global technological changes, local knowledge, regional infrastructures and traditional technologies set the stage for the long-term patterns of commodity production of these raw materials, including the exploitation of people and the environment. In these cases of the extraction of tropical natural resources, entangled histories of technology at different scales had clear – but often unintended – socio-economic consequences for indigenous peoples. At the same time, biological factors constrained the commodification, manufacturing and trade of such global commodities originated in this Maya-populated region.

Eve Buckley's contribution explores the ascent of an expert techno-class in the drought-ridden northeast area of Brazil during the mid-twentieth century. Here, the author further develops the conclusions of her recent book, *Technocrats and the Politics of Drought & Development in Twentieth-Century Brazil* (2017), incorporating other scholars' insights regarding techno-politics and agrarian reform, such as Mikael Wolfe's book on the influence of agronomists and water management on land distribution in the region of La Laguna during the Mexican Revolution.[64] Their respective books, at the intersection of political, environmental and technological history, clearly show the entanglements between institutional policies, middle-class technical professionals and natural conditions in years of nationalist and developmental policies. Irrigation technologies and projects mobilized by these technocrats not only failed to achieve their revolutionary promise but had pervasive environmental consequences.

Computers have received a great deal of attention in recent historiography. In her contribution, Debora Gerstenberger questions the role of the military as the spearhead of technological innovation. She deals with the example of the slow and odd introduction of computers in the Argentine military. The advantages and disadvantages of using computers were broadly discussed within several journals and gazettes related to the Argentinean armed forces. While a few civil institutions already had adopted this new technology early in the 1960s, several contributions to those military journals questioned the benefit of using this technology. They feared that it might challenge army habits and the position of men as soldiers. Gerstenberger shows that discussions within the military sphere also reflected the revolutionary potential of computers, the relationship between men and machines, and the future of warfare.

The entangled nature of technology and society is clear throughout this entire special issue, including in Josep Simon's article on education technologies. Simon reviews the long history of how technology has assisted learning in Latin America, from the pencil to the laptop. His article covers almost two hundred years of new educational technologies and questions the nature of technology in several pedagogical endeavours in Argentina, Mexico and Brazil. Furthermore, the role and impact of American educational projects on Latin American classrooms are considered and connected with national debates on new teaching and learning methods. His considerations of educational films, radio stations, new textbooks, audio-visual methods, and the export of Latin American educational inventions give a rich overview of historical developments in educational methods. The ephemeral nature of some of these technologies leaves important questions open for further research.

During the Cold War, political competition also extended to large scientific and technological projects like the Atacama observatory in Chile. Starting in the 1960s, Chilean astronomy developed in the international context of the space race. In her contribution, Bárbara Silva combines global with Latin American developments in politics, technology and science. Silva offers an account of astronomical observations since the early twentieth century regarding the competitive, international search for where to place new observatories in the 1950s. In this context, Chile became a place of competition between Soviet, European and American astronomers. The installation of international observatories involved Chilean and international political and academic institutions and opened an opportunity to train Chilean astronomers. This article, as with many others throughout this volume, reveals that the transfer of scientific knowledge and technology can only be studied through a consideration of a wide variety of factors, from political context to environmental conditions, from economic imperatives to knowledge structures.

The biotechnological industry has become a significant economic sector at the international level, including Latin America. Helen Yaffe studies Cuba's biotechnology miracle, placing her article at the interface of the history of medical science, business history and the history of the Cuban Revolution. To understand the rise of Cuba's biotech sector, she starts with an examination of the long tradition of applied scientific research and technological innovation in the island. The body of the article examines the rise of the Cuban biotech sector from the 1980s, with attention to the national policies and its distinctive characteristics. Through a study of Cuban biotech breakthroughs, this article provides a new entry point to the history of state socialism in Latin America.

State-led innovation in the petrochemical industries is a classic topic in Latin American history. Saul Guerrero's article in this special issue considers the patterns of technological change in the Venezuelan state-owned oil corporation (PDVSA) during the late twentieth and early twenty-first centuries. National innovation efforts, R&D programmes and product innovation were a definite success until 2005, primarily built on domestic expertise, as clearly shown by patent registration patterns. This enterprise managed to find technological solutions to particular problems, which then could be exported to other oil processing countries. Since then, the Venezuelan oil industry has entered a crisis resulting from many forces – domestic and foreign, technological and political.

## NOTES AND REFERENCES

1. Michael Lemon and Eden Medina, 'Technology in an Expanded Field: Review of History of Technology Scholarship on Latin America in Selected English-Language

Journals', in Eden Medina, Ivan da Costa Marques and Christina Holmes (eds.), *Beyond Imported Magic. Essays on Science, Technology, and Society in Latin America* (Cambridge, MA: The MIT Press, 2014): 111–138.

2. Juan José Saldaña, 'La Historiografía de la tecnología en América Latina: Contribución al estudio de su historia intelectual', *Quipu, Revista Latinoamericana de Historia de las Ciencias y la Tecnología* 15, 1 (2013): 7–26.
3. Pablo Kreimer and Hebe Vessuri, 'Latin American Science, Technology, and Society: A Historical and Reflexive Approach', *Tapuya: Latin American Science, Technology and Society* 1, 1 (2017): 17–37; Michael Lemon and Eden Medina, 'Technology in an Expanded Field'.
4. Edward Beatty, *Technology and the Search for Progress in Modern Mexico* (Oakland: California University Press, 2015); Alan Dye, *Cuban Sugar in the Age of Mass Production: Technology and the Economics of the Sugar Central, 1899–1929* (Stanford: Stanford University Press, 1998). Other economic histories of Latin America in which technology is extensively considered are Marvin D. Berstein, *The Mexican Mining Industry, 1890–1950: A Study of the Interaction of Politics, Economics and Technology* (Albany: State University of New York Press, 1964) and Aurora Gómez-Galvarriato, *Industry and Revolution: Social and Economic Change in the Orizaba Valley, Mexico* (Cambridge, MA: Harvard University Press, 2013).
5. Ramón Sánchez Flores, *Historia de la tecnología y la invención en México* (México: Fondo Cultural BANAMEX, 1980).
6. María Portuondo, 'Constructing a Narrative: The History of Science and Technology in Latin America', *History Compass* 7, 2 (2009): 500–522.
7. Ciro F. Cardoso and Héctor Pérez Brignoli, *Historia Económica de América Latina*, Vol. 2 (Barcelona: Editorial Crítica, 1979): 194.
8. Andre Gunder Frank, *Capitalism and Underdevelopment in Latin America* (New York and London: Monthly Review Press, 1967): 304 and 208–209.
9. See, for example, the recent re-edition in English of Víctor Figueroa, *Industrial Colonialism in Latin America: The Third Stage* (Leiden and Boston: Brill, 2013).
10. Leonardo Silvio Vaccarezza, 'Ciencia, tecnología y sociedad: El estado de la cuestión en América Latina', *Ciência e Tecnologia Social* 1, 1 (2011): 42–64; Antonio Arellano Hernández and Pablo Kreimer, 'Estudio social de la ciencia y la tecnología desde América Latina: Introducción general', in Antonio Arellano Hernández and Pablo Kreimer (eds.), *Estudio social de la ciencia y la tecnología desde América Latina* (Bogotá: Siglo del Hombre Editores, 2011): 21–55. See also Medina et al., *Beyond Imported Magic* and Portuondo, 'Constructing a Narrative'.
11. Luís Bértola and José Antonio Ocampo, *The Economic Development of Latin America since Independence* (Oxford: Oxford University Press, 2012).
12. This development can likewise be observed in the history of science; Cf. Jorge Cañizares-Esguerra, 'On Ignored Global Scientific Revolutions', *Journal of Early Modern History* 21 (2017): 420–432; Helge Wendt (ed.), *The Globalization of Knowledge in the Iberian Colonial World* (Berlin: Edition Open Access, 2016), http://edition-open-access.de/proceedings/10/toc.html

13. Juan José Saldaña (ed.), *Science in Latin America. A History* (Austin: University of Texas Press, 2006).
14. See, for example, Xavier Tafunell, 'Capital Formation in Machinery in Latin America, 1890–1930', *The Journal of Economic History* 69, 4 (2009): 928–950.
15. Bengt-Åke Lundvall, *National Systems of Innovation: An Analytical Framework* (London: Pinter, 1992); Chris Freeman, 'The "National System of Innovation" in Historical Perspective', *Cambridge Journal of Economics* 19, 1 (1995): 5–24.
16. See, for example, Jorge M. Katz and Nestor A. Bercovich, 'National Systems of Innovation Supporting Technical Advance in Industry: The Case of Argentina', in Richard R. Nelson (ed.), *National Innovation Systems: A Comparative Analysis* (New York: Oxford University Press, 1993); Maria Ines Bastos and Charles Cooper (eds.), *The Politics of Technology in Latin America* (New York: Routledge, 2005) and Mario Cimoli (ed.), *Developing Innovation Systems: Mexico in a Global Context* (London and New York: Continuum, 2000).
17. Bernhard Rieger, *The People's Car: A Global History of the Volkswagen Beetle* (Cambridge, MA: Harvard University Press, 2013): Chapter 7.
18. Julio Moreno, *Yankee don't go home!, Mexican Nationalism, American Business Culture, and the Shaping of Modern Mexico, 1920–1950* (The University of North Carolina Press, 2004).
19. Guillermo Guajardo, 'La tecnología de los Estados Unidos y la "americanización" de los ferrocarriles estatales de México y Chile, 1880–1950', *TST* 9 (2005): 110–129.
20. Diego Hurtado de Mendoza, 'Autonomy, Even Regional Hegemony: Argentina and the "Hard Way" Toward the First Research Reactor (1945–1958)', *Science in Context* 18, 2 (2005): 285–308; Emanuel Alder, *The Power of Ideology: The Quest for Technological Autonomy in Argentina and Brazil* (Berkeley: University of California Press, 1987).
21. Daniel Rood, *The Re-invention of Atlantic Slavery: Technology, Labor, Race, and Capitalism in the Greater Caribbean* (New York: Oxford University Press, 2016); Jonathan Curry-Machado, *Cuban Sugar Industry: Transnational Networks and Engineering Migrants in Mid-Nineteenth Century Cuba* (New York: Palgrave Macmillan, 2011); David Pretel and Nadia Fernández de Pinedo, 'Foreign Technology and Transnational Expertise in Nineteenth-Century Cuba', in David Pretel and Adrian Leonard (eds.), *The Caribbean and the Atlantic World Economy: Circuits of Trade Money and Knowledge, 1650–1914* (Basingstoke and New York: Palgrave Macmillan, 2015).
22. Gabriela Soto-Laveaga, *Jungle Laboratories: Mexican Peasants, National Projects, and the Making of The Pill* (Durham: Duke University Press, 2009).
23. Stuart McCook, *States of Nature: Science, Agriculture, and Environment in the Spanish Caribbean, 1760–1940* (Austin: University of Texas Press, 2002).
24. J. Justin Castro, *Radio in Revolution: Wireless Technology and State Power in Mexico, 1897–1938* (Lincoln, London: University of Nebraska Press, 2016). Among the most representative of this trend is Eden Medina, *Cybernetic Revolutionaries: Technology and Politics in Allende's Chile* (Cambridge, MA: MIT University Press, 2014) which examined Chile's socialist revolution from the lens of the history of computers.

25. Guillermo Guajardo, 'Tecnología y campesinos en la Revolución mexicana', *Mexican Studies/Estudios Mexicanos* 15, 2 (1999): 291–322.
26. Vera Candiani, *Dreaming of Dry Land: Environmental Transformation in Colonial Mexico City* (Stanford: Stanford University Press, 2014).
27. Ana María Otero-Cleves, 'Foreign Machetes and Cheap Cotton Cloth: Popular Consumers and Imported Commodities in Nineteenth-century Colombia', *Hispanic American Historical Review* 97, 3 (2017): 423–456.
28. Joel Wolfe: *Autos and Progress: The Brazilian Search for Modernity* (Oxford: Oxford University Press, 2010); Héctor Mendoza Vargas, 'El automóvil y los mapas en la integración del territorio mexicano, 1929–1962', *Investigaciones Geográficas* 88 (2015): 91–108.
29. For the case of Mexico see J. Brian Freeman and Guillermo Guajardo, 'Travel and Transport in Mexico', *Oxford Research Encyclopedia of Latin American History* (Oxford: Oxford University Press, 2018).
30. See in this special issue Christiane Berth's article 'Bringing Communication to the Countryside: Rural Telephony in Latin America, 1900–1985'. See also: Jonathan Hill, 'Circuits of State: Water, Electricity, and Power in Chihuahua, 1905–1936', *Radical History Review* 127 (2017): 13–38.
31. Regina Horta Duarte, 'Between the National and the Universal: Natural History Networks in Latin America in the Nineteenth and Twentieth Centuries', *Isis* 104, 4 (2013): 777–787.
32. Cf. Kenneth Pomeranz, *The Great Divergence: China, Europe, and the Making of the Modern World Economy* (Princeton: Princeton University Press, 2000) and Jürgen Osterhammel, *The Transformation of the World: A Global History of the Nineteenth Century* (Princeton: Princeton University Press, 2014).
33. Josephine Anne Stein, 'Globalisation, Science, Technology and Policy', *Science and Public Policy* 29, 6 (2002): 402–408.
34. Cf. Ramón Sánchez Flores: *Historia de la tecnología y la invención en México* (Mexico City: Fomento Cultura Banamex, 1980); Thomas Schott, 'World Science: Globalization of Institutions and Participation', *Science, Technology, & Human Values* 18, 2 (1993): 196–208. See also Hebe Vessuri, 'La actual internacionalización de las ciencias sociales en América Latina: ¿vino viejo en barricas nuevas?', in Antonio Arellano Hernández and Pablo Kreimer (eds.), *Estudio social de la ciencia y la tecnología desde América Latina* (Bogotá: Siglo del Hombre Editores, 2011): 21–55.
35. Manfred D. Laubichler and Jürgen Renn, 'Extended Evolution: A Conceptual Framework for Integrating Regulatory Networks and Niche Construction', *Journal of Experimental Zoology Part B: Molecular and Developmental Evolution* 324, 7 (2015): 565–577.
36. Lyn Carter, 'Globalisation and Science Education: Global Information Culture, Post-colonialism and Sustainability', in Barry Fraser, Kenneth Tobin and Campbell J. McRobbie (eds.), *Second International Handbook of Science Education* (Dordrecht: Springer, 2012): 899–912.
37. Ernesto Altshuler, *Guerilla Science. Survival Strategies of a Cuban Physics* (Cham: Springer, 2017).

38. Gail Davies, Emma Frow and Sabina Leonelli, 'Introduction: Bigger, Faster, Better? Rhetorics and Practices of Large-Scale Research in Contemporary Bioscience', *BioSocieties* 8, 4 (2013): 386–396; Jane M. Russell, Yoscelina Hernández-García and Mina Kleiche-Dray, 'Collaboration Dynamics of Mexican Research in Chemistry and its Relationship with Communication Patterns', *Scientometrics* 109, 1 (2016): 283–316.
39. Good examples of this line of research are: Irina Podgorny, 'Fossil Dealers, the Practices in Comparative Anatomy and British Diplomacy in Latin America, 1820–1840', *British Journal for the History of Science* 46, 4 (2013): 647–674; Cecilia Zuleta, 'Engineers' Diplomacy: The South American Petroleum Institute, 1941–1950s', in David Pretel and Lino Camprubí, *Technology and Globalisation. Networks of Experts in World History* (Cham: Palgrave Macmillan, 2018): 341–370; Manuel E. Contreras, 'Ingeniería y Estado en Bolivia durante la primera mitad del siglo XX', in Marcos Cueto (ed.), *Saberes andinos: Ciencia y tecnología en Bolivia, Ecuador y Perú* (Lima: Instituto de Estudios Peruanos, 1995): 127–158; J. C. Lucena, 'De Criollos a Mexicanos: Engineers' Identity and the Construction of Mexico', *History and Technology* 23, 3 (2007); 275–288.
40. Mariano Bonialian and Bernd Hausberger, 'Consideraciones sobre el comercio y el papel de la plata hispanoamericana en la temprana globalización, siglos XVI–XIX', *Historia Mexicana* 269, 1 (2018): 197–244.
41. Saul Guerrero, *Silver by Fire, Silver by Mercury: A Chemical History of Silver Refining in New Spain and Mexico, 16th to 19th Centuries* (Leiden, Boston: Brill, 2017).
42. Jürgen Renn and Malcolm D. Hyman, 'The Globalization of Modern Science', in Jürgen Renn (ed.), *The Globalization of Knowledge in History* (Berlin: Edition Open Access, 2012): 561–604, http://www.edition-open-access.de/studies/1/28/index.html#9
43. Chris Evans and Olivia Saunders, 'A World of Copper: Globalizing the Industrial Revolution, 1830–70', *Journal of Global History* 10, 1 (2015): 3–26.
44. Georg Fischer, *Globalisierte Geologie. Eine Wissensgeschichte des Eisenerzes in Brasilien (1876–1914)* (Frankfurt am Main: Campus, 2017).
45. Bartolomé Yun-Casalilla, 'Social Networks and the Circulation of Technology and Knowledge in the Global Spanish Empire', in Manuel Perez Garcia and Lúcio De Sousa (eds.), *Global History and New Polycentric Approaches* (Singapore: Palgrave Macmillan, 2018): 275–291; Eleonora Rohland, 'Hurricanes on the Gulf Coast: Environmental Knowledge and Science in Louisiana, the Caribbean and the U.S., 1722 and Beyond', in Patrick Manning and Daniel Rood (eds.), *Global Scientific Practice in an Age of Revolutions, 1750–1850* (Pittsburgh: University of Pittsburgh Press, 2016): 38–53; David Pretel and Adrian Leonard (eds.), *The Caribbean and the Atlantic World Economy.*
46. Angelo Baracca, Jürgen Renn and Helge Wendt (eds.), *The History of Physics in Cuba* (Boston: Springer, 2014).
47. Carlos A. Cabal Mirabal, 'Magnetic Resonance Project 35-26-7: A Cuban Case of Engineering Physics and Biophysics', in A. Baracca, J. Renn and H. Wendt (eds.), *The History of Physics in Cuba*: 315–322.

48. Lucía Lewowicz, *LEMCO. Un coloso de la industria cárnica en Fray Bentos, Uruguay / The Meat Industry´s Colossus in Fray Bentos, Uruguay* (Montevideo: INAC, 2016).

49. Stephan Haggard, *Pathways from the Periphery: The Politics of Growth in Newly Industrializing Countries* (Ithaca: Cornell University Press, 1990) and Alice Amsden, *The Rise of 'the Rest': Challenges to the West from Late-Industrializing Economies* (Oxford: Oxford University Press, 2001).

50. This combines statistics on schooling and life expectancy with income levels for all nations.

51. Figures are from *The Economist Pocket World Figures* (London: Profile Books, 2018), which summarizes the latest socio-economic data from, among others, the UN, World Bank and OECD.

52. Joseph Schumpeter, *The Theory of Economic Development* (Oxford: Oxford University Press, 1969 [first edition, 1934 by Harvard University Press]): chapters 2 and 3 but especially pp. 65–94; for a synoptic application, Ian Inkster, 'Inertia and Technological Change: An Elementary Typology', in Pascal Bye and Daniel Hayton (eds.), *Industrial History and Technological Development in Europe* (Luxembourg: European Commission, 1999): 343–348.

53. See Ian Inkster, *Science and Technology in History. An Approach to Industrial Development* (London: Macmillan, 1991). Also, the Special Issue, 'The World Exhibitions and the Display of Science, Technology and Culture: Moving Boundaries', *Quaderns D'Història de L'Enginyeria* 13, 2012.

54. Douglass C. North, 'Institutions', *Journal of Economic Perspectives* 5, 1 (1991): 97–112.

55. More realistic views of the Maoist component of this history have of course much modified the assertions of both actual economic surplus and the forces of inducement, see Barrington Moore Jnr., *Social Origins of Dictatorship and Democracy: Land and Peasant in the Making of the Modern World* (Boston: Beacon Press, 1966): 162–227; Ramon H. Myers, *The Chinese Peasant Economy: Agricultural Development in Hopei and Shantung 1890–1949* (Cambridge, MA: Harvard University Press, 1970).

56. Tonio Andrade, *The Gunpowder Age: China, Military Innovation, and the Rise of the West in World History* (Princeton, NJ: Princeton University Press, 2016); see also André Gunder Frank, *(Re)Orient: Global Economy in the Asian Age* (Oakland: University of California Press, 1998); Norman Jacobs, *The Origins of Modern Capitalism and East Asia* (Hong Kong: Hong Kong University Press, 1958); Daniel R. Headrick, 'The Tools of Imperialism: Technology and the Expansion of European Colonial Empires in the Nineteenth Century', *Journal of Modern History* 51, 2 (1979): 231–263, and his *The Tools of Empire: Technology and European Imperialism in the Nineteenth Century* (Oxford: Oxford University Press, 1981).

57. William Hutton, *The Writing on the Wall. China and the West in the 21st Century* (London: Little Brown, 2007). For a good iteration of the related issues see chapter 12 of Martin Jacques, *When China Rules the World* (London: Penguin, 2nd edition 2012).

58. Ian Inkster, 'Economy, Technology and the Huttonian Enlightenment: Approaches to

China in the International Political Economy since the Early Twentieth Century', *International History Review* 37, 4 (2015): 809–840.

59. John H. Coatsworth, 'Inequality, Institutions and Economic Growth in Latin America', *Journal of Latin American Studies* 40, 3 (2008): 545–569; See also the following article Kenneth L. Sokoloff and Stanley L. Engerman, 'History Lessons: Institutions, Factors Endowments, and Paths of Development in the New World', *The Journal of Economic Perspectives* 14, 3 (2000): 217–232.
60. David Pretel: 'El sistema de patentes en las colonias españolas durante el siglo XIX', *América Latina en la Historia Económica* 26, 2 (2019); Edward Beatty et al., 'Technology in Latin America's Past and Present: New Evidence from the Patent Records', *Latin American Research Review* 52, 1 (2017): 138–149.
61. Kenneth C. Shadlen, *Coalitions and Compliance: The Political Economy of Pharmaceutical Patents in Latin America* (Oxford: Oxford University Press, 2017).
62. T. Ono, 'The Industrial Transition in Japan', *Transactions of the American Economic Association*, 5 (1890): 1–12; Koichi Emi, 'Economic Development and Educational Investment in the Meiji Era', in M. J. Bowman et al. (eds.), *Readings in the Economics of Education* (Paris: UNESCO, 1968): 167–199; J. I. Nakamura, 'Human Capital Accumulation in Pre-Modern Japan', *Journal of Economic History* 41 (1981); David G. Wittner, *Technology and the Culture of Progress in Meiji Japan* (London: Routledge, 2008).
63. For some attention to what might be called the software of technology transfers see Noboru Umetani, *The Role of Foreign Employees in the Meiji Era in Japan* (Tokyo: Tokyo Institute of Developing Economies, 1971); special issue, 'Adaptation and Transformation of Western Institutions in Meiji Japan', *The Developing Economies* 15, 4 (1977); Hazel J. Jones, *Live Machines: Hired Foreigners and Meiji Japan* (Vancouver: University of British Columbia Press, 1980); Tessa Morris-Suzuki, *The Technological Transformation of Modern Japan; From the Seventeenth to the Twenty-First Century* (Cambridge: Cambridge University Press, 1994); Morris Low, *Science and the Building of a New Japan* (London. Palgrave Macmillan, 2005); Ian Inkster, 'Cultural Engineering and the Industrialization of Japan, circa 1868–1912', in Merritt Roe Smith, Leonard N. Rosenband and Jeff Horn (eds.), *Reconceptualizing the Industrial Revolution* (Cambridge, MA: The MIT Press, 2010): 291–308.
64. Mikael D. Wolfe, *Watering the Mexican Revolution. An Environmental and Technological History of Agrarian Reform in Mexico* (Durham: Duke University Press, 2017); Eve Buckley, *Technocrats and the Politics of Drought and Development in Twentieth-Century Brazil* (Chapel Hill: University of North Carolina Press, 2017).

# Centrifugal Capitalism: Struggles over Infrastructure in the Sugar Ports of Nineteenth-century Cuba

DANIEL B. ROOD

This article examines the role of infrastructural transformation in the history of nineteenth-century capitalism in Latin America, specifically the slave–sugar economy of Cuba. In the island colony, artisan sugar technology, together with early railroads, environmental transformation and a large slave trade helped the island become by far the world's biggest sugar producer by the 1830s. This part of the story is well-told.[1] However, the more chronologically delimited convergence of technological, political, economic challenges between 1837 and 1843 forced a deeper transformation of the production and transport of sugar. The most important of the innovations was the 'Derosne system', which was basically a factory in the fields of sugar plantations. It used steam power and vacuum enclosure to cheapen and accelerate the production of sugar, as well as to maximize yields of white sugar, which shipped better and brought higher prices in foreign markets. A key component of this new industrial sugar-making process was the steam-powered centrifuge. Centrifuges shortened the draining phase from a few weeks to a matter of hours, and more precisely separated different grades of sugar.[2]

Extending the angular momentum initiated by the centrifuges, a new class of merchant-planters used transport infrastructure to throw different grades of sugar into discrete levels of the economy starting in the 1840s. As expanding railroads transformed the countryside, these merchant-planters funnelled high-quality purged sugar to new wharf-warehouse complexes in Havana, while lower grades of molasses and muscovado were shipped directly out of secondary port towns; Matanzas and Cárdenas being the most important.[3] Havana, with its transnational financial connections, its deep-water harbour, its new warehousing complexes, its established reputation as an import destination, and its proximity to the halls of political power, hosted big ships and white sugar – the largest proportion of which went to Britain.[4] Matanzas sent locally produced white sugar to Havana by rail and sail, directly exporting muscovado to North America through the traditional means of small merchants. Cárdenas, which will be the focus of this article, had a comparatively shallow harbour; its merchants directly exported molasses on smaller US schooners. This secondary port also had access to the most modern plantations and most fertile soils on the island. However, due to the imperatives imposed upon the new transportation system by elite merchant-planters, 'plantation white' sugar traced a path around the town's warehouse district, going by coaster or steamship to the centralized remote warehouses ringing Havana.

In spite of the fact that most of Cuba's white sugar was produced around Matanzas and Cárdenas, Havana maintained its primary position as an export centre thanks to the city's new sugar warehouses, which would rank as the largest cast-iron structures in the world by 1860, their components pre-fabricated at a foundry in New York and assembled on-site in Cuba. Running its own railroad, and offering banking and insurance services to planters, the Regla Warehouse Company and others of its ilk would anchor the infrastructural reinvention of the island's sugar-making economy.[5]

Although encouraged by new technologies like the centrifuge and the cast-iron warehouse, as well as the topography and location of harbours, the divvying up of different parts of the sugar market among the three major ports was not a simple matter of technological or topographical determinism. Rather, rival merchant blocs struggled over infrastructural change, with an emergent clique of Spanish-born merchant-planters mobilizing technology and topography to put in place a hierarchical division of sugar labour among cities. Through an examination of events in the understudied port of Cárdenas, this article shows how technological change intersected with and was impacted by the competition between different blocs of merchants; what Marx called 'the division of capitals'. Class conflict has never only hinged upon a unified 'capital' versus a monolithic 'labour'. Groups of capitalists, tied together by familial, ethnic, religious or national affinities, engaged in competition among themselves.[6] With their 'littoral power' rooted in the different port towns of western Cuba, merchant blocs wrestled over sugar-related profits by attempting to reshape the colony's infrastructure.

They were in some ways responding to new patterns of maritime shipping that emerged in the Atlantic World after 1815. Due largely to shifting tariff policies in Spain and the United States, Havana deepened its relationship with European consumers of plantation white sugar, via tight associations with British capital. A less wealthy cohort of Cuban and North American merchants in Matanzas and Cárdenas increasingly supplied muscovado and molasses to the United States, primarily through New York merchant bankers like Moses Taylor. The port infrastructures that were built in each city reflected this divergence. While the Regla warehouses provided the kinds of large-scale, centralized, efficient infrastructure required for the white sugar business, bringing the requisite quantities of white sugar to Havana would require the transformation of mercantile life in Matanzas and Cárdenas as well.

Smaller sugar traders resisted these changes. In Cárdenas, a furore broke out in the 1850s over the plans of elite merchant-planters with control over a local railroad to move the freight terminal fifteen miles out of town and institute a discriminatory tariff that would concentrate white sugar in the hands of the well-connected individuals operating out of Havana. Recent episodes of slave resistance, however, enabled advocates of the new station to argue that secure containment of a large, non-white workforce, at a location far from centres of white population, was beneficial to all. For political reasons, then, the station was allowed to be removed, reinforcing the bifurcation of sugars.

The idea of the 'repeating island', or the 'plantation complex' common in histories of sugar can create the impression that an identical system replicated itself from island to island across the Atlantic over the centuries after 1400.[7] While accurate in one sense, this view hides the fact that the hemispheric economy structured around sugar only worked as a system because it was made up of very different parts organized into spatial divisions of labour at a series of geographic scales. Although 'sugar' is often treated as an undifferentiated historical unity, the centrifugal splitting of the commodity promoted the engineering of divergent pathways, with infrastructural change both enabling and responding to the ways in which profits tended to adhere to whiter grades of sugar.

## COUNTERINSURGENCY, MERCHANT CONFLICT AND A WAREHOUSE GRAB IN CÁRDENAS

In the 1850s controversy over infrastructure, the local merchants of Cárdenas offered a particular reading of local history to support their case. When they first arrived in town in the late 1830s, 'everything was desert, without more activities than loading Havana-bound schooners with boxes of sugar received from surrounding farms'.[8] In the fertile plains south of town, slaves were laboriously carving sugar estates out of this 'desert', but their ambitious masters were frustrated by the frequently flooded, primitive roads that made wagon transport virtually impossible during the rainy season. Planters paid $3 to $4 to get each box of sugar from their plantations to the Canímar River, from which the boxes could be floated down to the bay of Matanzas.[9] Due to these difficulties, in 1839 Creole planters heading the Railroad Company of Cárdenas (RCC) built a railroad southward into a quickly emerging neighbourhood of sugar plantations, adding a terminal and a deposit warehouse in the humble frontier town of two hundred residents.

The merchants of the nearby port town of Matanzas had until then assumed they would be the beneficiaries of the sugar expansion in the district of Cárdenas (as opposed to the port city of the district which went by the same name). They suddenly became concerned about all the major new plantations having access to the sea through Cárdenas, and tried to stymie construction of the new railroad. 'The businessmen of Matanzas are rising against our project', noted the RCC's chief engineer in 1839. Their efforts succeeded in delaying free port status to Cárdenas until 1843.[10] Trying to emerge from the shadow of established Matanzas merchants, scrappy sugar traders needed an infrastructure to rival that of Matanzas. A railroad, they would realize too late, was only the first step.

Just as the residents of Havana and Matanzas were discovering at the same moment, the building of the Cárdenas railroad also required ancillary transformations of urban space in order to weave together land and sea communications. The rough-and-tumble town had so few thoroughfares, that 'the warehouses that a few merchants had built upon the water did not have any way of reaching the new railroad stop'. Thus, 'for its own convenience', as well as its shared interests with the merchants of the city, the railroad company established interior railways to reach the merchant warehouses. The railroad firm planned to use oxen to pull 'freight cars to private warehouses of the city by means of rails'. Then, individual merchants, temporarily in control of the freight cars, loaded and locked the cars, the company only retaking possession of them 'in the moment they are hitched to the locomotives about to leave the station'.[11] Some of the merchants ran the rails to the ends of their private docks in order to transfer coasted cargo to the railroads or vice versa.[12]

Being a much older and larger city, Havana had an ingrained and expensive pre-railroad system that was subjected to heavy pressure by the sugar boom (even as Regla projects transcended the existing set-up, the *solar* system of hogsheads and boxes being dispersed to merchant homes throughout the city persisted alongside it). Cárdenas was different. Even though the first merchant warehouses built there preceded the railroad, the firms along Pinillos Street were from the start seaside sugar warehouses with private wharves, removed from the inland home of the merchant family – counting rooms and offices would likely have been at the warehouses as well. While Havana was blessed with a nearly perfect bay but hamstrung by the considerable costs sunk in an obsolete infrastructure, Cárdenas could start a railroad-specific infrastructure from scratch. Nevertheless, its merchants left themselves vulnerable by failing to coordinate in any systematic way.[13]

For a time, the RCC decided to leave the construction of warehouses to individual merchants, agreeing to limit its own small deposit warehouse to items other than sugar. As merchants' individual repositories multiplied along Pinillos, and their private wharves proliferated upon the bay, the railroad extended its line, and continued to transport sugar products at a reasonable rate. Linking the shoreline to some of the largest and most technologically advanced plantations on the island (newly established on fertile soils south of town), the town of Cárdenas grew from 200 to 4,000 inhabitants in the four years after the railroad began service. The robust commercial growth of the port city was recognized by the government in 1844, when it declared Cárdenas officially open for international trade, a legal development which confirmed the fears of Matanzas traders.[14]

In the meantime, however, the magnetic force of the railroad was subtly reorienting the commercial architecture of the city. New mercantile establishments concentrated themselves in the vicinity of the station, which forced the merchants to 'reduce the dimensions of their storefronts'. The increased density along Cárdenas's main thoroughfare sometimes left merchants 'incapable of dealing with the number of railroad cars that passed by'.[15] In other words, the very railway local merchants had supported sometimes made it difficult to conduct business because there just was not enough room in front of their narrow entrances when freight cars were present. Their buildings, although relatively new, were nonetheless architectural relics of a pre-railroad era.

However, an 1853 fire that began in the warehouse of US merchants Safford and Co. burned down many of the warehouses and destroyed the railroad. This destruction of sunk costs provided an opportunity for some, because the mayor was able to ask the merchants to rebuild closer to shore, which enabled a widening of Pinillos. Seeing the benefits for themselves, the local merchants paid for a good portion of the wharf reform undertaken in 1852–1853.[16] A large embankment and a masonry seawall replaced the old seawall and pushed the edge of town out into the bay. On top of this new surface was built Marina Street, meaning that Pinillos no longer had direct access to the water. At 180 feet wide and running the length of the warehouses parallel to Pinillos, Marina was the obvious choice for a new railroad branch.[17]

In fact, as the RCC's enslaved workers lugged heat-twisted rails and charred crossties from the roadway, the Júcaro Railroad Company began serving the individual merchant warehouses of Cárdenas by running its line from east of town over Marina Street.[18] A company shareholder and Havana-based merchant-planter called Pedro Lacoste seems to have been preparing for this moment for many years. Scion of a French family who first fled the Revolution and settled in New Orleans, and then moved on to Havana, Lacoste spent the 1830s and 1840s buying up waterfront property whose value would increase with the growing importance of the port of Cárdenas. In December 1852 alone he bought up fourteen contiguous lots. By 1858, Lacoste and the JRC had concentrated a solid mass of waterfront property.[19]

With one line running in front of the commercial district and the other line running in back, the two railroads duplicated roles, thereby 'doubling the disturbance' of urban space.[20] Therefore, all parties agreed, at least at first, that a merger between the two railroad lines would lessen this destructive form of competition, allowing the opposing companies 'to order their affairs with a considerable saving in the number of workers and of time'. Such economies, the local merchants acknowledged, were 'inappreciably necessary if one considers the peremptory quality that the labours of the sugar harvest require'.[21] Thus armed with the blessing of the city's existing merchants, the new railroad combination partnered with elite Havana-based merchant-planters to undertake an overhaul that would, unbeknownst to locals, suddenly make valueless local merchants' considerable infrastructural investments.

Once in operation, the combined Cárdenas-Júcaro Railroad Company (CJRC) purchased hundreds of bonded labourers, and sought to commandeer warehousing operations.[22] At the start, they had committed to building the new central freight station within city limits, on a lot they already owned (they went as far as purchasing an adjacent property to make a bigger site within city boundaries). However, there were problems with this plan. As the Board of Directors wrote in an 1858 report to shareholders, the proposed urban location 'occupied a space that was too small to hold a passenger house, a machine shop, a *barracon de los negros y asiáticos*, housing for various employees, and other buildings'. Furthermore, they noted, the shape of the property 'formed a long and narrow parallelogram, inadequate for the movement of the 400 cars that travel daily down both railway lines'. While these factors were discouraging to those who wished to build the new station in town, something else weighed more heavily on their minds. 'The most decisive condition as regards this issue', the Board wrote, 'was our examination of the topographic plan of Cárdenas'.[23]

To make their concerns clear, the Board asked its shareholders to envision a future urban geography, a layout of streets 'that today only exists as lines on a piece of paper'. The central station would be built on grounds that had already been planned as future roadways, and the Board feared that 'local authority' could mandate the actual construction of those roads at any time. They feared neighbouring landowners would demand free passage through areas that 'were supposed to have been reserved for the central station', running roads through it, building structures on it and so forth. An urban location also had other disadvantages, including the increased likelihood of trains running over the multitude of pedestrians, horses and wagons. The dense, multi-use environment of the city centre, the Board emphasized, 'does not allow for the easy, rapid, and orderly transfer of cars and locomotives from one set of tracks to another; nor does it permit the hitching, arrival, and departure of the trains, nor the thousand other tasks of interior service'.[24] A newly selected location well outside city limits would avoid all of these issues, while also exerting salutary effects on 122 slaves, 112 emancipados and 228 Chinese contract workers who worked for the railroad. They would 'commit fewer disorderly acts' because they would be under increased surveillance at a rural location.[25]

The memory of a slave uprising outside of Cárdenas in 1843 likely weighed on the minds of those crafting the policies. During that year, the town turned into a refugee settlement so crowded with white families who had fled the countryside that there were not enough houses to hold them. During the interrogations and torture of people of colour that came after the suppression of the uprising, it came out that two local free black artisans (a tailor named Segui and a butcher named Morejón) were the leaders. They were both executed by firing squad, along with four others. The transportation system played a central role in the mass castigations that followed: local planters who had decided rebellion could be beaten out of their captives sent scores of slaves on trains to *depósitos de negros* on RCC property. They were then transported in smaller groups to a warehouse belonging to a Sr. Carrerá, where they were tied face down to a ladder and beaten. If too gravely wounded to return to the plantation, they were kept in the railroad's infirmary overnight. The following morning the dead were loaded into freight cars and taken either to the cemetery or to the sea, 'where they were tossed in unceremoniously'.[26] Discussions over the location of the new terminal, in other words, took place within the context of racial terror.

With such compelling reasons against a city-centre freight terminal, the CJRC Board of Directors was able to carry out its plan to alter the location over the formal protests of local traders. Less than a year after the fire of 1853, the company completed 'a vast deposit warehouse' in the village of El Recreo, at 'a very central and commodious location', (the Board of Directors insisted) fifteen miles outside of town. Standing at the intersection of two key railway lines that ran through sugar country, the new warehouses at El Recreo handled seven locomotives and as many as 225 freight cars per day. The complex included 'a beautiful building with rooms for offices, the administrator and clerks; and another for iron-working tasks, carpentry, and accommodations for operatives'.[27] But there was more. The new railway line ran from El Recreo directly to the wharf belonging to a member of the railroad's Board: Pedro Lacoste. His wharf was one of the largest on the coastline, distinguished from its neighbours by the size of the ships that could dock there.[28] He also had a large cooper's shop in nearby Júcaro, as well as a fleet of five or six coasting schooners plying the route between his wharf and Havana, all staffed with between thirty and forty slaves.[29] Having been one of the major contractors in the project to modernize Havana's wharves in 1849–1850, Lacoste had deep knowledge of Regla's usefulness, so he sought to cultivate direct linkages to it.[30] In the following years, Lacoste purchased two large steamships from Merrick and Sons of Philadelphia, as well as a third from the Canada Marine Works in Montreal, and built a private steamship berth at the Luz Wharf in Old Havana to drop off passengers after its freight had been deposited in Regla.[31] The CJRC thus coordinated with Lacoste's steamship company to transport freight and passengers between Cárdenas and Havana.

Having sunk considerable capital into its intermodal system of sugar transport, the company now had to shift the flows of sugar over to its own system. To ensure the full freights that would cover construction costs, the company established what local merchants cursed as 'an absurd special tariff for urban service'. The CJRC's new ex-urban warehouse at El Recreo would of course be exempt from this 'urban service' surcharge, making it impossible for the urban merchants to compete with the new warehouse. Thus the unified company used discriminatory rates to enforce a new mercantile organization of urban space that echoed, at reduced scale, what had been accomplished by the Regla Warehouse Company outside Havana. Tripling the pre-existing rate for carrying sugar to the old warehouses within city limits, the fee structure guaranteed the sugar would flow to El Recreo and from there to Lacoste's wharf, and on to Havana, without passing through the hands of the Cárdenas traders. Local merchants had been betrayed by their own success. In establishing relationships with settler-planters whose slaves were cutting sugar estates out of forests in the 1840s, they 'naturally attracted the attention of speculators, of calculating people . . . who are trying to monopolize all of the operations of our great future', noted Cárdenas mayor José Zabala.[32] Rivals who had looked with envy upon the grassroots commercial growth had seized the opportunity provided by the 1853 fire (a moment of infrastructural flux) to transform the city's sugar trade.

The combined railroad–wharf–warehouse–steamship interest did not seem to have envisioned a complete replacement of the twenty-five urban merchant *solares* of Cárdenas with a single extramural clearing house. Instead they sought to bifurcate the sugar market by building a new and segregated pathway for the product that only elite planters were capable of producing: white sugar. In order to guarantee its monopoly of white sugar, the CJRC imposed a further concession on the town's smaller merchants, who now of course depended on the railroad to transport all of their products; city merchants agreed only to receive molasses and muscovado, leaving the purged sugar to elite merchant-planters like Lacoste.[33] With the new, centralized warehouse dedicated to white sugar, planters who used the railroad could store the product for as long as they liked at a much lower rate, and sell when the market was good, leaving the bulky, unwieldy and perishable hogsheads of molasses to the local merchants who had to unload it almost immediately, whether prices were soaring or sinking.

In their official complaint, local merchants tried to paint the Lacoste crowd as underhanded monopolists. The CJRC told a different story: the only reason carrying costs had previously been so low was the 'ruinous competition' between the two separate railroad companies. Countering local merchants' accusations of collusion, the Board claimed that

> 'the [small] warehouse owners of Cárdenas who promised always to maintain the two lines under their protection, due to the fact that the railroads lacked their own deposit warehouse in which accounts of the freights and returns could be stored, watched with anxiety while the railroads sought to do nothing more than establish a central station, with the object of freeing the company from the obstacles which it has suffered and in which its traffic suffers in the private warehouses'.[34]

The transport system that passed all sugar through the private wharves of local traders added several additional steps in the embarkation of goods, hampering the operations and cutting into the profits of the railroad. Local merchants, whose livelihood was based on the fees they collected from these extra steps, balked at the change because they would simultaneously be robbed of the market flexibility that came with access to white sugar. Lacoste had been fighting against these kinds of slowdowns and added costs for planters since at least 1847, when he opposed the hiring of a 'molasses inspector' of Cárdenas whom local merchants desired.[35] The CJRC, allied with the merchant-planter bloc headed by Lacoste, could portray their warehouse grab to wary government officials as a liberal and enlightened response to technocratic imperatives while engineering their own form of infrastructurally rooted market power.

The new generation of elite merchant-planters associated with Lacoste were the prime makers of white sugar in the region. Learning from a decade's experience that the railroad by itself was insufficient to break free from the grasp of the first-entry pioneer merchants, they successfully transformed the littoral infrastructure of Cuba, cancelling out the strategic advantage of certain coastline locations. These interlopers from Havana and Matanzas, through the mechanism of the CJRC, and the already-existing littoral power Pedro Lacoste, would pull off the construction of a multi-element system; only such a system would make their prized railroads bring the kind of benefits planters had dreamed of in the first shareholders meetings during the hungry 1840s.[36]

Although the government eventually sided with the smaller merchants by ruling against the CJRC's discriminatory rates for urban service, it appears that the overarching goals of the combined railroad–plantation–wharf–warehouse–steamship companies were achieved in Cárdenas. Perhaps the removal of the central station to El Recreo was enough. Two years after the new system went into operation, only seventeen merchant warehouses remained in crowded downtown Cárdenas, a decrease of 38 per cent. The local traders who survived, moreover, dealt mostly in molasses.[37] The bifurcation of molasses and purged sugar did not completely kill off the older generation of small merchants. They continued to build their own small jetties along the shoreline (that could probably accommodate one ship at a time), each being equipped with animal-powered rail spurs that intersected with the Cárdenas-Júcaro line. Clearly, however, the range of saccharine products to which they had access had narrowed. In the early days, Cárdenas's merchants had accepted molasses as payment for storage fees from cash-poor planters as long as those planters also purchased their containers from that merchant.[38] Thus, after Lacoste's 1858 warehouse coup, local traders depended almost exclusively upon what had formerly been an auxiliary activity.

Lacoste, meanwhile, enjoyed augmented market power in the infrastructure of flows he had helped engineer: acting as agent for the largest, perhaps most mechanized plantation on the island, Lacoste had 2,600 boxes of sugar in Havana's San José Warehouses in 1861. He first contracted with an associate of Moses Taylor, one of the brown sugar trade's most powerful Anglo-American merchants, then walked away from the deal when the agent declined Lacoste's unusual condition that he be paid in full as soon as the cargo left the warehouse. Lacoste sold the shipment to another party. This transaction shows the kind of flexibility a merchant-planter like Lacoste had with the warehouses. In this case, he saw fit to walk away from a deal with a powerful international commercial enterprise. Also, he was in a position to act on the Regla warehousers' promise to open up space for sellers to offer wares to multiple purchasers simultaneously. Indeed, he had taken this promise quite literally.[39]

With Cárdenas's sugar exports outperforming those of every city in the world save Havana in 1860, there were plenty of sugars to go around – they had simply been drawn apart into two separate commodity chains that are rarely noted in quantitative studies.[40] While the district of Cárdenas produced more sugar, molasses and rum than any other district on the island in 1852, it hardly exported any purged sugar directly. Most of it went to Havana. On the other hand, Cárdenas remained the island's number one exporter of molasses.[41] While this lopsidedness preceded the warehouse takeover, Lacoste and the CJRC took the white sugar out of local merchants' hands and monopolized what had formerly been divvied amongst them all: the commission trade through Havana. The relatively rapid emergence of discrete commodity pathways was the product of merchant-planter alliance, infrastructural capacity and technological capability. The CJRC, together with the deposit warehouse at El Recreo, and the deeper-draft wharf of Lacoste, with its schooners and steamship lines, took purged sugar to Havana. The remaining merchant wharves and warehouses in the port city of Cárdenas focused on molasses, which were exported directly to the US at lower profit margins, and with less control over when to sell. This reinforced the early emergence of colonial economic relationships that would be manifest by the 1890s.

Accumulators and deployers of littoral power, elite merchant-planters portrayed their actions as disinterested: merely detecting static in the system and smoothing it out. In reality, their infrastructural undertakings had a dual character: like water itself, they found tiny points of egress and widened them into avenues of traffic for commodities. Like sand, on the other hand, they clogged already-existing channels, forcing transport down new avenues that benefited themselves. Their attempts to cloak cutthroat merchant competition as disinterested technocratic duty served them well at a political moment where colonial treasurers thirsted for sugar revenues, and colonial officials hungered for racial order. However, the littoral zone, which they came to dominate in the railroad era, was only one crucial link to a wider Atlantic World over which they had less control.

## ATLANTIC MARITIME SHIPPING, TARIFF POLICY AND INFRASTRUCTURAL CHANGE

In the early decades of the nineteenth century Madrid legalized export–import activities at secondary port cities in Cuba in response to the eastward march of sugar cultivation away from Havana. Havana's continued role as a distribution point for increasingly distant plantations was premised upon infrastructural transformation that made the harbour attractive to large North American and British ships.[42] The most notable of these transformations began with the Regla warehouses, headed up by Havana-based merchants with agricultural interests in Matanzas and Cárdenas. Facilities in Havana help explain the continuing attraction of the city for larger ships and purged sugar export, while Matanzas specialized in raw sugar, and Cárdenas in syrups export. Each city developed a port infrastructure matched to the particular span of the sugar spectrum that was refracted in its direction.

The new Havana warehouse interests offered centralized financial services as well as large-scale bulk-breaking and cargo-shipping in the island's capital. Larger, more efficient ships took advantage of Havana's considerable facilities to bring in European and North American commodities and pick up purged sugar. In the first quarter of 1861, for example, the mean tonnage of vessels entering Havana was 352.5 tons. In Matanzas the average shrank to 264.7 tons, while those entering Cárdenas only measured 236.4 tons.[43] The shallow, winding and unpredictable channels that wormed in and out of the Bay of Cárdenas made hiring local pilots indispensable and much more costly than at Havana or even Matanzas. Even worse, multiple steps of unloading before arrival, and the inadequacy of ship chandler services in Cárdenas, ensured the port would play a very particular role in the system. Cárdenas took most of its imports by coasting schooners from Havana, so the shipping lines and the stop-off in the capital city were already in operation.[44] With buyers and shippers accustomed to the Havana market, and Havana long established as an import centre, merchant-planters had reason to maintain other activities in the colonial capital.

However, Havana-based merchant-planters did not monopolize all trade or even the trade of all products deriving from sugarcane. Rather, Havana played an ever-diminishing role in the export of molasses and muscovado, while pulling in far more than its share of 'boxed', 'purged' and 'clayed' sugar (which I have been grouping together under the name of 'plantation white' or 'white' for the sake of convenience).[45] Cárdenas and Matanzas, in fact, exported 62.7 per cent of all the molasses on the island, while the molasses export of Havana plummeted from a peak of 54 per cent in 1830 to 4.7 per cent in 1863–1864.[46] On the other hand, Cárdenas only exported 2.6 per cent of Cuba's boxed sugar in 1851, in spite of its surrounding plantations producing more boxed sugar than any other jurisdiction on the island. Over 50 per cent of boxed sugar exports went out of one port: Havana. In other words, while the jurisdictions of Matanzas, Cárdenas and Sagua la Grande sold most of the new sugar zone's muscovado and molasses out of their own ports, the purged sugars also made on nearby plantations were sold out of Havana warehouses because of the tendency of railroad and warehousing improvements to reshuffle commodity flows to the advantage of elite merchant-planters. If we envision Cuban sugar exports as a cone of sugar with white at the top, and progressively molasses-heavy grades towards the bottom, Havana would slice off the white cap and sell it to Europe. Matanzas would take the middle slice of brown and deliver it to the US. Cárdenas, finally, would spoon the mish-mash of leftovers into *yanqui* bottoms bound for *el norte*.

The centrifugal splitting of Cuban sugar complicates the usual story of increased US imperial influence over the island. While Matanzas and Cárdenas had indeed begun to look like US neo-colonies, as a centre of purged sugar export Havana actually had closer links with England until the 1860s. Tariffs had a profound impact. The 1830s were a hard time for Cuban planters, in part because Spain used its most valuable colonial possession as a pawn in an ill-considered tariff war with the US. Fought to protect Spain's struggling shipping industry, the tariff war came down hardest on Cuban consumers who paid inflated prices for North American imports. Retaliatory duties imposed by the US also stymied the young merchant marine of Cuba (deeply dependent on trade with the US), while Cuban sugar producers got cut out of the white sugar market in the US.[47] With steep duties protecting US refiners, little 'white, clayed, or powdered' sugar from Cuba found its way to North American markets. Nearly all of it was brown.[48] The tariff arrangement was not necessarily a raw deal for certain entrepreneurs in Cuba. In fact, Spain helped funnel cheap byproducts to the US by leaving open a duty loophole for molasses alone, exempting US vessels dealing in that product from otherwise burdensome tonnage charges.[49]

This incentive to North American traders to focus on molasses helped wealthy immigrants to the island like Lacoste exert continued control over the trans-Atlantic trade of white sugar to England, although US shippers made increasing inroads in the carrying trade between Cuba and European markets as well. After the British Parliament re-opened the empire's market to slave-grown sugar on equal terms with 'free labour' sugar in 1846, much of the expansion in Cuba's purged sugar exports went to the British Isles.[50] Contemporary with the famous repeal of the Corn Laws, the UK Sugar Duties Acts of 1846 and 1848 swept away all protections on sugar (both the penalties against slave-grown sugar, and 'the penal duty on superior plantation sugar').[51] The triumph of 'free trade' weakened the protectionist position of some abolitionists in Britain, and abetted the continued expansion of slavery in Brazil and Cuba, along with the illegal slave trade.

The newly permissive laws also helped the slaveholders of Cuba evade the mercantile grasp of the United States. Britain's imports of Cuban sugar increased from a low of 197,460 cwt (One cwt, or 'hundredweight' equals approximately 112 US pounds) in 1845, to a high of 1.64 million cwt in 1859, although the figures varied widely from year to year.[52] In 1849, twice as many boxes of sugar went to England as to Cuba's metropole Spain. England also received nearly four times as many boxes as the United States in that year.[53] The distinction between US and UK markets for different grades of sugar even overrode the control New York merchants exerted in Matanzas and Cárdenas. In 1861, for example, in spite of the prominence of North American merchants there, Matanzas shipped only 15 per cent of its boxed sugar to the United States. Eighty-five per cent went to Europe. However, the US was the destination for 81 per cent of the muscovado hogsheads, and 55 per cent of the molasses hogsheads that left Matanzas.[54] A US-centred, neo-colonial dynamic was emerging in the world of brown, but not of white, sugar.

A trader without significant wealth had an easier time getting a foot in the door of these second-tier port towns along the sugar frontier. They put themselves in a position to capitalize on a thoroughgoing change in Cuba's agricultural economy. Only a small minority of the wealthiest planters adopted versions of the Derosne system (that enabled their enslaved workers to produce more white sugar and less molasses per acre of cane planted). In the 96 per cent of non-elite sugar plantations, the steam-powered grinding mill was the main innovation adopted in the 1830s and 1840s. Relatively affordable and simple to maintain, these grinding mills brought a lot more cane juice into the pans, but with Jamaica trains still processing the juice, ratios of purged sugar to muscovado and syrups remained fixed: accelerated grinding did not provide a relative increase in white sugar; it merely multiplied the absolute yield of syrups. Cuba's total exports of syrup thus grew from 3.2 million gallons in 1815 to 43.7 million gallons in 1868. Most of this inexpensive byproduct found its consumers among residents of the United States, after having passed through the machinery of US-based refiners.[55] Elite merchant-planters' continued production of consumption-grade white sugar was premised on the strength of the newly opened British market. But competing with new producers in places like Java, Mauritius and India, which also supplied Britons with affordable sugar, required the endless depressing of production costs, which involved the ever-increasing exploitation of enslaved workers in the fields, combined with mechanization in the boiling house.[56] It also required pushing down shipping costs, which intensified the struggle among blocs of merchants in Cuba itself, and helps explain the at-first-glance quixotic and drawn-out struggles over the shape of sugar transport arrangements in and around the secondary port city of Cárdenas in the 1850s. Compelling logistical issues reinforced entrepreneurial rivalries in encouraging spatial segregation by grades of sugar.

## TRANSPORT LOGISTICS, CITY-SYSTEMS AND THE DIVISION OF CAPITALS

Molasses and syrups (*mascabádos* and *mieles finales*) had a limited shelf life. They had to be marketed almost immediately after they were packed.[57] Filtered, centrifuged and clayed sugar, having so little moisture, could be stored almost indefinitely. Furthermore, the lower quality '*mascabádos clásicos*', which were basically a sweet, yellowish-brown mush taken straight from the strike pan and placed in giant hogsheads (*bocoyes*), weighing over a thousand pounds when full, were costly and difficult to manoeuvre. Sometimes the hogsheads were partially purged of water content by drilling a hole in the bottom. Oftentimes, they were not drained at all.[58]

Because of this incomplete drainage, leakage continued long after the goods were shipped. Visiting a New York warehouse in 1877, the writer Ernest Ingersoll observed 'tier upon tier of hogsheads of sugar, perspiring molasses with the memory of the Cuban sun'.[59] Constant leakage of hastily drained shipments meant that many hogsheads of muscovado arrived in Europe or North America weighing 20 per cent less than they had at embarkation. One advantage of this phenomenon, a British sugar expert sarcastically pointed out, was that muscovado-laden vessels arrived at their destinations with a foot-and-a-half shallower draft than when they departed.[60]

Moreover, hogsheads were heavy in relation to the value of the product they contained. Unlike boxes of sugar, they could not be stored in tall stacks in a warehouse, meaning that the only way to store them effectively would be to construct a warehouse with many floors, or more likely a sprawling, single-storey edifice. Structures of this sort would have been costly, and only worthwhile if profits from sales considerably outweighed building expenses. Warehousers therefore charged far higher rates for hogsheads of molasses than they did for hogsheads of muscovado or boxes of purged sugar.

Because storing hogsheads of perishable, low-grade sugar in warehouses was cost-ineffective, sellers of this line of products had little freedom to arbitrate against the daily price shifts of the world sugar market. Their freedom of operation was further hampered by the fact that they may not have transported by rail and probably remained in the grasp of local sugar dealers. Regla customers mostly were big planters, according to company head Eduardo Fesser.[61] The marketing agility made possible by imperishable white sugar, combined with the price spread between grades of sugar, and merchant-planters' unique access to European markets through the Regla warehouses, helps explain why elite merchant-planters wanted to monopolize Cuba's supply of white sugar. Logistical differences among different kinds of cane products made warehousing an effective strategy for disaggregating white grades from other sweet byproducts. In a sense, durable white sugar, combined with railroads and warehouses, gave the new generation of merchant-planters the power to cope with price shifts over which they had no control. In an increasingly unified global sugar economy commodity prices could change rapidly through the fortunes of a sugar harvest halfway around the world, or the activities of speculators in London and New York.

The elite merchant-planter cohort used the railroad and warehousing systems to realign the circuits of sugar. They used transport infrastructure like a centrifuge to throw different grades of sugar (that had formerly been jumbled together on the streets of Havana, on the government docks of the old city, and in the home storehouses of *solares*) into discrete levels of the economy. In order to pull off this bifurcation, however, they needed to do more than conquer the Havana waterfront: they needed to exert their will in the port which had most direct access to the major producers of white sugar in the world at the time: the port of Cárdenas. It was just too important a business to run through such a port which had such obvious disadvantages, they might have said. And they operated in a variety of ways to make sure white sugar found cheaper, faster channels of egress.

In the 1840s and 1850s, a new geographic division of labour among Cuban port cities was founded on the separations initiated by the centrifuge. The various saccharine products emitted by the sugar mill thus lit up different pathways on the Atlantic switchboard, inscribing a set of discrete economic relationships on the ocean's surface. Technical and organizational change on plantations, in particular ports, and between ports, helped lay the circuits of the larger system. There was no 'global empire of sugar' in the age of Atlantic slavery to parallel the 'global empire of cotton' controlled from a single place.[62] In fact, the competing centres of England and the United States reflected and completed the bifurcation of sugars which began in the centrifuge bowls of the boiling houses of Cuba's sugar plantations, but not without being further refracted through a revolutionized transport infrastructure engineered by elite merchant-planters attempting to navigate tariff barriers and retain flexibility – a requirement for the success of capitalists operating far from the centres of global economic power in the nineteenth century.

## NOTES AND REFERENCES

1. Manuel Moreno Fraginals, *El ingenio: complejo socio-economico Cubano de azúcar* (Havana: Editorial Ciencias Sociales, 1978) 3 vols; Dale Tomich, 'World Slavery and Caribbean Capitalism: The Cuban Sugar Industry, 1760–1868', *Theory and Society* 20, 3 (1991): 297–319; José Guadalupe Ortega, 'Machines, Modernity, and Sugar: The Greater Caribbean in a Global Context, 1812–50', *Journal of Global History* 9, 1 (2014): 1–25; and Leida Fernández Prieto, 'Islands of Knowledge: Science and Agriculture in the History of Latin America and the Caribbean', *Isis* 104, 4 (December 2013): 788–797.
2. For more on sugar mill technology in Cuba, see Daniel Rood, *The Reinvention of Atlantic Slavery: Technology, Labor, Race, and Capitalism in the Greater Caribbean* (New York: Oxford University Press, 2017), chapters 1–2.
3. Names and types of sugar were notoriously variable. For our purposes, 'purged' sugar referred in markets to white and yellow sugars that had been carefully drained of molasses and water content. A rough guide to British classifications is as follows: in 1845 'two classes of sugar were adopted: one called "equal to white clayed by any process" and other "not equal to white clayed or muscovado." The sugars came to be referred to as "yellow muscovado" and "brown muscovado." The former paid duty at the rate of sixteen shillings four pence per hundred weight and the latter at fourteen shillings. . . . The production of a good class direct consumption sugar continued to be handicapped.' Noel Deerr, *The History of Sugar*, 2 vols, vol. 2 (London: Chapman and Hall, 1949), 466, 468.
4. While the exports of Cuba were divvied up among many ports, Havana nearly monopolized imports. See statistical table in Jose Maria de la Torre, *The Spanish West Indies. Cuba and Porto Rico: Geographical, Political, and Industrial* (New York: Colton, 1855), 124. Havana was uniquely convenient for the marketing of imports since only it had bonded warehouses where imports could temporarily be stored without duties being levied. At ports like Matanzas, importers had to pay duties immediately upon disembarkation of goods. *Letter of the Secretary of State Transmitting a Report on the Commercial Relations of the United States with Foreign Countries* (Washington, 1863), Part One, 232.
5. See Rood, *Reinvention of Atlantic Slavery*, third chapter.
6. For a recent discussion of Marx's idea of a 'division of capitals', see Scott Reynolds Nelson, 'Who Put Their Capitalism in My Slavery?', *The Journal of the Civil War Era* 5 (2015): 289–310.
7. Philip Curtin, *The Rise and Fall of the Plantation Complex: Essays in Atlantic History* (New York: Cambridge University Press, 1998). The idea that a particular crop (whether sugar, corn or potatoes) gives rise to particular forms of civilization, which

clearly informs the sugar literature, is known as 'staple theory'. For example, see Philip Morgan, *Slave Counterpoint: Black Culture in the Eighteenth-Century Chesapeake and Lowcountry* (Chapel Hill: University of North Carolina Press, 1998).

8. '*Expediente relativo a la reclamacion de la Sociedad Almirall y Hermano contra el estab del rrd proyectado por Dn Rafael Ferrer y Vidal para el trafico en varias calles de la Ciudad de Matanzas*', 1857. Archivo Nacional de Cuba (ANC), Fondo Real Sociedad y Junta de Fomento (JF), Leg 137, exp 6724.
9. Abiel Abbot, *Cartas escritas en el interior de Cuba: entre las montañas de Arcana, en el este, y las de Cusco, al oeste, en los meses de febrero, marzo, abril y mayo de 1828* (La Habana: Consejo Nacional de Cultura, 1965), 54.
10. Old-line Creole family names predominated on the Board of Directors and principal investors of RCC: Juan Montalvo y O'Farrill, Anastasio Carrillo, el Conde de Jaruco, Angel Pérez, Carlos Drake, Antonio Mariátegui, Joaquin Pedroso Echevarría, Joaquín de Peñalver, José Joaquín Carrera, Juan Vázquez, Juan Ignacio Echarte and Domingo del Monte. Eduardo Moyano Bazzani, *La nueva frontera del azúcar: El ferrocarril y la economía cubana del siglo XIX* (Madrid: CSIC, 1991), 146. The Júcaro board had more of the Catalán group presented. These included Joaquín de Arrieta and Pedro Diago. But the separation between the two groups was never absolute, since they collaborated on so many projects. Moyano Bazzani, *La nueva frontera del azúcar*, 152. They were opposed by the local Cárdenas merchants Tomas Fernandez Cossio, Francisco Jimenez and Vicente Medina. For the Cárdenas–Matanzas rivalry, see Oscar Zanetti and Alejandro García, *Sugar and Railroads: A Cuban History, 1837–1959* (Chapel Hill: University of North Carolina Press, 1998), 41.
11. '*Para evitar que se conduzcan clandestinamente armas al interior por los ferrocarriles*', 1854. ANC, JF, Leg 137, exp 6695.
12. These wharves had to be quite sturdy in order to support the freight cars. One of them cost 12,300 pesos for the rails, the crossties, and the 9'-thick planks which the weights required. '*A los autos que sigue D. Juan Giraud contra D. Juis Bourdaut y Timeolon Bartemeli sobre acreditar que es socio des estos en unos almacenes . . . contra D. Pedro Lacoste*', 1847. ANC, Escribanías de Pontón, Leg 119, exp 1.
13. Author conversation with Ernesto Alvarez Aramis Blanco, historiador de la ciudad de Cárdenas, 21 June 2014.
14. Moyano Bazzani, *La nueva frontera*, 147–148.
15. '*Expediente relativo a la reclamacion de la Sociedad Almirall y Hermano contra el estab del rrd proyectado por Dn Rafael Ferrer y Vidal para el trafico en varias calles de la Ciudad de Matanzas*', 1857. ANC, JF, Leg 137, Número 6724.
16. '*Sobre construccion de un muelle en el puerto de Cárdenas*', 1848. ANC, JF, Leg 84, exp 3430; '*Documento incomplete sobre un muelle en Cárdenas*', 1852–1853. ANC, JF, Leg 162, exp 7855. When the wharf plan was first announced in 1847, the projected dock was to depart from the Plaza of Pinillos, but the rapid completion of 540 feet of new embankment, largely 'to obtain a greater quantity of water' in the problematically shallow Bay of Cárdenas, left the plaza too far from the water.

17. '*Mocion hecha por el Sor Alcalde ordinario D. José Zabala a consecuencia del proyecto que tienen los Sres Noriega Olmo y Cía y el Crédito Mobiliario sobre construir un muelle general que atraviese de este a oeste la bahía de Cárdenas*', 1857. ANC, OP, Leg 11, exp 289.
18. The Júcaro Railroad Company (JRC) was a new kind of railroad on the Cuban scene. While the Havana Railroad Company (HRC) and the RCC had been built to service existing plantation districts, the JRC was a speculative railroad, preceding and anticipating future development. It did not simply respond to existing geography – it governed the evolution of new plantation neighborhoods. Zanetti and García, *Sugar and Railroads*, 43.
19. '*Relación de los solares que la Real Hacienda posee en la Villa de Cárdenas*', 1859. GSC, Leg 42, exp 2827. For a map showing Lacoste's and the JRC's *solar* buy-up, see '*Expediente propuesto por D. Juan Vidal para hacer un muelle con su almacen en esta bahía*', 1858. ANC, OP, Leg 11, exp 244.
20. ANC, JF, Leg 137, exp 6724.
21. ANC, JF, Leg 137, exp 6724.
22. On CJRC slave ownership, see Zanetti and García, *Sugar and Railroads*, 118.
23. '*Exposicion que hace la Junta Directiva de la empresa de los caminos de hierro de Cárdenas y Júcaro a la general de accionistas que ha de celebrarse el día 14 de Abril de 1859*' (Habana: Imprenta 'La Habanera', 1859). ANC, JF, Leg 164, num 7919.
24. '*Exposicion que hace la Junta Directiva de la empresa de los caminos de hierro de Cárdenas y Júcaro a la general de accionistas que ha de celebrarse el día 14 de Abril de 1859*' (Habana: Imprenta 'La Habanera', 1859). ANC, JF, Leg 164, num 7919.
25. '*Exposicion que hace la Junta Directiva de la empresa de los caminos de hierro de Cárdenas y Júcaro a la general de accionistas que ha de celebrarse el día 14 de Abril de 1859*' (Habana: Imprenta 'La Habanera', 1859). ANC, JF, Leg 164, num 7919. After the slave and free black conspiracy of 1843 (remembered as 'La Escalera'), it was thought good policy to remove as many people of colour from Cuba's cities as possible.
26. Carlos Hellberg, *Historia Estadística de Cárdenas* (Cárdenas: 1893), 46–47.
27. Jacobo de la Pezuela, *Diccionario geográfico, estadístico, histórico de la Isla de Cuba*. 4 vols (Madrid, 1863), vol. 2, 344.
28. Pezuela, *Diccionario*, vol. 1, 307. 'The largest coasting schooners' only, not oceangoing ships.
29. '*A los autos que sigue D. Juan Giraud contra D. Juis Bourdaut y Timeolon Bartemeli sobre acreditar que es socio des estos en unos almacenes . . . contra D. Pedro Lacoste*', 1847. ANC, Escribanías de Pontón, Leg 119, exp 1.
30. Rolando García Blanco, *Francisco de Albear: un genio cubano universal* (Havana: Editorial Científico-Técnica, 2007), 52–53.
31. 'Propellers for Cuba', *Journal of the Franklin Institute* 3rd ser., 35 (1858), 350.
32. '*Mocion hecha por el Sor Alcalde ordinario D. José Zabala a consecuencia del proyecto que tienen los Sres Noriega Olmo y Cía y el Crédito Mobiliario sobre construir un*

*muelle general que atraviese de este a oeste la bahía de Cárdenas*', 1857. ANC, OP, Leg 11, exp 289.

33. '*Expediente relativo a la reclamacion de la Sociedad Almirall y Hermano contra el estab del rrd proyectado por Dn Rafael Ferrer y Vidal para el trafico en varias calles de la Ciudad de Matanzas*', 1857. ANC, JF, Leg 137, exp 6724.
34. '*Exposicion que hace la Junta Directiva de la empresa de los caminos de hierro de Cárdenas y Júcaro a la general de accionistas que ha de celebrarse el día 14 de Abril de 1859*' (Habana: Imprenta 'La Habanera', 1859). ANC, JF, Leg 164, exp 7919.
35. '*Expediente relativo a la solicitud de varios vecinos de Cardenas para que se nombre un fiel de mieles en aquel puerto*', 1847. ANC, JF, Leg 76, exp 2986.
36. An overlapping group, including the coolie importer Rafael Rodriguez Torices, pulled off a similar hostile takeover of the lucrative Sagua la Grande Railroad in the late 1850s, part of a general Catalán conquest of Cuba's sugar infrastructure.
37. Pezuela, *Diccionario*, vol. 1, 307.
38. Not always, though. Merchants had to keep an eye on the rapid ups and downs of syrups prices because, as one merchant lamented, 'it is not easy to dispense with storage fees'. '*A los autos que sigue D. Juan Giraud contra D. Juis Bourdaut y Timeolon Bartemeli sobre acreditar que es socio des estos en unos almacenes . . . contra D. Pedro Lacoste*', 1847. ANC, Escribanías de Pontón, Leg 119, exp 1.
39. Francisco Lastres, *Contratacion sobre efectos publicos de los corredores de comercio y de los agentes de bolsa* (Madrid, 1878), 323. The man was Drake's/Taylor's consiglieri Jose Maria Morales. For more on this network of entrepreneurs, see Roland Ely, *Comerciantes cubanos del siglo XIX* (Havana, Cuba: Editorial Liberia Martí, 1960), 10–11.
40. While Laird Bergad, *Cuban Rural Society in the Nineteenth Century: The Social and Economic History of Monoculture in Matanzas* (Princeton, NJ: Princeton University Press, 1990), citing no sources, claims that Cárdenas 'never developed into a major Cuban export center', a more recent work, based on Carlos Rebello's census, puts the 1860 exports of the city at 126,185 metric tonnes, those of Havana at 144,726, and those of Matanzas at 93,841. Alberto Perret Ballester, *El azúcar en Matanzas y sus dueños en La Habana* (La Habana: Editorial Ciencias Sociales, 2007), 271, 459. The key here is that Rebello simply recorded 'puntos de embarque' for the products of the island's *ingenios*. Many of these sugars may have been shipped coastwise to Havana before being sold to export merchants.
41. Of thirty districts in the island, Cárdenas produced nearly 25 per cent of total rum and 25 per cent of molasses, as well as 30 per cent of its sugar (purged and muscovado included). But the city only exported 2.6 per cent of Cuban sugar, while Havana exported nearly half of the island's total. Torre, *Spanish West Indies*, 122–124.
42. Oscar Zanetti Lecuona, 'La capital del azúcar', in Bernardo García Díaz and Sergio Guerra Vilaboy (eds.), *La Habana/Veracruz Veracruz/La Habana: Las dos orillas* (Veracruz: Universidad Veracruzana, 2002), 262. Linda Salvucci, 'Supply, Demand, and the Making of a Market: Philadelphia and Havana at the Beginning of the

19th Century', in Peggy Liss and Franklin Knight (eds.), *Atlantic Port Cities: Economy, Culture, and Society in the Atlantic World, 1650–1850* (Knoxville: University of Tennessee Press, 1991), 43.

43. 332 vessels entered Havana, with an aggregate tonnage of 33,090. Figures taken from *Letter of the Secretary of State Transmitting a Report on the Commercial Relations of the United States with Foreign Countries* (Washington, 1863), Part Two, 146–147, 159, 164.

44. On these conditions, which seem largely unchanged in the 1920s (and even today), see Herminio Portell Vilá, *La decadencia de Cárdenas* (Havana, 1929), 19, 21, 75. For a similar division of maritime operations among ports in the UK, see Kenneth Morgan, 'Bristol and the Atlantic Trade in the Eighteenth Century', *The English Historical Review* 107, 424 (Jul. 1992), 650.

45. While Havana 'pretty much monopolized the exports of clayed sugar', one maritime historian explains, Matanzas evolved into the 'headquarters for muscovado'. Robert Albion, *The Rise of New York Port, 1815–1860* (New York: Charles Scribner's Sons, 1939), 183.

46. In 1851 the district of Cárdenas produced almost 25 per cent of the island's total, while 28.7 per cent of the lower-cost byproduct went directly out of Cárdenas. Torre, *Spanish West Indies*, 122–123, and Leví Marrero, *Cuba: Economia y sociedad,* 14 vols, vol. 12 (Madrid: Editorial Playor, 1984), 126.

47. One expert writes, 'Not infrequently consumers in Cuba paid more than double the original market price for US imports. One midcentury estimate calculated that approximately $15 million worth of North American textiles, agricultural goods, furniture, and tools was resold in Cuba for more than $30 million. A total of $73,000 in duties was paid on $91,000 worth of U.S. flour.' Louis Pérez, Jr., *Cuba: Between Reform and Revolution*, 2nd edn. (New York: Oxford University Press, 1995), 16–17.

48. 'Cuba', *Hunt's Merchant Magazine* 28 (1852), 153. In 1852 the US took 196,485 boxes of Cuban sugar (about 28.5 per cent of Cuba's total) while taking 100 per cent of the island's molasses exports. 275 million lbs. of brown sugar were imported into the US from Cuba in 1851. 'Commerce of Havana', *Hunt's Merchant Magazine* 28 (1852), 480. While most of this was used to make rum, North American refiners also re-worked Cuban molasses into 12,000 tons of brown sugar in 1860. Ramón Sagra, *Cuba en 1860, o sea cuadro de sus adelantos en la población, la agricultura, el comercio y las rentas publicas. Suplemento a la primera parte de La Historia Política y Natural de la Isla de Cuba por D. Ramón de la Sagra* (Paris, 1862), 131. In spite of all the barriers to exchange, writes one historian of Cuban–North American relations, 'North American traders shrewdly and effectively pursued the Cuban market, providing slaves and manufactured goods at reasonable prices, often extending generous credit arrangements, and accepting in return sugar and molasses as payment.' Merchants, bankers, and shippers from northeastern states established houses in the port cities of Havana, Matanzas, Cienfuegos, Cárdenas, Sagua la Grande, Trinidad, and Santiago de Cuba. Perez, *Cuba and the United States*, 16–17.

49. US vessels were charged $1.50 per ton for merely entering the port. 'Of Navigation between the United States, Cuba, Etc. Etc.: Circular Instructions to the Collectors and Other Officers of the Customs', *Hunt's Merchant Magazine* 27 (1852), 238.

50. David Murray, *Odious Commerce: Britain, Spain, and the Abolition of the Cuban Slave Trade* (Cambridge: Cambridge University Press, 1980), 208–210.
51. Deerr, *The History of Sugar*, vol. 2, 430.
52. Murray, *Odious Commerce*, 243, http://www.convertunits.com/from/cwt/to/pounds
53. 'Exports of Produce from Havana and Matanzas', *Hunt's Merchant Magazine* 22 (1850), 662.
54. The city also exported 16,261 hogsheads of muscovado, 13,204 of which went to the US, 2958 to Europe (81 per cent to US). When it came to molasses, the US took 17,640 of a 32,147-hogshead total. Europe took the remainder. *Letter of the Secretary Of State, Transmitting a Report of the Commercial Relations of the United States with Foreign Nations, for the Year Ending September 30, 1861* (Washington: Government Printing Office, 1862), 167–168.
55. Fernando Charadan Lopez, *La Industria Azucarera en Cuba* (Havana: Editorial Ciencias Sociales, 1982), 43. Charadan discusses how the excess of syrups was the source of the great push to reform the processing train. While the older Jamaica train technology, with its inexact temperature control, its slow rate of evaporation, and its lack of both filters and centrifugal dryers, could not 'work the syrups', the modern trains drew more granulated sugar out of molasses that had already been processed once, largely from investment in carbon filtration systems, as well as revivification.
56. For this story, see Rood, *Reinvention of Atlantic Slavery*.
57. Perhaps to extend the shelf life of syrups, some wealthier merchants had large syrup tanks built on their property. In Cárdenas, a foundation of piles, plus the three tanks, and a shelter built around the tanks, cost 5,270 pesos in the early 1840s. '*A los autos que sigue D. Juan Giraud contra D. Juis Bourdaut y Timeolon Bartemeli sobre acreditar que es socio des estos en unos almacenes . . . contra D. Pedro Lacoste*', 1847. ANC, Escribanías de Pontón, Leg 119, exp 1. Extreme heat and humidity, temperature fluctuation, and the lack of completely airtight containers shortened the shelf-life of molasses, which can last for years under the right conditions, http://www.bgfoods.com/int_faq.asp
58. Manuel Moreno Fraginals, *El ingenio: complejo socio-economico Cubano de azúcar* (Havana: Editorial Ciencias Sociales, 1978) 3 vols, vol. 3, 12–13.
59. Ernest Ingersoll, 'The Lading of a Ship', *Harper's New Monthly Magazine*, Sept 1877, 481–493. Quoted in Dara Orenstein, unpublished manuscript, 27.
60. John Scoffern, *The Manufacture of Sugar in the Colonies and at Home, Chemically Considered* (London, 1849), 79.
61. Escribanías de Pontón, Leg 119, 1, 1847. Nevertheless, the largest Matanzas warehouse, which 'serves the same function for Matanzas that the Warehouses of Regla fulfill for Havana', was constructed so as to hold 80,000 boxes of sugar and 20,000 hogsheads of molasses, so at least some unpurged sugar was warehoused, probably because molasses played such an important role in Matanzas's export fortunes. Pezuela, *Diccionario*, vol. 4, 40, quoted in Bergad, *Cuban Rural Society*, 170. '*Expediente sobre aumento del capital de la empresa de Almacenes de Regla*', 1857. ANC, Gobierno Superior Civil, Leg 1571, exp 81322.
62. Sven Beckert, *Empire of Cotton: A Global History* (New York: Knopf, 2014).

# *Machucados* and *Salvavidas*: Patented Humour in the Technified Spaces of Everyday Life in Mexico City, 1900–1910

DIANA J. MONTAÑO

## INTRODUCTION

*'What's dripping blood in your basket?' asked the old lady.*

*'My poor mother! It's all I managed to rescue,' answered the young girl, adding 'Fortunately, I found the head and the heart, even though they were left as* tortillas.
*In this bag, I have the eyes and the teeth.'*

*'An eléctrico, right?' confidently queried the woman.*

*'Yes!' replied the grieving girl explaining, 'A leg, a hand and the entrails were flattened out on the tracks. This foot is missing toes and the shoes were nowhere to be found.'*

*Oh lord! exclaimed the bewildered woman commanding,*
*'Leave the basket in the sink so the cat doesn't eat it. Go buy the casket,' warning,*
*'Listen up! There is a tram coming, be careful it doesn't run you over because I don't have time to pick up your remains!'*

*'That's the tram that ran over my mother!' shouted the girl,*
*'I hope I can catch it!'*

Mexico City welcomed the twentieth century as a thriving metropolis. Similar to other major urban centres around the world, its increasing technification and higher population density oftentimes proved a deadly combination. Since the late nineteenth century, the concentration of manufacturing, the introduction of new fuels, and the beautification and improvement projects placed lives and property at risk of fire.[1] On the streets, the intersection of increased traffic, new modes of transportation and higher speeds substantially altered the navigation of urban space. The opening exchange fictionalized the carnage *eléctricos* (electric streetcars) brought to the capital. 'To that degree,' concluded its writer, 'we have become habituated to the slaughter'.[2] The sketch encapsulates how faster transportation changed streets and the rise of industrial deaths; however, it also stands as evidence of humour as a response.

President Porfirio Díaz had been in power for two decades by the turn of the century, and the capital's material transformation was boasted as the consequence of the country's political stability, heavy foreign investment and the growth of the export-based economy as well as light manufacturing. Technological progress served as regime legitimization. Inaugurated in January 1900, *eléctricos* embodied the tangible notions of modernity, complementing the ideal modern cityscape of paved and lighted boulevards and avenues criss-crossed by telegraph, telephone and power lines. Introduced in the context of rapid population growth and far-reaching sanitation, public health, and urban renewal programmes, trams connected the crowded commercial and business city centre to the affluent western districts, to the country's chief religious centre in Villa de Guadalupe to the north, and to the suburban areas to the south.

These transformations neither encompassed all areas of the city nor went without glitches. Electricity replaced animal and steam traction at a fast pace, powering half of the city's railway network within two years from its introduction.[3] Besides more frequent and faster service, trams struck down pedestrians and animals, often crushing them under the wheels and undercarriage. Sensational reporting exploited their public and quotidian nature. Vividly described, scenes of gruesome maimed bodies and scattered bodily remains underlined the physical and psychological strains of modern life. Critics inveighed against pedestrians and drivers, as well as towards the tepid responses of the authorities and the foreign-owned Mexico Electric Tramways (MET).

Mexicans sardonically grappled with the quotidian mishaps resulting from rapid urbanization, nascent industrialization and the technification of public spaces. Humour as part of the human condition offers important insight into how new technologies were lived in everyday life, the ideas and values that were challenged and threatened, and the ensuing negotiation. Scholars of laughter see humour as key to the cultural codes of the past that 'can take us to the heart of a generation's shifting attitudes, sensibilities and anxieties'.[4] Thus, humoristic production offers rich veins to historians of technology.[5]

Wit and ridicule became preferred combat weapons in covering and exposing the pretences of Porfirian Mexico City. In particular, *humor negro* (gallows humour) and satire channelled the horror, frustration and anger felt towards the *eléctricos*. As disorderly public spectacles, accidents exemplified the darker side of an electric metropolis and directly defied the Díaz government mantra of order and progress. With its multifaceted and shifting role as both symbol of civilization and the 'site of danger, destruction and death in the public mind', the streetcar undermined what

William Beezley has defined as the 'Porfirian persuasion' by demonstrating the fragility of technological modernity.[6] However, if accidents threatened the regime's mantra, the public discussion and trial of safety devices promulgated the promise of a technical fix.

Understanding that a machine's social reality is constructed, emerging not only through its use as a functional device, but also through the ways in which collective human experience shape its meaning, this article examines how humour was employed to digest accidents and to scrutinize technical solutions.[7] It brings together two strange bedfellows: humour and patents. Their intercourse demonstrates a sceptical faith in a technological fix. An analysis of the vernacular employed in the discussion of safety devices known as *salvavidas* (lifeguards) evinces it. *Salvavidas* were fenders attached to the motorcars' front end that varied material, design, operation and, ultimately, performance. The quest for a useful *salvavidas* illustrates how ordinary citizens lived and shaped the everyday realities of urban technification. The glitches of the modern city and stress on technical solutions reveal the ways in which life and technology were reevaluated. A technological fix was not passively dreamed; instead, social critics, wordsmiths, engravers and inventors embarked on an experimentation quest. They scrutinized different *salvavidas*' models. In doing so, this article argues, these voices echoed faith in a technical solution, making evident a sort of technological folklore, 'a word-of-mouth culture of technological possibility'.[8]

*Salvavidas* patent applications, formulaic in nature, provide scant information on the applicant and their devices' social biography. Humoristic production becomes a rich illustrative resource for the latter. Mining humour from the written record has its limitations. Despite its elusive and ephemeral nature, humoristic production entered the written record through the filtered pens of urban chroniclers. Comical illustrations, jokes, sonnets, sardonic verses and fictional stories survived nestled in the pages of the capital's dailies. Compendia of the work of wordsmiths are rare. Satirical broadsides lay scattered in various collections. These are the sources that serve as raw material for the following discussion. Although their circulation via print subtracts the richness of their verbal version, tone, tempo and gesticulation, the essence of a technological folklore survives. Linda Degh in her defence of folklore disseminated via print reminds us that talented narrators 'always combined material from oral and print sources . . . [and] that the notion that folklore was properly the domain of illiterate cultures [is] inaccurate and short-sighted'.[9]

After a word on humour and the Porfirian press, the article summarizes the city's physical transformation to set the stage for the analysis of accidents, their comical coverage as well as the debate on *salvavidas*. The etymological characteristics of the latter demonstrates how Mexicans critically embraced, appropriated and 'made their own' new technologies. Accidents were also tactically framed to bolster or criticize the casting of the lower classes as an obstacle to progress. Thus, an intimate engagement with trams developed both symbolically and physically.[10]

## *HUMOR NEGRO*

As part of the language of everyday life, humour scales down major processes such as modernization and urbanization to ordinary individuals on the ground, 'to lives as lived, crisis suffered, [and] technology encountered'.[11] Across temporal and spatial boundaries, serious and shocking topics have proven fertile soil for humoristic reactions with jokes serving as socially sanctioned outlets for expressing taboo ideas and subjects.[12] Technological mishaps joined oppression, war and catastrophe as another source of threatening or frightening sentiments.[13] Big-city traffic at the turn of the century became a popular source not only for humour, but also for sensationalist reporting. Urban dailies fixated on the sensory intensity of modernity with its frenzied life tempo.[14] The combination of rapid transportation, industrial time regimes, the disciplinary demands of the assembly line on the body as well as the bombardment of commercial advertisements in public spaces resulted in what many described as a horrific assault on the senses.[15] In this context, the violence, suddenness and randomness of accidental deaths in the streets of New York, Paris or Mexico City translated into a fondness for lurid descriptions of mangled bodies and monstrous technologies. Editorialists, cartoonists and engravers resorted to an urban imaginary of danger in their construction of a public culture of violence.[16] Ben Singer posits that their painstaking attention to the physical details of accidental deaths is not only justified by the fact that grotesque sensationalism sold copies, but it evinces 'a distinctive hyperconsciousness of environmental stress and physical vulnerability in the modern city'.[17]

Modernity with its thrills and chills placed increasing demands on the body. The new dangers of urban life, whether on the street or in the factory, called for physical readjustment and alertness.[18] Danger assessment was considered a rational act; those who lacked it were by extension irrational, backward, unfit for city life.[19] Railroads, part of the Porfirian restructuring of the economy, with their mechanized killings contributed to what Claudio Lomnitz calls 'the massification of death'.[20] Accidental injury and death resulting from technical malfunction or undisciplined bodies opened the gates for comical renditions. Heirs of a robust satiric press and rich graphic tradition, Porfirian wordsmiths and engravers mastered *humor negro* to document the capital's transformation.[21]

Laughter cannot be rightly interpreted unless situated in its historical context and the utility of its function determined. To 'get the joke' forces a cultural reading of public space and the technological transformation it endured, a journey into the world of symbols and memory, 'into the core of what [made] the city a site of both dreams and nightmares'.[22] Mexican scholars of humour claimed that national laughter often appears 'stained by pain, mockery or aggression',[23] therefore, its analysis must go beyond recognizing its apparent cruel nature to grasp its function 'as a resource to adapt to a specific cruel reality. An intellectual exercise seeking to overcome a certain grief.'[24]

## SETTING THE STAGE

Mexico City changed tremendously in the late nineteenth century. A city of 250,000 residents in 1880 claimed 350,000 by 1900, and close to 500,000 by 1910.[25] Its urban footprint grew from 8.5 square kilometres in 1858 to 40.5 square kilometres in 1910 with thirty-four new residential areas.[26] Contemporaneously, electrification of factory work allowed the city's industrial life to develop.[27] The capital's large consumer market along with its sizeable, relatively experienced and cheap workforce encouraged the consolidation of corporations and the advent of larger factories south of the capital.[28] By the 1890s, mule-drawn trams that had serviced the city's for decades were deemed unsuited to its needs.

*Eléctricos* were promoted as the answer for faster and reliable transportation, but anxieties ran high once plans to electrify streetcars were announced in 1896.[29] *El Imparcial* forecasted that trams would not lend themselves to the city's passenger demands.[30] Undesirable habits included dodging cars, jumping on platforms and acrobatically exiting cars while in motion. The persistence of these *imprudencias* (reckless actions) along with the trams' higher speed would bring forth numerous tragedies. Echoing this, *El Universal* made a harsh appraisal: 'Civilization has and will continue to cost many *pellejos* (lives).Those unwilling to die were to step aside!'[31]

Accidents took place as expected. A 1901 report revealed an increase in railway accidents for the previous year. *Eléctricos* were involved in a staggering 80 per cent of the total 283, with half resulting in death.[32] Deaths were likely higher since courts registered casualties at the scene of the accident, not those who perished from injuries. Thus, nowhere else does the statement that railway accidents 'introduced an unprecedented form of carnage into everyday life' take a greater dimension than in Mexico City.[33] Figure 1 compares tram accidents to three national railroad lines. The MET's numbers closely resemble those of the Mexican Central Railway. Accidents on the latter occurred along its five-million kilometres, while trams were confined to the two-hundred plus kilometres operated within the capital and suburban area. From 1904 to 1906, accidents yearly surpassed 300 with over half of the injured having no relation to the *eléctricos*' operation or service.[34] Over fifty people were killed annually, with pedestrians accounting for 80 per cent. These official statistics are quite conservative. Dr. Orvañanos placed the annual number of accidents above 600.[35] He reported a higher mortality rate for trams than the leading causes of death (tuberculosis and yellow fever). In comparison, the county of London, at the lower end of the spectrum, claimed ten killed in 1903 (pop. at over 4.5 million), while 227 were killed within the limits of Greater New York in 1906 (pop. at over 4.7 million).[36]

*Motoristas* (drivers) and pedestrians were allocated greater responsibility for accidents. Drivers were chastised for travelling at outrageous speeds, not making full stops and taking off too fast, not ringing the bell when approaching intersections, and not doing enough when an accident was eminent. They were also harangued for driving inebriated. Enticed to discover the typical driver's 'nature', newspapers, magazines, songs, poems and broadside illustrations analysed their actions.[37] Passengers and pedestrians were labelled *imprudentes* (imprudent) who committed *imprudencias* (imprudent actions). Within months of the *eléctricos*' inauguration, the report '*Imprudencias*' cited the public's carelessness as the major source of tragedies.[38]

The fixation on *imprudentes* centred on the proper use of streets, a concern with important colonial precedents. Enlightened urban planners privileged the tenet of free circulation of goods and people. Understood as one of the triumphs of individualism, they favoured the unencumbered flow of commercial life and an individual's movement through streets and sidewalks.[39] This characteristic of the modern city merged with the Porfirian public health discourse that deemed circulation one of the most effective strategies to confront social pathologies. Pedestrian and vehicular flow was privileged over social interaction.[40] The city

| Injuries and fatalities in railway accidents[1] | | | | | | | | | | | | | | | | | | |
|---|---|---|---|---|---|---|---|---|---|---|---|---|---|---|---|---|---|---|
| | | | | Injured | | | | | Deaths | | | | | | | | | |
| Year | Railway Company | Km. in operation | Total no. of passengers in all lines | Passengers | Company and government employees | Persons unrelated to the tramway's service and operation | Total number of injured individuals | ratio of injured passengers to the total number of passengers | Passengers | Company and government employees | Persons unrelated to the tramway's service and operation | Total number of fatalities | ratio of fatalities to the total number of passengers | Total number of personal injuries | Ratio of personal injuries to the total number of passengers | Ratio of personal injuries of passengers to the total number of passengers | Ratio of number of accidents in general to the kilometers in operation |
| 1904 | Tranvías Eléctricos de México | 241.006 | 42,602,194 | 63 | 55 | 196 | 314 | 1:676,225 | 2 | 3 | 52 | 57 | 1:21,301,097 | 371 | 1:114,831 | 1:655,418 | 1: o.650 |
| | Interoceánico and all its lines | 1,062,164 | 1,141,655 | 3 | 84 | 26 | 113 | 1:380,551 | ...... | 20 | 13 | 33 | ...................... | 146 | 1:7,819 | 1:380,551 | 1: 7,276 |
| | Nacional de México | 2,448,252 | 1,325,847 | 14 | 106 | 55 | 175 | 1:94,703 | 2 | 14 | 18 | 34 | 1:662,923 | 209 | 1:6,343 | 1:82,864 | 1: 11,714 |
| | Central and its lines | 5,055,422 | 2,941,998 | 60 | 174 | 86 | 320 | 1:49,033 | 13 | 34 | 48 | 95 | 1:226,307 | 415 | 1: 226,307 | 1:40,301 | 1: 12,182 |
| 1905 | Tranvías Eléctricos de Mexico | 241.54 | 47,757,440 | 88 | 48 | 218 | 354 | 1:542,698 | 5 | 6 | 46 | 57 | 1:9,551,488 | 411 | 1:116,198 | 1:513,520 | 1: 0,588 |
| | Interoceánico and all its lines | 1,020,693 | 1.481,743 | 25 | 63 | 24 | 112 | 1:59,269 | 2 | 17 | 18 | 37 | 1:740,871 | 149 | 1:9,810 | 1:54,879 | 1: 6,850 |
| | Nacional de México | 2,527,344 | 1,885,069 | 31 | 95 | 45 | 171 | 1:60,808 | 3 | 22 | 26 | 51 | 1:628,356 | 222 | 1:8,491 | 1:55,443 | 1: 11,384 |
| | Central and its lines | 5,056,028 | 3,371,281 | 70 | 198 | 92 | 360 | 1:48,161 | 14 | 33 | 57 | 104 | 1:240,805 | 464 | 1:7,266 | 1:40,134 | 1: 10,897 |
| 1906 | Tranvías Eléctricos de México | 249.986 | 54,562,725 | 123 | 40 | 203 | 366 | 1:443,599 | 6 | 7 | 55 | 68 | 1:9,093,787 | 434 | 1:125,720 | 1:422,966 | 1: 5,076 |
| | Interoceánico and all its lines | 1,188,664 | 1,483,629 | 6 | 67 | 22 | 95 | 1:247,271 | ...... | 15 | 9 | 24 | .................. | 119 | 1:12,467 | 1:247,271 | 1: 9,987 |
| | Nacional de México | 2,527,344 | 1,782,802 | 20 | 101 | 34 | 155 | 1:89,140 | 2 | 17 | 29 | 48 | 1:891,401 | 203 | 1:8,782 | 1:81,036 | 1: 12,449 |
| | Central and its lines | 5,134,728 | 3,806,536 | 38 | 247 | 79 | 364 | 1"100,172 | 2 | 34 | 118 | 154 | 1:1,903,268 | 518 | 1:7,348 | 1:95,163 | 1: 9,912 |

[1] Source: Put together by author based on data from *Reseña Histórica y Estadística de los Ferrocarriles de Jurisdicción Federal* (DF, México: Topografía de la Dirección General de Telégrafos Federales, 1907).

FIGURE 1: Injuries and fatalities in tram accidents from three national railroad lines, 1904–1906.

council echoed this when it stated, 'one of [its] first obligations . . . [was] to assure the cleanliness and elegance of the *vía pública* (public thoroughfare), and freedom of circulation for all'.[41]

The capital was notorious for its lively street life, whose lively and profuse activity was partially due to the diversity of residents who earned livelihoods on the streets. By the early twentieth century, revised traffic regulations and etiquette manuals called for constrained public conduct. Improper behaviour included outrunning cars, stopping mid-street and taking over sidewalks. Individuals attempting to outrun *eléctricos* were often run down given the toil to prevent close-range accidents. On paper, traffic regulations granted trams precedence over animal-drawn cars, coaches, carts and all other types of vehicles as well as pedestrians. Article 15 stipulated the *eléctrico* was to be given the right-of-way when *motoristas* rung the bell. Article 16 ordered *motoristas* or gendarmes to arrest violators and direct them to the district's government office to receive a penalty. In practice, their right-of-way was contested.

## *IMPRUDENTES* VS *MACHUCADOS*

Accidents became raw material for the production of jokes, songs, sardonic sonnets, cartoon strips and comical illustrations. These fed off two decades of verse and popular lyrical production that considered, questioned and engaged the railroad's tremendous social, cultural and political transformation.[42] Wordsmiths and engravers had mastered the art of reversing the symbols of elite discourse to challenge the in-kind price being paid for technological progress. Patriotic concerns pervaded humorous diatribes against foreign railway companies and American conductors. This narrative predominated *corridos* of train wrecks. This nationalist rhetoric was partially muted in the case of *eléctricos* as accidents were not only traced back to 'a rapacious foreign-owned company', but also to pedestrians and native *motoristas*. This shared responsibility placed the country's modernization and industrialization at risk.

Accidents were portrayed as a spatial dispute between pedestrians and *eléctricos*. Elitist commentary on accidents centred on the reckless individuals that took over the streets while popular outlets mocked such a portrayal. Conservative outlets such as *El Imparcial* and *El Universal* censured the lower classes' averseness to forgo 'uncultivated' behaviours. In contrast, the popular dailies such as *El Popular* and the Catholic *El País* sensationalized accidents satirizing alongside the classed gaze of the regime-friendly press.

Mexicans used language as a weapon in their battle to 'retain, or establish, legitimate title to the streets'.[43] Accident reportage labels not only exposed class prejudice but served to allocate responsibility. Conservative outlets upheld the *eléctricos*' right to the streets to the public's detriment. Employing statistics, *El Imparcial* alleged that after 'lengthy observations', the public's *imprudencias* accounted for at least 90 per cent of accidents.[44] This mirrored a federal district government's report released earlier that year. *Gendarmes* journeyed on streetcars' platforms and, after two months, their valuations clearly demonstrated that the public's *imprudencias*, and drivers who disputed the tram's right-of-way, were liable for 90 per cent of the tragedies.[45] Both sources shared their methodology to legitimate estimates. After all, statistics was the language of science, of modernity.[46]

The '*imprudentes*' these reports refer to belonged to the lower classes. The semiofficial *El Imparcial* characterized *imprudentes* as being in 'a deep state of mental engrossment and digression, so unique of the country's indigenous and poor classes'.[47] Occupying streets improperly included holding social gatherings *right on* the railways; mothers letting their *prole* (brood) romp over the rails; and drunkards and the blind falling under the cars' wheels.

With the consolidation of the Díaz regime in the 1880s, newspapers considered the social effects of modernizing and capitalist policies newsworthy items. Pro-government dailies developed a discourse of the lower classes and the issues of poverty, alcoholism, crime, marginalization, social inequalities and the maintenance of public order.[48] These also dealt with the growing presence of the lower classes on the streets. The greater visibility of the social question in the national press, nonetheless, lacked both self-criticism and discussion of state responsibility. A classed gaze at the consequences of industrialization and urbanization combined with close association to the interests of the regime's economic project blinded a true inquiry into the roles of the state and industrialists in the making of and solutions to the social question. *La gente de pueblo* did not promenade along lighted and paved boulevards. They invaded streets with their *prole*! Lacking 'education, good taste, civility, and constituting the negation of modernity, of the progress that reigned among the elites', they obstructed a modern and cosmopolitan Mexico.[49]

Popular dailies viscerally assessed the bloodshed on the streets and used humour to contest the characterization of *imprudentes*.[50] The 'disemboweling streetcars', alluding to the horrible aspect of crushed bodies, had apparently 'made a bet with the public over who would get tired first: individuals of being made *guacamole* or the *eléctricos* of making human *guacamole*'.[51] Most of the avocados pretended to occupy streets as they had previously taken over sidewalks 'taking over the space and obstructing [the tram's] way'. It was clear *mataristas* believed it was enough to ring the bell frantically without even saying 'out of my way!'. The popularity of the term *matarista*, a wordplay of *motorista* and *matar* (to kill), was evinced a few years later in the song *Los Motoristas,* which held that was their common name among unrefined people.[52]

The scientific language of statistics was dismissed in their narrative by using in-kind quantification. A note graphically estimated, 'if the flesh torn from the *machucados* was placed in a pile, it would be the height of the *Peñon*', a rocky hill northeast of the city.[53] A closer reference asked readers to imagine a pot of seafood salad and guts the size of the *Zócalo*, the city's main square. By rejecting statistics, the writer provided readers a mental map populated with recognizable guideposts to visualize the extent of the trams' carnage.

A wide spectrum of professionals and technocrats had increasingly marginalized, pathologized, criminalized and caricaturized the city's lower classes.[54] Commercial ads used visual and linguistic tropes that echoed their conclusions. For example, accidents entered the advertising cartoon strips of the tobacco company El Buen Tono, famous for mastering popular themes and the lower classes' common linguistic techniques.[55] A set of cartoon strips in 1904 cautioned against risky behaviour on streets and the cigarette's role in equipping individuals to navigate them (Figure 2). These cartoons ridiculed the manliness and technical naiveté of the *gente de pueblo*.

FIGURE 2: El Buen Tono cartoon, 1904.

## *SALVAVIDAS/SALVA-NADA!*

Mexican inventors tasked themselves with delivering a workable *salvavidas*. From 1900–1912, more than forty individuals filed patent applications, joining homegrown inventors who identified opportunities in the city's technification.[56] Much of the upstream and downstream activities of inventive work relating to *salvavidas* remain obscure. For instance, there are reports of inventors who announced public trials, trials that did not materialize or went undocumented. Thus, a granted patent does not tell much. Take the case of Don Gamboa, who in 1904 received a patent for his '*Salvavidas Mexicano*' and paid 300 pesos to Enrique Álvarez for its construction.[57] In 1909, Gamboa accused him of neither delivering it nor returning the money. For unstated reasons, a substantial number of patent applications went unprocessed.[58] Therefore, conferred patents should not be taken as evidence of inventive/innovative activity as these 'measure investment in patent rights', making them mere snapshots of a wider inventive landscape.[59] After 1900, foreigners captured about 80 per cent of the granted patents in Mexico.[60] This is reversed in the case of *salvavidas*. National inventors dominated this niche, an outlier since foreigners typically monopolized certain economic activities, such as railways.[61]

The examination of *salvavidas* adds a unique opportunity to analyse technological change on the ground and elucidate how Mexicans embraced, appropriated and invented new technologies. Its dynamics in Latin America remain relatively *terra incognita*. Limited availability and accessibility of national patent records partially explains indifference to the systematic use of patents. The perspective that the region's overall contribution has been 'largely marginal, derivative or mimetic' also explains this indifference.[62] The view that technologies 'found little engagement with local society' emanates from the understanding that imported technologies primarily served the interests of foreign investors and local associates.[63] This distortion erases 'dynamics of local invention, adaptation, and engagement'.[64]

## FOOD FOR THOUGHT

Safety devices were commonplace by the turn of the century. Cities around the world passed ordinances requiring their use. The precautions section of the MET's 1896 concession demanded protective devices, *salvavidas* or wheel guards.[65] The first cars to hit the streets in January 1900 rode without these. After a few accidents, the municipal council reminded the MET of its requirement to meet all modern improvements.[66] The railways commission pressed its compliance to guarantee public safety.[67] The company manager claimed cowcatchers were ordered, but the foreign shop's work overload prevented delivery. He promised to adapt them upon arrival.[68]

*Salvavidas* appeared in mid-April 1900. Introduced on a trial basis, the inverted 'S' shape lifeguard knocked down *obstacles* on the rails but failed to place them inside its net or brush them aside.[69] Their poor performance was promptly scorned using food imagery. On their inauguration day, it ran down driver Carlos Alcocer reportedly turning him into an '*albondiga*' (meatball).[70] The death of Felipe Estrada, among the first pedestrians to test it, was also reported in a sarcastic tone. Although the car was to 'make *picadillo* (mincemeat) out of him, the lifesaver worked, with such precision, that in less than a second, it tore his leg to pieces'.[71] Undoubtedly, it proved to be 'good . . . good for nothing!'.

Although technical language (trial basis, performance and results) permeated the debate, wordsmiths sceptically interpreted the situation in the vernacular through biting satire and speech play. *El Popular* applauded the device's 'superb results', but announced yet a better model. The model offered the advantage of 'snatching pedestrians from the street and placing them inside the rails for an instantaneous death!'.[72] Instead of lessening casualties, the fictitious model, as the real one, facilitated them. The exaggeration of the trams' physical reach speaks to the anxiety that nobody stood at a safe distance.

Speech play defined as the manipulation, conscious or unconscious, of elements of language and speech, in relation to one another and/or in relation to the social and cultural contexts of their use, serves as a tool to analyse how Mexicans expressed and lived their own language and culture.[73] References to human *guacamole* and meatballs denote the indescribable nature of mangled and disembowelled bodies. Culinary metaphors, in particular, allude to the limits of everyday speech to come to terms with the spectacle of industrial deaths.

Food metaphors ought to be considered not as mere rhetorical artefacts. Cognitive linguists argue that these should be analysed 'as a reflection of the more general human tendency to categorize abstract concept ideas in terms of the more concrete bodily experience of food'.[74] Figurative culinary uses belong 'to a system of metaphors which bring to the surface a mental mechanism that allows people to understand mental processes in terms of foodstuffs'.[75] In the case of accidents, the source domain of food allows making sense of another experiential target domain – injured bodies.[76]

The selection of meatballs, *guacamole* and *tortillas* as figurative substitutions for the bodies of victims reveals much more. First, victims were compared to avocados and maize as ingredients being transformed, ground in a way that radically altered their physicality. Throughout the city, *molcajetes* and *metates*, time-tested stone implements, technologies particular to the country's indigenous cultures, transformed these ingredients on a daily basis. The comparison was facilitated by the labelling of accidents as *machucamientos* and victims as *machucados*. *Molcajetes* and *metates* are used to *machacar/machucar* (to grind/mash) ingredients. Second, these choices were not only native products, but were representative of an indigenous diet that was under assault. Positivists such as Francisco Bulnes identified the maize-based diet as the source of the indigenous and peasants' backwardness.[77] *Eléctricos* as representative of modern technology were simply eradicating those ingredients in a more succinct manner. Third, the symbolic replacement of victims with avocados, maize and *picadillo* framed the issue via a familiar visual vocabulary. Readers would have no trouble imagining, likely repulsively so, bodily harm via familiar dishes.

## *REATA*! (A ROPE!)

Humorous language scrutinized technical solutions, ridiculing rudimentary models. It exposed the faith in *salvavidas*, but also the fact that these would not be accepted uncritically. Growing criticism of the S-shaped model led to its suppression within two months. The town council called for the immediate implementation of a new fender.[78] Within a week, San Ángel line cars donned a new fender. It consisted of a folded-up iron grill that would open by *motoristas* pulling a rope.[79] When extended, its semicircular shape covered the rails' width resembling a spoon that would pick up obstacles. Once the obstructing object was lifted, the device folded halfway preventing it from falling.

Jocular material centred on its rudimentary nature, results and selective implementation. *El País* underlined that if until then lives had been at the *motoristas*' mercy, now they literally hanged on a thread or on whether he felt like pulling it on time.[80] The fact that nothing of its mechanism was automatic was also highlighted. *El País* commented, 'were it made of iron, it would at least work to brown meat'. The invention was dated to Noah's Ark times and its half shell was useless as a body perfectly fitted under it.

Benito Muñoz-Serrano, who versified editorials for the Catholic newspaper *El País* under the pseudonym Khit, articulated a counternarrative of accidents and advocated for a technical fix. Khit criticized the *reata* model since it required the *motorista*'s intervention. He believed drivers would likely not pull the rope, as in the past they failed to brake. Thus, the cure would prove worse than the original problem. He also questioned the term *salvavidas*, a better name would be dumb killer given the proverb 'have faith in the *salvavidas* and do not run!'.[81] *El Popular* sarcastically applauded the model as 'it bounced pedestrians like balls, leaving them as a sack of bones'.[82] One reporter proposed an improved model: a pair of perfectly padded steel arms attached to the car's front end. Similar to the fictitious model that would snatch pedestrians from sidewalks, the padded-steel hands would courteously and gently pick up the individual, pay the tram's ticket, sit and light up a cigarette, saving both lives and money! 'Let no one claim,' it concluded, 'we don't have inventors such as Edison, or with Jules Verne's imagination'.[83]

Khit energetically advocated for *salvavidas* in two particular verses. *Barley on the tail of the dead donkey!* adapted a Hispanic short story whose moral held that no amount of food would revive a dead animal. It held that an inventor was 'a fool that assured he "had the secret," to stop cars from making mincemeat out of individuals'. It pondered, 'If the invention was ready/ and its efficiency had been proven/ and everyday translated into fatalities/ what explained the delay?'. The verse concluded, 'It would be a shame/ if it's successfully implemented/ when there isn't a single citizen left/ whose life to save!'. Khit also ridiculed the *reata* model's selective adoption. *One with and two hundred without* adapts the children's fable *Two frogs* by the eighteenth-century Spanish neoclassical fabulist Félix María Samaniego. The adaptation replaced the wagon that killed a frog with an eléctrico. Two scoundrels standing on their fixed spots began a conversation. The first one asked, 'Why is it that only certain trains carry *the sublime invention*? Those will save the machucado, while the rest will continue making mincemeat out of the distracted ones.'[84] Concerned about where his friend stood, he warned, 'I am afraid trams on that side don't carry them. Don't be stupid and come next to me!' The second scoundrel laughed making crude remarks. All of the sudden, a Tlálpan car passed leaving him as a *tortilla*. He cried in agony, 'by the time they adopt them in the other lines, *ya estaremos muchos masticando arcilla* (many of us we'll be chewing mud)'. This verse again calls for a prompt technical fix.

The *reata* model gave poor results. In one case, it picked up Ciriaco Cedilla, but when the rope was pulled, it forcibly threw him fracturing both arms.[85] *El Tiempo* called it a 'bloody mockery'.[86] *El Popular* celebrated its prompt removal as they 'were completing the work of the destructive wheels, leaving pedestrians in the miserable *albondiga* state'.[87] It acerbically applauded that rather than being carved with the device that made escalope out of ribs, people could go on to enjoy a more comfortable death.

José Guadalupe Posada, the period's most famous printmaker and engraver, offered his own '*Practical Salvavidas for Mexico*'. Posada dedicated several broadsheets to *eléctricos* developing his own representational schema of chaotic streets, uncontrolled technology and horrified crowds. The broadsheet below illustrated his *salvavidas* (Figure 3). A tram dominates the action at the moment when a man on the railways is about to be struck. Onlookers underline the incident's

FIGURE 3: José Guadalupe Posada's *Salvavidas* cartoon from broadsheet.

spectacularity: attentive well-dressed individuals held back on the sidewalk by ropes and a few others observing from the adjacent buildings' balconies. The tram itself has been modified with a frock-coat top-hat-wearing man replacing the missing trolley on top of the car. He holds a short sword on the air labelled *LEY* (law) and the car's reins with his other hand. The personification of the trolley, the section that powered the streetcar, speaks to the human control over the technology. A second individual stands inside the driver's cabin holding a club over a third individual knelt on the car's front end, a human fender. Held back by the reins, the latter reaches out to pick up a person that fell on the rails. The symbolism of the labelled sword is clarified by the text. Responsibility for deaths went unpunished. If captured, the *motorista* was sent to a *bartolinas* (a dark and narrow cell), only to be taken out on the third day by the company. The text offered a solution, 'let's place a *motorista* on the front end with a girdle and brake close to the ground so he can pick up whoever falls on the tracks. If he fails to and both died, it would be *justicia seca* (immediate and rigorous punishment).'

While Posada was not alone in proposing the intervention of the legal system, it is remarkable that both the image and the text advocate for holding the driver accountable, not the company. While his model failed to gain traction, three-dimensional models hit the streets.

## STREETS INTO LABS

The accidents' public nature and the ritualistic culture of Porfirian Mexico made the discussion of different models a public affair. Ceremonies to memorialize technological achievements operated as 'festivals of progress', which accidents undermined by exposing the fragility of technological modernity.[88] The *salvavidas*' quest converted streets into laboratories through public trials: rituals whose spectacularity restored, to a certain extent, the faith in technology.[89] The following four trials illustrate this.

In May 1900, Ciriaco Garcillán and the MET's general manager greeted President Díaz at the Calzada Belén, near the Indianilla powerhouse.[90] A large crowd alongside the railway witnessed an *eléctrico* arriving with Garcillán's '*Salvavidas Porfirio Díaz*' attached. Electricity was not employed for the demonstration; rather, a handful of men pushed the car until it reached considerable speed. As it approached the crowd, the confident inventor set aside the dummy prepared for the test, throwing himself across the tracks. Spectators held their breath horrified as the approaching car gained speed. A collective sigh of jubilation escaped as the *salvavidas* picked and tossed him aside unharmed. Unsatisfied, the inventor prostrated across the rails again, successfully repeating the audacity for the audience's delight. A third test run was made with the body lying over a single rail. The device violently struck the body throwing it off the rails. Fortunately, the dummy had taken Garcillán's place. The MET filed the results away.

Faustino Cervantes tested his '*Salvavidas Porfirio Díaz*' at the same location two years later.[91] The device's snout took a protruding angle made of metallic net. The edge of the angle would throw obstacles found on the rails. The wheels fractured one of the bran-filled dummy's feet on the first run. It escaped the wheels on the second run, but the front running board harmed the other foot. Cervantes made a few modifications before the final run. The dummy was picked up and carried for a certain distance. Given its poor results, *El Tiempo* offered instead an infallible lifeguard: heavy damages![92] Once it realized 'the cost of making human fricassee, [the MET] would move on its own to prevent accidents'.

Juan Vallejo filed a patent application for his own '*Salvavidas Porfirio Díaz*' in 1904 (Figure 4). Vallejo claimed it was 'the true *Salvavidas*' and although it would run down the individual, it did not injure since 'the individual would endure a soft blow on the feet'. The device's ropes were to secure the individual. If the individual failed to fall on the device, the *vigilante* (guard) would use his tongs to save him/her. The illustration accompanying the patent application depicts the individual being run down with a wide-brimmed hat, a characteristic in the representation of lower class men. *La Patria* concluded, 'no longer will there be victims. Those knocked down will walk out uninjured.'[93] It invited readers to the official trials in front of the presidential balcony.[94]

A sensational public trial brought close to five hundred curious individuals to the Indianilla in 1908. The multitude constrained the car's movement, thus engineer Leopoldo Villareal moved the trial inside the warehouse grounds.[95] The Ministry of Communications and Public Works had ordered Villareal to test all the models presented by the public. The first one, identified as 'the defender', consisted of wood boards and wire screens forming an acute angle. The former to be placed on the

**3459, provisional. Juan Vallejo,** *mexicano.*—3.ª *calle Vanegas* 3, *México, D. F.*—*Enero* 12 *de* 1904.—*Expediente* 5252.

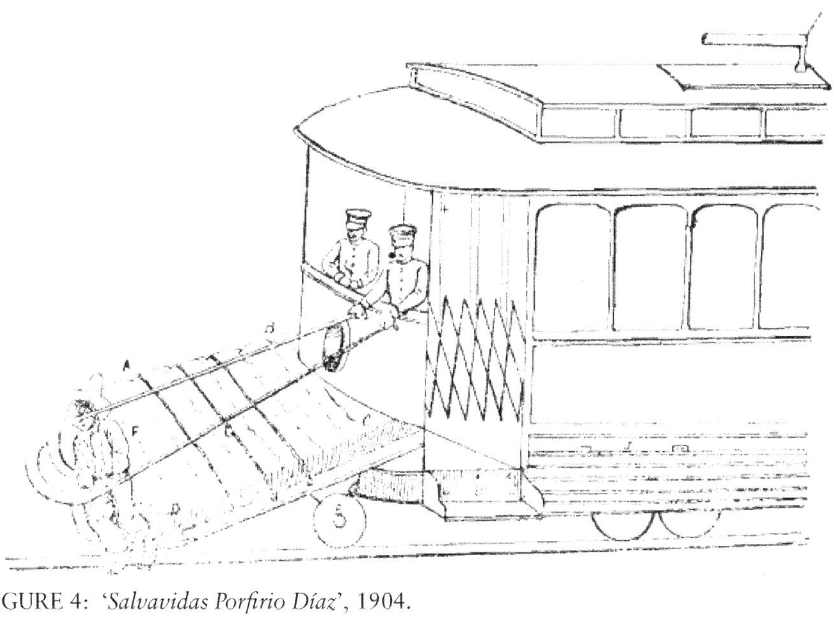

FIGURE 4: '*Salvavidas Porfirio Díaz*', 1904.

front and back end of the car, the latter covered the wheels. The sawdust-filled dummy was thrown from the rear platform, where most passengers fell. Although *El Imparcial* claimed all three tests had been successful, *La Patria* held the dummy was left in pieces after each run.[96] Demonstrating familiarity with trials, the shouting crowd called for the inventor to replace the dummy. The second device gave better results. It consisted of a wood semicircle that spun over an axle very close to the ground.[97] It gave three-quarter turns ejecting the 'future victim' to either side of the rails. The driver would lower it by pushing a button. Baptized 'Human Defense' by its inventor Francisco Nava, the device brushed objects aside while its rubber-covered circumference promised to soften the impact's blow. The first two runs left the dummy under the wheels. Nava believed the driver had not lowered it properly. After coaching the driver, the third run gave satisfactory results. Although Villareal reserved his opinion, *El Imparcial* articulated its own: '"The Defender" was better than the "Human Defense", its construction was simple and, more importantly, did not require the driver's intervention. The second model demanded the driver's alertness to lower the axle when he spotted a person on the railway. The latter to be done while trying to bring the car to a halt. It was too complicated!'[98]

What was being tried were not models but rather the faith in technological progress. Each trial brought witnesses back to the streets for a dramatic spectacle. The president himself along with technocrats, company employees, the inventor and the crowd gathered in hope for a technological fix. The irony of the handful of *salvavidas* named after Díaz is that neither the models were ever adapted nor was his regime able to soften the blow the adopted modernization model delivered on the bulk of the population. The quest for an efficient *salvavidas* continued. In 1909, the Ministry of Communications and Public Works requested particulars of the devices adopted in New York.[99] A year later, the Porfiriato's socio-economic model was put on trial.

## CONCLUSION

The spectacle of deathly machines challenged the mantra of order and progress. *Salvavidas*' public trials and their discussion restored the conviction in technological modernity: a technical fix was only a model away. This faith afforded advocates a better future by 'imagining [that a] new technology will rescue [them] from [their] persistent social ills'.[100] These fixes have an exculpatory function allowing to momentarily forgo harder questions, helping 'abrogate the responsibility of both [those] involved in the problem and the designers of the technology themselves'.[101] As long as the conversation centred on the responsibility of the *machucados* and the possibility of a technological fix, alternate conversations failed to gain traction.[102] As products of their socio-cultural context, *salvavidas* reveal the extent to which faith in technological progress was shared and how cracks to its edifice were plastered over.

The use of humour reveals ingenuity to denounce the *eléctricos*' human toll, but also a collective search for a technical solution. The company's and government's slow and unsuitable responses along with the discourse that had rationalized casualties as the price for progress generated indignation among certain sectors.

*Salvavidas* were crafted as decisive answers to the tumultuous present. Employing a recognizable grammar, comedic criticism became an ideal venue to harshly expose and ridicule this violence. This criticism, however, adhered to a technification outlook not far removed from that of the *científicos*. As in the case of Nicolas Zuñiga y Miranda, a wise fool who in the late nineteenth century paraded through the streets with his far-fetched inventions and earthquake predictions; humour and the *salvavidas* quest reminded Mexicans that not all inventions performed as advertised and that technology did not have to be imported.[103] Cautioning against questionable behaviour, or acting as a mechanism of protest or ridiculing rudimentary models, laughter renewed the 'courage to live with reality rather than instilled the bravery to dream new worlds'.[104]

The fender quest both exposed the fragility and worked to reassert the legitimacy of technological progress. Turning streets into laboratories, public trials serve to reinforce the unmitigated hope that a device could be out there, that it was achievable, that Mexicans were busy perfecting it.

## NOTES AND REFERENCES

1. Anna Alexander, *City on Fire: Technology, Social Change, and the Hazards of Progress in Mexico City, 1860–1910* (Pittsburgh: University of Pittsburgh Press, 2016).
2. *El Popular*, 27 August 1903.
3. About 116 kilometres of 240.
4. Vic Gatrell, *City of Laughter: Sex and Satire in Eighteenth Century London* (New York: Walker & Co., 2006): 5.
5. In calling for the study of cartoons, Jon Agar underlined that more people will see a cartoon than will ever read a treatise on technology, 'Technology and British Cartoonists in the Twentieth Century', *Transactions of the Newcomen Society* 74, 2 (2004):181.
6. Anton Rosenthal, 'Spectacle, Fear, and Protest: A Guide to the History of Public Space in Latin America', *Social Science History* 24, 1 (2000): 38. Rosenthal places trams in the region's urban imaginary in 'The Streetcar in the Urban Imaginary of Latin America,' *Journal of Urban History* 42, 1 (2016): 162–179.
7. David Nye, *Electrifying America: Social Meanings of a New Technology, 1880–1940* (Cambridge: MIT Press, 1992): ix.
8. Lisa Gitelman, *Scripts, Grooves, and Writing Machines: Representing Technology in the Edison Era* (Stanford: Stanford University Press, 1999): 77.
9. Linda Degh, *American Folklore and the Mass Media* (Bloomington: Indiana University Press, 1994). Also see Trevor J. Blank, *Folk Culture in the Digital Age: The Emergent Dynamics of Human Interaction* (Boulder: University Press of Colorado, 2012).
10. Peter Soppelsa considers both the material and symbolic appropriation in his 'Reworking Appropriation: The Language of Paris Railways, 1870–1914', *Transfers* 4, 2 (2014): 104–123.

11. William Beezley, 'Foreword', in Stephen Neufeld and Michael Matthews (eds.), *Mexico in Verse: A History of Music, Rhyme, and Power* (Tucson: University of Arizona Press, 2015): viii.
12. Alan Dundes, *Cracking Jokes: Studies of Sick Humor Cycles & Stereotypes* (Berkeley: Ten Speed Press, 1987).
13. For nuclear era humour, see Daniel Wojcik, *The End of the World as We Know It: Faith, Fatalism, and Apocalypse in America* (New York: NYU Press, 1997).
14. Ben Singer, *Melodrama and Modernity: Early Sensational Cinema and Its Contexts* (New York: Columbia University Press, 2001): 66. Also Gregory Shaya, *Mayhem for Moderns: The Culture of Sensationalism in France, c. 1900* (PhD diss., University of Michigan, 2000).
15. Singer, *Melodrama and Modernity*: 67.
16. For this commodification of violence, see Rielle Navitski, *Public Spectacles of Violence: Sensational Cinema and Journalism in Early Twentieth-Century Mexico and Brazil* (Durham: Duke University Press, 2017).
17. Singer, *Melodrama and Modernity:* 74; for sensational reporting and suicide, Kathryn Sloan, *Death in the City: Suicide and the Social Imaginary in Modern Mexico* (Oakland: University of California Press, 2017).
18. Scott Bukatman discusses its prevalence among American cartoonists in *The Poetics of Slumberland: Animated Spirits and the Animating Spirit* (Oakland: University of California Press, 2012).
19. For neurasthenia and suicide, see Sloan, *Death in the City*.
20. Claudio Lomnitz, *Death and the Idea of Mexico* (New York: Zone Books, 2005): 377–381.
21. Maria Elena Díaz, 'The Satiric Press for Workers in Mexico, 1900–1910: A Case Study in the Politicisation of Popular Culture', *Journal of Latin American Studies* 22, 3 (1990): 497–526; Fausta Gantús, 'La ciudad de la gente común. La cuestión social en la caricatura de la Ciudad de Mexico a través de la mirada de dos periódicos: 1883–1896', *Historia Mexicana* 59, 4 (2010): 1247–1294. For a discussion of how the satirical press worked to educate and validate the sentiments of working-class men see Robert M. Buffington, *A Sentimental Education for the Working Man: The Mexico City Press, 1900–1910* (Durham: Duke University Press, 2015).
22. Rosenthal, 'Spectacle, Fear, and Protest', 38.
23. Martha Munguía, *La Risa en la Literatura Mexicana* (Mexico: Bonilla Artigas Editores, 2012).
24. Rafael Barajas, El Fisgón, *Sólo me rio cuando me duele: La cultura del humor en México* (Mexico: Editorial Planeta Mexicana, 2009).
25. For the city's transformation, see Claudia Agostoni, *Monuments of Progress: Modernization and Public Health in Mexico City, 1876–1910* (Mexico: UNAM, 2003).
26. Agostoni, *Monuments of Progress.*

27. Tony Morgan, 'Proletarians, Politicos, and Patriarchs: The Use and Abuse of Cultural Customs in the Early Industrialization of Mexico City, 1880–1910', in William Beezley, Cheryl Martin and William French (eds.), *Rituals of Rule, Rituals of Resistance: Public Celebrations and Popular Culture in Mexico* (Wilmington: SR Books, 1994): 151–171. For early industrial development, see Gustavo Garza Villareal, *El proceso de industrialización en la Ciudad de México (1821–1970)* (Mexico: Colegio de México, 1985).
28. John Lear, *Workers, Neighbors, and Citizens: The Revolution in Mexico City* (Lincoln: University of Nebraska Press, 2001): 59.
29. For a social history, see Georg Leidenberger, *La historia viaja en tranvía: el transporte público y la cultura política de la ciudad de México* (Mexico: UAM, 2011). For the jurisdictional conflict between municipal and federal authorities, see Ariel Rodríguez Kuri, *La Experiencia Olvidada: El Ayuntamiento de Mexico: Política y gobierno, 1876–1912* (Mexico: Colegio de Mexico, 1996).
30. *El Imparcial*, 3 October 1899.
31. *El Universal*, 16 January 1900.
32. Based on the Federal District's courts' statistics, *El Imparcial*, 18 January 1901.
33. Michael Matthews, *The Civilizing Machine: A Cultural History of Mexican Railroads, 1876–1910* (Lincoln: University of Nebraska Press, 2014): 146.
34. Matthews finds this pattern for railway accidents. *The Civilizing Machine:* 145.
35. *Gaceta Médica de México*, Vol. V, 2a Series, No. 1 (Mexico: Tipografía y Litografía de Juan Aguilar-Vera y Co., 1905): 12–14.
36. Accident data from John Fox, 'The Needless Slaughter by Street-Cars', *Everybody's Magazine*, March 1907. Population data from *Quarterly Publications of the American Statistical Association* XIII, 97 (1912): 105.
37. For a thorough discussion, see Diana Montaño, *Electrifying Mexico: Cultural Understandings of a New Technology, 1880s–1960s* (PhD diss., The University of Arizona, 2014).
38. *El País*, 17 April 1900.
39. Mario Barbosa, *El trabajo en las calles: Subsistencia y negociación política en la Ciudad de Mexico a comienzos del siglo XX* (Mexico: Colegio de Mexico, 2008): 55–56.
40. Barbosa, *El trabajo en las calles.* A mechanism to prevent crowd formation and thus confront the fear of the masses.
41. *Boletín Municipal*, September 1902, cited in Susie Porter, '"And That Is Custom Makes It Law": Class Conflict and Gender Ideology in the Public Space, Mexico City, 1880–1910', *Social Science History* 24, 1 (2000): 121.
42. See Michael Matthews's chapter in *Mexico in Verse.*
43. Peter Norton, 'Street Rivals: Jaywalking and the Invention of the Motor Age Street', *Technology and Culture* 48, 2 (2007): 332.
44. *El Imparcial*, 30 October 1901.

45. *La Voz de México*, 15 March 1901.
46. Mauricio Tenorio-Trillo, *Artilugio de la nación moderna: México en las exposiciones universales, 1880–1930* (Mexico: FCE, 1998): 173–184.
47. *El Imparcial*, 7 August 1901.
48. Gantús, 'La ciudad de la gente común': 1252.
49. Gantús, 'La ciudad de la gente común': 1272.
50. For an examination of how satiric penny press editors capitalized on the tension between proper and improper behavior, see Buffington, *A Sentimental Education*.
51. *El Popular,* 27 August 1903. Also referred to *guacamole* of Christians. *El Popular*, 7 June 1903. An 1898 dictionary defined *guacamole* as a stew of avocado and *mole*.
52. Antonio Vanegas Arroyo, *Canciones Modernas* (1906): 13. On the city's linguistic transformation, Tenorio-Trillo posits 'the language of the city was, as was the city itself, corrupt, promiscuous, uncontrollable, unpredictable, rich, creative, and, at times, virtuous', where 'the middle- and upper-class language was itself transformed by inevitable contacts with the vulgar eloquence of the street', in *I Speak of the City: Mexico City at the Turn of the Twentieth Century* (Chicago: University of Chicago Press, 2012).
53. *El Popular,* 27 August 1903.
54. For the criminalization of the working class' use of public space, see Pablo Piccato, *City of Suspects: Crime in Mexico City, 1900–1931* (Durham: Duke University Press, 2001), for public health concerns, see Agostoni, *Monuments of Progress.*
55. Steven Bunker, *Creating Mexican Consumer Culture in the Age of Porfirio Díaz* (Albuquerque: University of New Mexico Press, 2012): 44–45.
56. For advertising, see Bunker, *Creating Mexican Consumer Culture*; for fire prevention/control, see Alexander, *City on Fire*.
57. *El Tiempo*, 10 February 1909.
58. *Lista dispuesta por orden de clase y subclases de las patentes* (Mexico: Imprenta y Fototipia de la Secretaria de Fomento, 1912): 35. Nine were also abandoned.
59. See Edward Beatty, 'Patents and Technological Change in Late Industrialization: Nineteenth-Century Mexico in Comparative Context', *History of Technology* 24 (2002): 133.
60. Beatty, 'Patents and Technological Change': 136.
61. Beatty, 'Patents and Technological Change': 138–140.
62. A current project seeks to address this http://www.ibcnetwork.org/project.php?id=46
63. Edward Beatty, Yovanna Pineda and Patricio Saiz, 'Technology in Latin America's Past and Present: New Evidence from the Patent Records', *LARR* 52, 1 (2017): 139.
64. Beatty et al., 'Technology in Latin America's Past and Present': 139.
65. *Diario Oficial,* 24 February 1900.
66. *Diario Oficial,* 24 February 1900.
67. Archivo Histórico del Distrito Federal (AHDF)/Ayuntamiento/Ferrocarriles/box 1044/file 367, 23 January 1900.

68. AHDF/Ayuntamiento/Ferrocarriles/box 1044/file 367, 31 January 1900.
69. *El Imparcial*, 16 April 1900.
70. *El Popular*, 18 April 1900.
71. *El Popular*, 13 May 1900.
72. *El Popular*, 18 April 1900.
73. Joel Sherzer, 'On Puns, Comebacks, Verbal Dueling, and Play Languages: Speech Play in Balinese Verbal Life', *Language and Society* 22, 2 (1993): 217.
74. Irene Lopez-Rodriguez, 'Are We What We Eat? Food Metaphors in the Conceptualization of Ethnic Groups', *Linguistik Online*, [S.l.], v. 69, n. 7, Sep. 2014. ISSN 1615-3014. Verfügbar unter: <https://bop.unibe.ch/linguistik-online/article/view/1655/2798>. Date accessed: 12 Feb. 2018. doi:http://dx.doi.org/10.13092/lo.69.1655
75. Lopez-Rodriguez, 'Are We What We Eat?'.
76. For a thorough discussion, see Antonio Barcelona (ed.), *Metaphor and Metonymy at the Crossroads* (Berlin: Mouton de Gruyter, 2003): 3–7.
77. Jeffrey Pilcher, *Que vivan los tamales! Food and the Making of Mexican Identity* (Albuquerque: University of New Mexico Press, 1998).
78. AHDF/Ayuntamiento/ Ferrocarriles Urbanos/1044/391/1901. Fines were established for *motoristas*' violations, but not the company's. *El Popular*, 15 December 1901.
79. *El Imparcial*, 20 December 1901.
80. *El País*, 5 January 1902.
81. *El País*, 22 December 1901.
82. *El Popular*, 5 January 1902.
83. *El Popular*, 5 January 1902.
84. *El País*, 21 December 1901.
85. *La Patria*, 4 January 1902.
86. *El Tiempo*, 4 January 1902.
87. *El Popular*, 11 March 1902.
88. Term coined by Matthews.
89. For the power of rituals see Beezley et al., *Rituals of Rule, Rituals of Resistance.*
90. *La Voz de México*, 2 June 1900.
91. *El Popular*, 9 June 1902.
92. *El Tiempo*, 27 July 1902.
93. *La Patria*, 24 November 1904.
94. No record has emerged. *La Patria*, 24 August 1904.
95. *El Imparcial*, 9 May 1908.
96. *La Patria*, 10 May 1908.
97. *La Patria*, 10 May 1908.
98. *El Imparcial*, 9 May 1908.

99. AGN/SCOP/Box 144/reg. 25/file 530/792 (1909).

100. Tarleton Gillespie, *Wired Shut: Copyright and the Shape of Digital Culture* (Cambridge: MIT Press, 2007): 2.

101. Gillespie, *Wired Shut*: 2.

102. Physical conditions (traffic density, trams' size and speed, sharp curves, poor grading on roads), managerial reforms or the company's legal responsibility.

103. William Beezley, 'Mexican Sartre on the Zócalo: Nicolas Zuñiga y Miranda', in William Beezley and Judith Ewell (eds.), *The Human Tradition in Latin America* (Wilmington: SR Books, 1997): 68. The history of technology in Mexico has demonstrated the importance of following technologies through a cultural lens. In addition to Alexander, *City on Fire* and Matthews, *The Civilizing Machine*, see Rubén Gallo, *Mexican Modernity: The Avant-Garde and the Technological Revolution* (Cambridge, MA: MIT Press, 2005); Justin Castro, *Radio in Revolution: Wires Technology and State Power in Mexico, 1897–1938* (Lincoln: University of Nebraska Press, 2016); and Araceli Tinajero and Brian Freeman (eds.), *Technology and Culture in Twentieth Century Mexico* (Tuscaloosa: University of Alabama Press, 2013).

104. Beezley, 'Mexican Sartre': 74.

# Bringing Communication to the Countryside: Rural Telephony in Latin America, 1900–1985

CHRISTIANE BERTH

This article analyses how rural telephony developed in Latin America through local demand, national modernization schemes and global development plans. I argue that each change in large technological systems created a need to readapt telecommunication systems to isolated regions with little traffic volume. This process led to interactions between foreign experts, Latin American engineers, government bureaucrats, local technicians and telephone users. In this story, we find histories of technological dependency and inventive appropriation. As a result, a new kind of 'entangled knowledge' evolved based on bringing together sets of knowledge from different origins.[1]

Despite power imbalances between the Latin American nations and foreign donors, the history of rural telephony is not just a story of technological diffusion from north to south. Beginning in the 1970s, Latin American actors organized to develop alternative systems based on low-cost equipment. In particular, larger nations, such as Mexico, Colombia and Brazil built up research centres and specialized training through national science and technology development programmes.

Although the first explicit programmes for rural telephony originated in the 1950s, the history of rural telecommunications dates back to the early twentieth century. During this period, Latin American governments built up lines to administrative centres as well as to regions important for export trade. With an expanding telephone service, the demand for lines increased by the 1930s. However, investing in rural lines was not profitable for private enterprise. The price for the installations was high and only small groups of solvent users existed. Hence, the firms demanded community participation. Villages had to build up local infrastructure, mobilize their workforce and appeal to government institutions for financial support. This model persisted throughout the twentieth century.

With urbanization and international migration, the demand for a telephone service in rural areas expanded. I argue that rural telephony was both a means to extend state power and also a means of empowerment for local communities. A government institution's telephone call could order police actions against rural guerrillas or request data on population structure. At the same time, citizens could also call for help, organize resistance against powerful landowners, or reduce their dependency on local merchants.

Research on rural technologies in Latin America is still scarce and has concentrated on the agricultural sector.[2] In this article, I propose to write the history of rural technologies as an entangled history. Rural communications formed part of global development programmes but also played a role in internal civilizing missions. State actors from the centre frequently considered rural people as backward, a recurrent theme also in scholarly depictions of villages opposed to modern technologies.[3] By contrast, I argue that there was considerable activity at the local level to mobilize to telecommunications. People founded pro-telephone committees, initiated telephone cooperatives and petitioned to the state authorities. In their attempts to convince higher authorities, rural politicians upheld the narrative of technological progress.

The existing research on Latin American telephone systems has not taken into account these local activities. A volume edited by economist Eli M. Noam and communication scientist Cynthia Baur provides an overview of the different historical pathways of the telephone sector in Latin America focusing on large enterprise. The editors highlight inequality in access and deficient networks as the main characteristics of Latin American telecommunications. The few historical articles on Latin American telephone history mainly deal with the relationship between national governments and foreign telecommunication multinationals.[4] Beyond that, some institutional histories of state firms and private companies exist.[5] To conclude, studies on smaller firms and local telephone systems are rare as well as social and cultural histories of telephone technology in Latin America.

Research on rural telephony in Latin America is nearly non-existent. Most studies arose in the context of foreign development cooperation. In addition, contemporary telecommunication engineers have written about technological systems and network planning, which I utilize as historical sources. This article also draws on documents from Mexican and Chiapanecan state archives. The sources from the Mexican National Archives illustrate how the rural communities petitioned authorities for access to the telephone system. Documents from German development cooperation introduce the activities and perceptions of foreign experts on rural telephone systems in Colombia and Paraguay. I also consider official reports of foreign donors, such as the International Telecommunication Union (ITU) and the World Bank, analysing rural telecommunications from a comparative perspective. Finally, I include reports of Latin American state programmes for rural telephony. Most of these archival documents look from a top-down perspective at a rural telephone service and exclude the perceptions of local operators or users.

I begin the article with an examination of an early telephone network built in Chiapas beginning in the late nineteenth century without foreign technological expertise. This network is an interesting case: a region marginalized by Mexican development plans that implemented its own system. The Chiapanecan example demonstrates how state administration, export business and local firms were the first users of rural telephony. By contrast to all other systems analysed in this article, state institutions in Chiapas mainly used the telephone for written correspondence. They sent around written messages, so-called *telefonemas*, to communicate important events, government decrees and administrative instructions. Next, I show how rural demand for telecommunications increased in the 1930s. Third, I analyse different paths to rural telephony throughout Latin America. With the turn to public telecommunication monopolies, state programmes for rural communications developed. In some countries, telephone cooperatives developed as an alternative model. Rural citizens initiated these cooperatives, which adapted better to rural users' needs than centralized state programmes.

Next, I discuss international donors' motivations for supporting telecommunications on the continent. Using the examples of Colombia and Paraguay, I conclude that externally financed projects failed to benefit rural communications when donors competed or national governments changed priorities. In the fifth section, I discuss the impact of satellite technology on rural telecommunications. Although the technology promised cheaper access to isolated regions, it took over a decade until experts developed viable solutions for low-cost earth stations. In the last section, I evaluate the few existing developmental studies on telephone use in rural areas. These show that community representatives highlighted the telephone's usefulness for emergencies and agricultural activities, whereas statistics reveal a majority of calls for social purposes.

## A STEP TO REGIONAL INDEPENDENCE: RURAL TELEPHONY IN CHIAPAS, 1896–1930

The local telephone network in Chiapas was exceptional for two reasons. First, it was constructed and maintained without strong technical expertise. Second, the state administration used the telephone mainly as a medium of official written communication. The independent system survived more than fifty years until it was connected to the national telephone network in 1958. Although historians have often characterized early telephone history as a history of urban networks serving a small elite and middle-class clientele, this interpretation overlooks the history of local companies and early rural networks that existed in different Mexican regions. As neither the Mexican government nor the foreign telephone companies were interested in investing in telephony in Chiapas, the local government took the initiative.

For the Mexican government, Chiapas was a marginal region in the far south without strong economic potential. Hence, it assigned low priority to expanding national infrastructures to this area, except for border security, for which the telegraph network was sufficient. The foreign telephone companies considered Chiapas as no promising investment. When expanding their long-distance networks, they concentrated on expansion to the industrial north and the US border, which promised higher revenues. Hence, the Chiapanecan government decided to launch its own telephone system. This was a step to more independence: first, independence from external interferences through the national government and second, independence from the companies' commercial priorities for new lines in profitable areas with high traffic volume. While the Chiapanecan government had to negotiate each change in telegraph lines with the national authorities, it could determine the telephone system autonomously.

Within only twelve years, the network in Chiapas expanded significantly and covered all administrative districts with at least one station. It is still unclear who launched the initiative for an independent telephone network as archival documents for the early years are scarce. The government started to install lines in roughly 1896 and 1897. The first lines connected the most important cities, San Cristóbal de las Casas and Tuxtla Gutiérrez, with smaller settlements of around 2,500 to 3,500 inhabitants.[6] By 1909, the network covered sixty-three official telephone stations and thirty-nine private connections. The total length of lines had expanded to 1,507 km.[7] At first, the new network served mainly official purposes. But soon local plantation owners also requested access, which was generally conceded for a monthly rate of four Pesos. In addition, local entrepreneurs and hotels were connected to the telephone network. While the government relied mainly on written messages, private clients also used the medium for oral communication.

The early network was based on simple constructions with heterogeneous technical equipment. It consisted of modest telephone stations connected by open-wire lines. It is unclear who advised the local government on the technology for the first installations. Sources from the 1920s indicate heterogeneous equipment from different suppliers. Among others, the government ordered articles from Ericsson and Siemens as well as General Electric.[8] As the transmission capacity was still limited, intermediary stations were necessary to transmit the messages over long distances. This benefited small settlements that otherwise probably would have remained without a telephone service. However, these had a very low traffic volume of ten to twenty messages a month.[9]

Correspondence on technical problems and maintenance shows that the staff's technical knowledge was rudimentary. Most problems occurred as a result of interrupted lines or broken telephone sets. A central telephone office in the state capital Tuxtla Gutiérrez administered the network and ensured maintenance. Generally, the documents on technical matters were not written in technical jargon. For example, the state government lacked knowledge of the technical models of telephone equipment. Once it ordered four telephone exchanges from Siemens without any specifications. Filled with consternation, the firm representatives requested more information on the network in order to provide adequate equipment. Instead of providing further explanations, the government sent an example and asked Siemens to supply the same type. Hence, government authorities lacked the adequate technical vocabulary to order equipment and replacement parts.[10] In a similar way, local telephone operators reported on broken equipment without providing technical details. As these operators were unable to fix technical problems, they sent the defective apparatus to the central telephone office, where it was repaired. In 1926, the central telephone office's deputy director commented that repair attempts by local telephone operators were generally fruitless.[11] Before starting their work, these operators had received no technical training. Hence, the maintenance of the network was a process of learning by doing with expertise centralized in the Tuxtla Gutiérrez office. It was mainly the office's telephone inspectors that improvised solutions with the heterogeneous equipment. By doing so, they created a knowledge adapted to the specific local circumstances, or in the words of historian Leida Fernández Prieto, an 'island of knowledge'.[12]

In addition to untrained staff, insufficient financial resources, the tropical climate, frequent robberies and local resistance also affected the telephone network. Politicians often complained that it was impossible to send urgent communications as the lines were frequently interrupted.[13] Heavy rains and strong wind frequently made telephone lines fall from the stakes, which interrupted the network until an attendant put it up again. Due to geography and scarce roads, line maintenance was difficult, time-consuming work. Sometimes, the local authorities required local workers to participate in the repair works; sometimes they were unable or unwilling to provide a workforce.[14] Moreover, frequent robberies of wire caused interruptions. By the mid-1920s, the problem had become so large that the governor's secretary general suggested the establishment of a reward for any leads on thefts.[15] Historical research has interpreted the robbery of wire as an expression of resistance against the telephone.[16] While this might be true, archival sources provide no explanations for why people in Chiapas took the wire from the lines. It remains unclear if they wanted to interrupt the network, use wire for other purposes, or rejected the telephone as such. Overall, maintenance was like Sisyphus' mythic punishment: once a problem was fixed, another problem showed up. Throughout the year, the telephone inspector, attendants and local labour force worked on reparations while complaints about inadequate telephone operators never ceased.

The government tried to find trustworthy people for the position of telephone operators as confidential government correspondence and private communications ran through their office. Customers accused them of abandoning their workplace, ignoring messages, making errors in written correspondence, and even drunkenness. The local authorities had difficulty in finding good operators as they only offered low salaries. In the early 1920s, a telephone operator received eight Pesos a month and had to work multiple jobs.[17] Both technical problems and unreliable operators caused severe delays in official communication.

During the years from 1910 to 1930, the 1910 Mexican Revolution resulted in serious destruction, which downsized the telephone system. Frequently, the military or rebels took over the telephone stations and destroyed lines to cut their opponents off from information. The state authorities tried to reconstruct the network as soon as possible but lacked financial resources. Hence, in 1926 a local politician concluded that the 'demonstration of progress' remained only as a memory.[18] In fact, statistics show a reduced size for 1930 with 814.5 km of lines connecting eighty-six telephones.[19] The reduced telephone network became a symbol of the general instability during political transition. During the 1930s and 1940s, more communities requested access to the network although its technical problems continued. This was also the trend in other Mexican regions.

## THE TELEPHONE AS 'CONTRIBUTION TO LOCAL PROGRESS': MOBILIZING FOR ACCESS IN MEXICAN RURAL COMMUNITIES

Rural communities from all over Mexico demanded telephone lines beginning in the 1930s. By then, the majority of telephone lines were concentrated in Mexico City. There as well as in the rest of the country, two multinationals operated separate networks: Ericsson and ITT (International Telephone and Telegraph Corporation). In the early twentieth century, these companies expanded to other large cities but also covered smaller towns. With expanding service, rural communities became aware of the new medium and petitioned the Mexican president to provide them with access.

The political context of the 1930s favoured rural mobilization for telephones. The petitioners were inspired by President Lázaro Cárdenas' (1934–1940) stronger attention to rural Mexico, which raised hopes on access to land and better living conditions. In addition, Secretary of Communication Francisco J. Múgica strengthened the idea of the telephone network as a public service that had to serve people's needs. His public statements encouraged rural people to petition for network expansion.

The requests show that local community leaders perceived access to the telephone network as an indicator for belonging to the modern Mexican nation. Beginning in the mid-1930s, petitions from all over the country reached the presidency. For example, representatives of the distant chamber of commerce in Acuitzio del Canje, Michoacán, insisted that they needed communication with the country's centre. In that sense, they perceived the access to the network as a 'contribution to local progress'.[20] In a similar way, many petitions referred to the telephone as an expression of 'civilization' and modernity.[21] Other communities stressed the need to save money and time as well as to combat isolation.[22] In other cases, the communities felt threatened by local landowners. To defend their land rights, they needed a direct communication to the political centre, argued *ejido* members from Omitlán, Guerrero in their petition.[23] In some regions, the inhabitants founded local pro-telephone committees. These associations campaigned with political authorities for a telephone service and organized community work for the installations. It is difficult to determine the extent of support for the new medium in local communities. Some petitions, however, included many signatures, which meant local political leaders might have forced community members to sign.

As rural telephony did not generate large profits, the telephone companies did not invest in line installations and hence required communities to mobilize workers for the installations. Ericsson, for instance, let the communities construct their own line connecting to its network. However, local people lacked sufficient resources to pay for wire and telephone sets, which was another motive for petitioning the president. For example, the community Concordia Vicente in Guerrero offered the 'enthusiastic cooperation' of the local workforce but requested material support for the installation.[24]

In general, Ericsson offered two different services to rural places: the rural subscriber service with a public phone or general service. For general service, the company installed a telephone exchange that allowed connecting private clients and required formal authorization by the Ministry of Communication. Ericsson charged the communities a rate for connection to the next telephone exchange. Petitions from the National Archives show that communities frequently perceived these rates as being too high.[25] The inhabitants lacked sufficient income to pay for such expensive telephone calls.

In most cases, it is impossible to conclude from existing documents why the government supported certain communities and denied support in other cases. Most likely, political favouritism played an important role as well as budget concerns and distance to the telephone network. Unfortunately, the statistics for the 1930s only distinguish between Mexico City and the rest of the country. Consequently, it is impossible to analyse if the number of telephone lines in rural areas increased significantly as the statistics do not differentiate between urban and rural areas. Given the high installation costs and lack of solvent clients, it is fair to assume that companies concentrated on expanding in urban areas.

Nonetheless, the petitions indicate that there was a strong rural demand, at least among the communities' leadership. Local leaders interpreted the telephone as symbolizing 'civilization'. Being part of the telephone network meant being part of the modern Mexican nation. Whether or not the rest of the community shared this vision is questionable. While local elites embraced the idea of modernization through technology, the local workforce built the basic infrastructure. Expensive rates, however, made it almost impossible for most people to make calls. Nonetheless, some rural inhabitants mobilized for access to telecommunications under favourable political conditions as they hoped better communication with national authorities would improve their living conditions.

The tide turned from policies favouring rural areas to policies benefiting Mexican cities with President Manuel Ávila Camacho (1940–1946) who put stronger emphasis on urban Mexico and industrial development. In this new political environment, urban users felt encouraged to request telephone lines from the presidency while petitions from rural zones decreased. Similarly to Mexico, early Latin American statistics provide no data on rural telephone density. In the next section, I will identify general trends in telecommunications and describe initiatives for expanding the service to rural areas in the 1950s.

## UNEVEN EXPANSION: STATE PROGRAMMES AND COOPERATIVES FOR RURAL TELEPHONY IN LATIN AMERICA, 1940s TO 1980s

Throughout the twentieth century, the expansion of Latin American telephone networks was a slow, uneven process interrupted by short growth periods. In the larger nations, such as Mexico, Argentina and Brazil, service expanded significantly in the 1920s and 1930s. During the 1940s, the growth of telephone networks slowed down but picked up again from the 1950s onward. Part of the uneven development were sharp interregional differences between Brazil, Argentina, Mexico, Venezuela and Colombia and the rest of the continent, where the telephone service expanded at a slower pace.[26]

From the 1940s to the 1980s, large state monopolies dominated Latin American telephone networks although in some countries local enterprises survived. Both were important for expanding networks to rural zones. In the 1940s and 1950s, many governments decided to create state monopolies for telecommunications and appointed political allies as managers. These managements sometimes lacked specialized skills and suffered from frequent personnel changes. In consequence, decision-making on network expansion became more political. In general, many of these enterprises failed to fulfil demand for a telephone service. Network expansion also decelerated when governments used telecommunication revenues for other state projects.[27] Simultaneously, local companies persisted in many countries, whose history has not been explored. Evidence from Mexico, Argentina and Bolivia indicates that local companies remained important for telecommunication in rural areas throughout the twentieth century. The local firms also contributed to the technological heterogeneity characterizing many rural telephone systems.

The performance of rural telephony depended on the technological equipment, attention by central telecommunication authorities, and the local operators. The typical rural telephone infrastructure between the 1950s and the 1980s was either a small manual exchange with twenty to thirty subscribers or a HF radio connection.[28] Radio telephony connected isolated areas where a physical line was difficult to install. Nevertheless, it suffered from many deficiencies, such as spectrum overcrowding. For example, in a Peruvian project the stations with VHF radio telephones remained out of service during one-third of the year throughout the early 1980s. If central administrations did not ensure regular maintenance, technical problems could persist for long periods.[29] Telephone stations out of service or with deficient call quality turned from symbols of modernity into symbols of neglect. In addition to technical problems, urban operators frequently gave calls from rural telephone offices lower priority. Also, the telephone operators determined the quality of the telephone service. If people considered operators as unreliable or invading their privacy, they hesitated to make calls. In regions inhabited by different ethnic groups, communication problems also occurred when the operator did not understand indigenous languages.[30]

The first rural telephone projects in Latin America originated in Mexico and Argentina but under totally different premises. In Argentina, beginning in the late 1950s, a movement of local cooperatives offered a telephone service in rural communities. Although there is no systematic historical analysis of this phenomenon, case studies permit the following preliminary conclusions: first, many cooperatives benefited from Italian immigrants' knowledge of cooperative organization. Second, telephone users financed the service through fees and their own labour. Third, the cooperatives were more aware of local needs and adapted their service and rates accordingly. Last, they sometimes introduced new technologies faster than the large national telecommunication firms. By 1975, there were already 175 cooperatives in Argentina with around 17,000 telephone subscribers.[31] Hence, cooperative organization encouraged the spread of 'entangled knowledge' in telecommunications as cooperatives exchanged among each other and with telephone users.

The Mexican case, by contrast, is an example of state-led expansion of rural telephony with centralized planning. In 1959, a presidential decree assigned the Ministry of Communication the responsibility for a rural telephone service. This measure was the starting point for the first initiatives. The state commissions for rural telephony followed a tripartite agreement model. Three actors – the Ministry of Communication, the state government and the local governments – signed agreements for the establishment of telephone lines. Generally, the national and regional government supported the construction financially, whereas the communities contributed labour and poles to the effort. During 1966 and 1967, the Commission for Local Telecommunications installed twenty-seven new telephone lines. The Ministry of Communication contributed with more than 1.4 million Pesos, the state governments with 2.3 million Pesos, and private persons with 579,000 Pesos.[32]

Official state sources and local politicians presented the network in the 1960s and 1970s as a success story of modernization. At the same time, urbanization increased the demand for a telephone service in rural areas. In 1966, the telephone company's magazine *Voces de Teléfonos de México* introduced its readers to a modern landscape in the countryside characterized by microwave stations and telephone poles now unifying the Mexican population. Once a new telephone line was constructed, an inauguration ceremony took place. Generally, a local representative made a symbolic first call to a higher political authority. In 1970, the politician Mario Ruiz Aburto from Tarímbaro in Michoacán called the state government to inaugurate long-distance service. During the conversation he stated: 'Now, civilization has arrived in my community.'[33] In 1967, Telmex managers, the state governor and the local priest inaugurated a long-distance service in San Miguel Canoa, north of Puebla. In his speech, the priest highlighted a telephone service as a solution for local emergencies as well as an opportunity for better communication within Mexico and to other countries.[34] Although global connectedness was probably not so important for local telephone users, with increasing migration to large cities or the United States, long-distance communication became more urgent. This was also true for other Latin American nations, such as Colombia, Ecuador and Peru, where the governments launched plans for rural telecommunications. These programmes generally benefited from loans by international donors.

## BETWEEN PLANNING EUPHORIA AND LOW-COST EQUIPMENT: THE CONTRIBUTIONS OF DEVELOPMENT COOPERATION

Beginning in the 1960s, international donors provided credits and foreign experts for the development of telecommunications in Latin America. A global planning euphoria and post-war development ideas shaped these projects. At the same time, the differences between 'developed' and 'undeveloped' or 'First World' and 'Third World' defined new roles in international cooperation.[35] Based on the firm conviction that technological modernity was a precondition for economic development, the World Bank and the Inter-American Development Bank (IADB) provided loans for telecommunication projects. For example, the World Bank signed over ninety-three loan agreements for telecommunications between 1962 and 1983. In Latin America, it provided the first loans to Costa Rica, El Salvador, Venezuela and Colombia.[36] Initially, rural areas had no priority in these funding schemes, but rising global attention in the 1970s to rural development led to more financial resources for rural telephony.

According to the World Bank's focus on economic growth, the bank's telecommunications projects emphasized three principles: first, telecommunications should operate in a profitable way; second, they should provide business clients and international investors with good conditions for their transactions; and third, national telecommunication firms should take over smaller local companies to create strong enterprises. To make telecommunications more efficient and profitable, the World Bank suggested higher rates, investment in new technologies and staff reduction at telecommunication enterprises. These measures favoured business clients over private users and urban areas over rural communities.[37] However, in the 1970s, the World Bank renewed its rural development policy, which opened opportunities for rural telecommunications.

With donors' rising awareness of rural development efforts, the telecommunication loans expanded to include rural telephone programmes. At the World Bank, its president Robert McNamara (1968–1981) promoted changing the bank's strategy. In his vision, growth alone would not lead to a significant reduction of poverty. Hence, bank projects would have to benefit the poor directly, for example, through providing credits to small farmers.[38] As a result, the telecommunication loans increased attention to rural areas. However, these programmes only made up a small proportion of the entire loan.[39] At the Inter-American Development Bank a similar turn occurred: early IADB projects in Latin America had supported long-distance microwave networks while those of the 1970s brought rural telephony to the fore.[40] The IADB established the following criteria for its telecommunication projects: connection to national development plans, use of low-cost solutions, establishment of public phones in rural communities, and meeting the demands of the productive sector. Beyond IADB and the World Bank, the ITU and bilateral donors were also involved in telecommunication development projects.

The ITU provided experts for telecommunication projects and, although hesitantly, acted as a centre for knowledge about rural telecommunications. In addition, the organization supported telecommunication training centres in Global South countries in the 1960s. During the 1940s and 1950s, the organization had been reluctant to support development projects. For decades, African, Asian and Latin American delegates had demanded that ITU respond to their specific needs for low-cost equipment and technological solutions for serving rural, isolated areas with an extreme climate.

Early on, Latin American nations demanded more support for rural telephony projects. At the 1970 meeting of the Inter-American Telecommunication Commission, for example, members asked ITU for a regional seminar on rural telecommunications. The event took place in Managua that same year.[41] At these types of meetings, experts exchanged knowledge within the region but also with international organizations and firm representatives. The ITU experts contributed as a new group to the 'entangled knowledge' of rural telecommunications. Their knowledge complemented and coexisted with Latin American engineers' expertise. At the next Latin American workshop in 1974, the sixty-six representatives urged ITU to provide more experts and publish a regular bulletin on rural telecommunications.[42]

In addition, participants argued for equipment without expensive royalties as a step towards 'more technological independence'.[43] Probably not all delegates shared this vision given the political heterogeneity within the region at that time, but the concern at the workshop was strong enough to enter the ITU report.

From the 1960s to the 1980s, new technologies changed Latin American telecommunications networks significantly. Microwave networks, satellite systems and digital technologies offered new opportunities for designing cheaper rural telephone networks. Whenever new technologies were introduced, foreign technicians played a key role. For example, a German expert mission between 1963 and 1968 supported the design of a long-distance network in Colombia based on a microwave system. Such a system transmitted telephone and television signals through microwave transmitters and receivers. Prior to its installation, a detailed geometrical analysis and test measurements of hops were necessary.[44] The German experts conducted a survey for twenty-two hops over a distance of 1,500 km. Their work contributed to a larger telecommunication project financed by the World Bank. In the course of fieldwork, local conditions challenged the experts' original calculations.

During their mission, the experts faced problems with measuring instruments and vague data. First, the expert group struggled with inexact maps. Sometimes, the maps and real terrain differed by 500 metres in altitude, which forced the group to adapt their routes. Second, the high temperatures as well as the transport on mules affected gauges, sometimes leading to inexact results.[45] Finally, some hops had already been occupied by other projects, which was a result of insufficient coordination by the Colombian telecommunication authorities.[46] If a network was constructed based on erroneous data, this affected transmission and service quality later. In the end, the Colombian government assigned the network construction to a Japanese firm that worked with rougher measurements than the Germans. Hence, the German Embassy in Bogotá doubted if the measurements, produced during the first five years of intense work, would ever be used.[47] As a result of German–Japanese competition, Colombian engineers incorporated different sets of knowledge into their network planning. Next, experts adapted plans designed in urban headquarters in the field to local realities, which added to the 'entangled knowledge' of telephone systems. The expert team's thirty-seven reports demonstrate that they had to change their original plans designed in Bogotá headquarters frequently. Finally, each expert group had its own approach to network planning and measurements, which complicated cooperation. International organizations were aware of these conflicts, but silenced them in their official reports.

International donors generally stick to neutral, apolitical language when promoting the development effects of telecommunications. However, establishing telephone communications with remote communities was a political act. Consequently, governments considered not only economic, but also political and security factors when making decisions about network expansion. This aspect remains at the margins of planning documents but needs to be considered in any analysis of rural telephone projects. Especially in societies characterized by political violence, such as Colombia, Guatemala or Peru, security concerns were an important motivation for connecting isolated zones to telecommunications. For example, a project for rural telephony in

Eastern Colombia mentioned as one aim to assure 'control and security'.[48] However, rural telecommunication could also contribute to local empowerment. As shown in my analysis of early Mexican rural telephony, communities perceived the telephone network as improving their self-protection. Take the case of communities reporting on human rights abuses, asking for support in case of robberies, and using the telephone for social mobilization.[49] Further research needs to address how Latin American governments used telecommunications for extending political vigilance against opposition groups. This is also relevant for Paraguay, where an autocratic government commissioned a national telecommunications plan in the 1960s.

In 1963, the Paraguayan government requested United Nations Development Programme (UNDP) support for the National Telecommunication Plan. At that time Paraguay was governed by dictator Alfredo Stroessner (1954–1989). His regime was characterized by political repression, corruption and social control through the official party. After an initial ITU mission, German development cooperation supported the plan with credits and technical experts. Between 1969 and 1977, the Paraguayan telecommunication firm ANTELCO received loans of a volume of 44.1 million German marks. In addition, German development cooperation supported the project with more than 4.4 million German marks.[50]

Ultimately, externally financed telecommunication projects in Paraguay failed to benefit rural communities. While the Paraguayan side officially claimed that it would provide a telephone line to all population groups, it used the credit mainly to expand urban telephone networks. Over the years, the Paraguayan government manipulated the donors successfully: step by step it changed expansion goals from interior regions to the capital, which German development cooperation approved without follow-up questions. At that time, German telecommunication projects clearly supported Siemens business interests. The firm had already supplied the Paraguayan telephone network from 1955 onwards and benefited substantially from the loan. By the late 1970s, however, the situation had changed.

When German development cooperation established sustainability criteria, prior funding was revised. In June 1978, the German Ministry of Development Cooperation reported on the Paraguayan project with disastrous conclusions. According to the survey, the project had mainly served German export interests. Mostly Paraguayan elites had benefited from the new installations in urban areas.[51] Moreover, the installations were oversized. Paraguay had spent more money on telecommunication equipment than necessary, especially on new telephone exchanges, which the report characterized as too grand. In fact, the government documentation of the National Telecommunication Plan mainly consisted of colour photos showing new, spacious buildings.[52] Finally, the German side recommended to end support.[53] The example demonstrates how some Latin American governments only paid lip service to development goals while following their own agenda for network expansion. After the end of German support, the Paraguayan government obtained several Japanese credits for telecommunications. Among others, Japanese firms built an earth station that connected Paraguay to the global satellite network in 1978.[54] As satellite technology offered new opportunities for the integration of rural zones, Latin American governments intensified rural telecommunication plans in the late 1970s.

## NEW OPPORTUNITIES, BUT FEARS OF TECHNOLOGICAL DEPENDENCY: THE IMPACT OF SATELLITE TECHNOLOGY

At first, the United States dominated the technological development and international agreements for satellite technology. Communication satellites were developed in the early 1960s. The satellites communicated via electromagnetic signals with earth stations and hence could transmit television signals and telephone calls over long distance. At the height of the Cold War in 1962, the US designed a new global communication system bypassing international organizations. The Satellite Communications Act created a new private enterprise, COMSAT, which played a leading role in establishing Intelsat, the International Telecommunications Satellite Consortium. It achieved a global coverage in the late 1960s and focused in the beginning on international telecommunications.[55]

Intelsat members owned a part of the satellite infrastructure according to their investment shares and traffic volume. They paid for satellite communications according to their traffic volume. From the mid-1970s on, Intelsat leased special capacities for domestic services.[56] By 1970, the large Latin American economies had joined Intelsat, such as Colombia, Argentina, Peru, Brazil and Mexico. Smaller countries followed in the 1970s and 1980s, when the costs for leasing circuits had decreased. However, Latin American feelings about the new system remained ambiguous: they needed satellite access to expand broadcasting facilities but also feared technological dependency.[57]

On the one hand, the satellite system created opportunities for telephone networks in isolated regions. To connect a community by an earth station was cheaper than to build a telephone line, especially when geographic conditions made access difficult. On the other hand, Intelsat mainly provided technology for high traffic countries and put regions with medium or low traffic intensity at a disadvantage. In its early stage, Intelsat required large expensive earth stations, which made access for poor countries difficult. For instance, the equipment ordered for a USAID project in rural Peru in the early 1980s turned out to be $600,000 more expensive than initially calculated. By renouncing a turn-key agreement and modifying earth stations' design, USAID and ENTEL (Empresa Nacional de Telecomunicaciones del Peru) managed to reduce the cost during the negotiations with the supplier. As a consequence, they could install fewer stations than planned and based their budget on a cheap Hungarian radio telephone system that caused many technical problems later on.[58] Again, the combination of different technologies contributed to an 'entangled knowledge' although in this case technicians found no convincing solution to ensure compatibility. Nonetheless, this knowledge prevented other projects from making the same mistakes and inspired experts to search for technical alternatives.[59]

Contemporary researchers concluded that a telephone network for rural communities needed small, cheap earth stations with low electricity consumption.[60] From the late 1970s onwards, different research institutions and enterprises worked on viable solutions for rural telephone networks, among them the Stanford Communication Satellite Planning Center. They were particularly successful in adapting earth station design to rural systems' needs.

Over the years, other industrial countries acquired knowledge about satellite communication, which led to a diversification of foreign expertise and increased competition. Although initially the US dominated technological knowledge, European nations and Japan quickly caught up. In the 1970s, other countries had worked on satellite systems to become independent from US dominance. For example, Canada introduced its Telsat system in 1972, followed by the regional ARABSAT system in 1976 and the European EUTELSAT in 1977.[61] Although the systems competed against each other, knowledge on satellite communication became increasingly entangled when applied in rural Latin America. Foreign advisers brought their technological solutions to the region while some Latin American nations launched their own research and education programmes.

In particular, European and Japanese experts advised Latin American administrations on satellite technology. In the 1980s, German experts worked in Colombia on a proposal for a telephone network in Eastern Colombia based on satellite technology. They highlighted the need to adapt small earth stations to high temperatures and high humidity. In addition, they recommended to develop solar energy supply solutions for the stations. After submitting their proposal to the government, the German engineers feared that Japanese experts would take on the project.[62] In a situation where technological solutions for rural systems were still in development, foreign experts embarked on different solutions and competed against each other. In the background was a fierce competition among firms to win over market segments in the satellite sector. Even Intelsat offered by 1983 a special service for low volume telecommunications and cheaper earth stations.[63]

In Latin America, again the large countries were at the forefront of satellite initiatives. Mexico and Colombia, for example, first joined Intelsat, but worked simultaneously on their own solutions. The Colombian SATCOL failed after four years. Designed in 1978 to save the high Intelsat rates, the new president Belisario Betancur (1982–1986) did not take on the initiative. By contrast, Mexico launched its first two Mexican satellites (Morelos I and II) in 1985. In this country, the telecommunication administration was concerned about dependency on foreign experts and had started its own education programme for space communication beginning in 1979.[64] Mexico also successfully developed expertise for rural telephony. In 1972, the Ministry of Communication founded the CIDET (Centro de Investigación y Desarrollo de Telecomunicaciones), which official sources declared as a step towards technological independence.[65] The magazine of Dirección Nacional de Telecomunicaciones, *Teledato* published regular reports on the technological advances.[66]

With the spread of satellite technology and increasing international attention to rural development, the planning of rural networks in Latin America advanced significantly. Initially, foreign expert knowledge was important, but over time Latin American experts generated more knowledge to serve the technological needs of rural communities. The larger countries with their own research facilities particularly contributed to this 'entangled knowledge'. In Mexico, international meetings influenced the rural telephone programme. During the 1970s, Mexican experts participated in international workshops on rural telecommunications and brought back new ideas on system planning.

The Mexican rural telephone programme grew significantly in the early 1970s for two reasons: first, *Teléfonos de México* had become a state enterprise in 1972, which facilitated cooperation on state development projects. Second, the government of President Luis Echeverría dedicated more attention and financial resources to the Mexican countryside.[67] Between 1973 and 1975, the government invested 36.2 million Pesos in the rural telephone programme that focused on areas with a large indigenous population: the Chiapanecan highlands and the Sierra Tarahumara and the huicot region located in Northern Mexico.[68] Héctor Arellano, Director of the Commission for Rural Telecommunications, characterized these regional programmes in 1978 as a positive exception with strategic planning. In general, the rural programme had created a map with islands of telephone service instead of a coherent network. Hence, the selection of communities for a telephone service lacked long-term planning. As already outlined above, political criteria sometimes dominated network expansion.[69]

During the next period of rural telephone planning in the late 1970s, expert debates at international conferences and foreign planning tools transformed Mexican rural telecommunications. Héctor Arellano had participated in a 1978 ITU conference on telecommunications in isolated and underprivileged regions. In his contributions for the magazine *Teledato*, he referred to Canada as a model for bringing a telephone service to rural communities, as the country had established its own satellite system for connecting the large number of isolated rural communities within the country. In addition, he reported that Mexican engineers had worked with a modified version of planning software developed at Stanford University.[70] This was another example of 'entangled knowledge' as Mexican engineers improved on Stanford's software. If Stanford researchers followed-up on these modifications is still unclear.

*Teledato* also served as a forum for expert knowledge. It reported on new technical solutions for rural networks, such as a measuring device for rates in rural public phones. Sometimes, engineers from other Latin American telecommunication administrations contributed articles. Hence, we can see that Mexican actors drew on foreign tools for network planning and technical solutions developed in other Latin American countries but adapted them to their own needs.

Official publications presented the network expansion as a success story for incorporating isolated regions into modern Mexico. For example, the magazine *Voces de Teléfonos de México* regularly published reports on the establishment of new telephone offices with headlines such as 'Telephone service is reaching all corners of Mexican provinces' or 'Our aim: to unite and serve Mexico'.[71] Hundreds of photos of first calls, inauguration ceremonies and new buildings in the countryside represented this version of the story of telephony in Mexico. In contrast to this triumphant narrative, Héctor Arellano reported in 1978 that service had remained deficient in many areas. Hence, there was also a story of neglect that would produce a different narrative if told from the communities' perspective. Unfortunately, little evidence on use and the appropriation of rural phones is available for the Mexican case. This was different for other rural telephone projects started in the late 1970s in Latin America.

## SURVEYS ON TELEPHONE USE IN RURAL LATIN AMERICA, 1970s TO 1980s

For the first time, projects in the late 1970s and 1980s started to collect data on telephone use. These surveys indicate that rural people perceived the telephone as important during emergencies, for access to information and for staying in contact with family members. For example, twenty prominent personalities of Salto de Guairá in Paraguay explained to German experts in 1978 that the telephone permitted fast access to prices, foreign exchange rates and the reliability of bank checks.[72] Agricultural settlers in Eastern Paraguay raised similar arguments: the latest price information made them less dependent on local merchants. However, the German report remarked critically that a rural user with low income would have spent one-third of his monthly income for a three-minute call to Asunción.[73] As foreign experts often interviewed local elites and telephone users, the perspective of the poorest sector remained unconsidered. They also fell short in analysing long-term social changes after the introduction of a telephone service.

Overall, the surveys documented intensive use for social and business purposes. However, some sectors still perceived no social necessity for telephone calls while others abstained from calls due to high prices and poor quality. In 1974 and 1975, the University of Costa Rica and the World Bank conducted a survey among 674 telephone users in three rural villages. The research team found out that people made more than 50 per cent of calls for conversations with their relatives and friends. Other important reasons to make phone calls were commerce and agriculture as well as government administration. The team also analysed the social composition of users and concluded that public call offices were beneficial for all social groups.[74]

However, other studies showed a social imbalance.[75] A 1985 Peruvian investigation is important, as the team conducted its survey over a period of two years. By contrast to the Costa Rican study, it revealed social differences: the average user had a higher income, was well educated and had a professional position or owned a business. In addition, the team distributed a household survey in two places to include non-users with the following results. In general, awareness of the telephone service was widespread: 97 per cent of all interviewed persons were aware of the telephone service. In total, 58 per cent of all interviewees had made a telephone call. Those who had not, mentioned as a main reason that they lacked relatives or friends in other communities. They also saw no reason for calling friends in other places. Interviewees mentioned the high price and poor service quality as well.[76] Unfortunately, the authors of the surveys did not reflect on the telephone's impact on communication habits and the long-term social changes it caused in the communities. The experts were mainly interested in data to request more funding or facilitate decision-making on network expansion.

## CONCLUSION

In the early twentieth century, the telephone reached rural communities mainly for political and administrative reasons. Early telephone networks formed part of Latin American nations' ambitions to stimulate export-led growth and foster national integration. The technological systems were still basic and faced challenges related to climate, untrained staff and robberies. For private companies, the expansion to rural areas was not particularly lucrative, which is why they required communities to construct their own lines and pay high rates.

With telephone networks in the larger Latin American economies expanding significantly in the 1930s, more people in small towns and rural regions became aware of the telephone service. Local elites perceived the new medium as a symbol of technological modernity and an indicator for belonging to a 'civilized' nation. Despite growing demand, only a minority of rural areas gained access during this period. During the 1940s, many governments prepared for the nationalization of telecommunications. With national monopolies, it became easier to establish state rural telephone programmes. These contributed significantly to knowledge production on rural telecommunications.

Another important initiative for connecting the countryside came from below: in several nations, local cooperatives filled the vacuum left by private telecommunication firms. These cooperatives were in close contact with users and hence better served their needs. Hence the knowledge generated was more entangled with users' experiences than the knowledge of state programmes. Parallel, microwave technology allowed for the expansion of long-distance telephony so that local networks could connect to the national system. With urbanization and international migration rural people needed to communicate with relatives over long distances. However, high rates and service deficiencies impeded long, regular conversations.

Beginning in the 1960s, international donors provided loans for telecommunications projects to Latin American countries. In the first decade of international cooperation, the telephone became part of national development plans and was perceived as a means to economic development. With foreign loans, foreign experts also arrived in the region, working on measurements, staff education, planning or system design. Their reports frame access to the telephone as a neutral development mission, but decision-making related to network expansion was highly political. Instead of systematic planning, political considerations as well as economic priorities created a system of small connected islands in large seas of villages without telecommunications.

With donors' turn to rural development in the 1970s, more financial resources became available for the expansion of communications to the countryside. At the same time, satellite technology offered new opportunities for connecting isolated places. In the beginning, the satellite caused a stronger technological dependency on the US and foreign expertise. Only the large nations were able to launch their own satellite projects and develop technological expertise to substitute the foreign specialists. During that process, the engineers adapted technologies to local conditions and tried to find low-cost solutions. The financial crisis of the 1980s limited ambitious expansion plans while the first digital telephone exchanges reached the countryside.

The telephone connected actors at the global, national, regional and local levels. Multinationals and international organizations interacted with Latin American politicians and state bureaucracy. Sometimes together, sometimes in opposition they framed ideas of development, planning and control. These ideas shaped the design of rural telephone systems. The expansion plans involved technicians and engineers from diverse backgrounds. ITU missions intermingled with local engineers, the staff of foreign multinationals, national monopolies' bureaucracy and technicians. These actors met at telecommunication schools, international conferences, training workshops and in the field. There, they also encountered telephone operators and users but at least the foreign advisers frequently limited their conversations to local elite members. The adaptation of foreign technologies occurred in the capitals' laboratories and telecommunications administration offices as well as in isolated villages.

Climate, geography conditions and spontaneous changes in political priorities prompted improvisation and learning by doing. Through the international organizations the knowledge travelled within Latin America and to other world regions. Hence, the knowledge on rural telecommunications can be understood as 'entangled knowledge'. With the large technological systems (microwave and satellite), a stronger dependency on foreign technologies went hand in hand. As companies promoted competing solutions, Latin American engineers had difficulties ctaching up with knowledge and dealing with heterogeneous systems. Tensions occurred when expert teams competed against each other and governments acquired equipment based on foreign policy criteria instead of technical needs. Also, the acquisition of expensive foreign equipment strained limited state budgets. Especially early bilateral cooperation ignored social development aims and acted as market support for national telecommunication firms.

Petitions and demand from rural communities continued throughout the whole twentieth century. However, the telephone also faced non-use and resistance. Telephone service was expensive for rural inhabitants and frequently obliged locals to work in line construction. Future research should locate more historical sources illustrating the perspectives of rural operators, users and non-users.

In official discourse, the telephone was a symbol of technological modernity and progress. It promised integration into the nation through being connected. However, an inconsistent service and technological problems undermined this ideal. A radio telephone, being one-third of the year without service, could not fulfil the high expectations generated by politicians and engineers. Instead these telephones became symbols of neglect. Many rural communities remained disconnected and had to wait until the mobile phone filled the gaps in rural telecommunications in the late twentieth and early twenty-first century.

## NOTES AND REFERENCES

1. Gesa Mackenthun and Klaus Hock, 'Introduction: Entangled Knowledge, Scientific Discourses and Cultural Difference', in Klaus Hock and Gesa Mackenthun (eds.), *Entangled Knowledge: Scientific Discourses and Cultural Difference*, Cultural Encounters and the Discourses of Scholarship 4 (Münster: Waxmann, 2012): 7–28. The authors provide no definition of 'entangled knowledge' but argue that a social history of intercultural knowledge needs to acknowledge the 'plurality and multiplicity of knowledges' (p. 9). In a similar way, historian Harald Fischer-Tiné introduced the notion of pidgin-knowledge as a contact knowledge in continuous transformation. Harald Fischer-Tiné, *Pidgin-Knowledge: Wissen und Kolonialismus* (Zürich: Diaphanes, 2013): 13.

2. Yovanna Pineda, 'Farm Machinery Users, Designers, and Government Policy in Argentina, 1861–1930', *Agricultural History* 92, 3 (2018): 351–379; Billie R. DeWalt, 'Appropriate Technology in Rural Mexico: Antecedents and Consequences of an Indigenous Peasant Innovation', *Technology and Culture* 19, 1 (1978): 32–52.
3. For an overview on Mexican debates, see Edward Beatty, *Technology and the Search for Progress in Modern Mexico* (Berkeley: University of California Press, 2015): 20–21.
4. Rosanna Ledbetter, 'ITT: A Multinational Corporation in Latin America During World War II', *The Historian* 47, 4 (1985): 524–537; Arturo D. Grunstein, 'In the Shadow of Oil: Francisco J. Múgica vs. Telephone Transnational Corporations in Cardenista Mexico', *Mexican Studies* 21, 1 (2005): 1–32; Marcelo Bucheli and Erica Salvaj, 'Reputation and Political Legitimacy: ITT in Chile, 1927–1972', *Business History Review* 87 (2013): 729–755; Marcelo Bucheli and Erica Salvaj, 'Adaptation Strategies of Multinational Corporations, State-Owned Enterprises, and Domestic Business Groups to Economic and Political Transitions: A Network Analysis of the Chilean Telecommunications Sector, 1958–2005', *Enterprise and Society* 15, 3 (2014): 534–576.
5. Asociación Hispanoamericana de Centros de Investigación y Empresas de Telecomunicaciones, *Las telecomunicaciones en Hispanoamérica: Pasado, Presente y Futuro* (Madrid: Comprint, 1993); Oscar Szymanczyk, *Historia de las telecomunicaciones en la República Argentina: Servicios de comunicación audiovisuales; radiodifusión y teledifusión; tecnologías y políticas, estatización, privatización, oligopolio, nacionalización y competencia* (Buenos Aires: Dunken, 2011); Juan Balsevich, *Historia de las telecomunicaciones en el Paraguay (1864–2002)* (Asunción: AGR S.A, 2011); Teléfonos de México S.A., *Historia de la telefonía en México 1878–1991* (México, D.F: Scripta, 1991).
6. *Anuario Estadístico del Estado de Chiapas 1908* (Tuxtla Gutiérrez: Tipografía del Gobierno, 1909): 9–12.
7. *Ley de Ingresos y presupuesto de Egresos que regirán en el ejercicio fiscal de 1896* (Tuxtla Gutiérrez: Imprenta del Estado, 1895); Fernando B. Corzo Nájera, 'El desarrollo de los medios de comunicación en Chiapas durante el Porfiriato: los telégrafos y los teléfonos 1875–1911' (Tesis de Licenciatura: UNICACH, 2006): 68–70.
8. Secretario General a Daniel Villegas, 9 January 1926, Exp. 617, Tomo X, 1926, Archivo Histórico de Chiapas, Universidad de Ciencias y Artes de Chiapas (AHC).
9. See, for example, Cortes de Caja, Tomo XIII, Exp. 318, Sección Fomento, AHC.
10. Siemens Schuckert al Gobierno del Estado, 6 October 1909, F. 423; Secretario General a Siemens, 20 October 1909. Exp. 20, Sección Fomento, AHC.
11. Oficial Mayor Encargado al Telefonista de Colgante B. Domínguez, sin fecha, 1926, F. 2. Exp. 642. Sección Fomento, AHC.
12. She introduced the term for analysing agricultural knowledge in Latin America and the Caribbean. The regions of agricultural production or 'islands of knowledge' created, adopted and applied scientific methods from different traditions. Leida Fernández Prieto, 'Islands of Knowledge: Science and Agriculture in the History of Latin America and the Caribbean', *Isis* 104, 4 (2013): 788–797.

13. Presidente Municipal de Villa de Acala al Secretario General, 14 August 1923, Foja 6. Tomo X, Exp. 102, Sección Fomento, AHC.
14. For the 1940s, documents prove active resistance against the work recruitments.
15. Secretario General al Inspector General de Teléfonos, 4 January 1926, F. 71, Exp. 642, 1926, Tomo X, Sección Fomento; Francisco Coutiño al Secretario General, 28 January 1925, F. 14. 1925, Tomo I, Exp. 3, Sección Fomento, AHC.
16. Jörg Becker (ed.), *Fern-Sprechen: Internationale Fernmeldegeschichte, -soziologie und -politik* (Berlin: VISTAS, 1994): 166–172.
17. L. Estrada al Secretario General, 1 November 1921, F. 17. Exp. 222, Tomo VII, Sección Fomento, AHC.
18. Presidente de la C. Permanente del 3er Congreso de Ayuntamientos al Gobernador, 21 April 1926. Exp. 657, Sección Fomento, AHC.
19. *Informe rendido por el gobernador constitucional de Chiapas C. Ing. Raymundo E. Enríquez: ante la XXXIII legislatura del Estado el 1° de nov. de 1930* (Tuxtla Gutiérrez: Talleres Tipográficos del Gob. del Estado): 29–30.
20. Cámara Nacional de Comercio e Industria, Acuitzio del Canje al Secretario de Comunicaciones y Obras Públicas, 11 August 1939. 512.2/315. Fondos Presidenciales, Lázaro Cárdenas del Río, Archivo General de la Nación (AGN).
21. Junta Popular de Mejoras Materiales, Paracho al Presidente, 31 May 1937. 512.2/110; Heliodoro Hernández al Presidente, 3 June 1937. 512.2/130. Fondos Presidenciales, Lázaro Cárdenas del Río, AGN.
22. Extracto: Jesús N. Arratia al Presidente, 26 May 1939. LCR 512.2/288; Modesto Zarza al Presidente, Xoconuzco, 13 January 1935. 512.4/1. Fondos Presidenciales, Lázaro Cárdenas del Río, AGN.
23. In Mexico, the term *ejido* was used for communal land ownership, abolished in the mid-nineteenth century. After the revolution, land reforms again assigned beneficiaries communal plots of land. Presidente del Comisariado Ejidal, Omitlán al Presidente, 16 November 1936. 512.42/36. Fondos Presidenciales, Lázaro Cárdenas del Río, AGN.
24. Raúl Castellanos al Secretario de Comunicaciones y Obras Públicas, 7 June 1938. 512.2/202. Fondos Presidenciales, Lázaro Cárdenas del Río, AGN.
25. Secretaría de Comunicaciones y Obras Públicas a Teodoro Martínez, 11 May 1939. 512.2/255, Ernesto Flores, Unión Autónoma de Peluqueros, Jalapa al Presidente, 12 July 1939. 512.2/304. Fondos Presidenciales, Lázaro Cárdenas del Río, AGN.
26. In many countries, statistical data on the early twentieth century is scarce. See, for an overview the Latin American Economic History Database (http://moxlad-staging.herokuapp.com/home/es). By 1960, average telephone density was still low with 4.4 per 100 inhabitants in Argentina, 1.7 in Colombia and 1.6 in Venezuela. ITU, *World Telecommunication/ICT Database 2017* (Geneva: ITU, 2017).
27. Eli M. Noam and Cynthia Baur, 'Introduction', in Eli M. Noam (ed.), *Telecommunications in Latin America*, Global Communications Series (New York: Oxford University Press, 1998): xi–xxviii.

28. High frequency (HF) radio covers the frequency range between 3 and 30 Megahertz. Also, VHF/UHF connections were used frequently. VHF frequencies range from 30 to 300 Megahertz, UHF from 300 to 3,000 Megahertz.
29. Ricardo A. Criscolo, 'Rural Telecommunications in Latin America', *IEEE Transactions on Communications* 24, 3 (1976): 325–329; John K. Mayo, Gary R. Held and Steven J. Klees, 'Commercial Satellite Telecommunications and National Development: Lessons from Peru', *Telecommunications Policy* 16, 1 (1992): 67–79.
30. Robert J. Saunders, Jeremy J. Warford and Björn Wellenius, *Telecommunications and Economic Development*, 2nd edn, A World Bank publication (Baltimore: Johns Hopkins University Press, 1994): 246–250.
31. Criscolo, 'Rural Telecommunications'; Susana Finquelievich, 'Las cooperativas de telecomunicaciones y la democratización social: TELPIN, un estudio de caso de organización comunitaria de la sociedad de información', *Revista de Estudios Sociales* 22, diciembre (2005): 37–47; Celeste de Marco and Talia Gutierrez, 'Las cooperativas de servicios y el medio rural, estudios de caso, Saladillo y Colonia Urquiza (Buenos Aires), 1970–2010' (XII Jornadas Nacionales y IV Internacionales de Investigación y Debate: Economía social y cooperativismo en el agro hispanoamericano: Territorio, actores y políticas públicas, Buenos Aires, 2015), http://jornadasrurales.uvq.edu.ar/media/public/Ponencia_De_Marco_-_Gutierrez.pdf
32. Secretaría de Comunicaciones y Transportes, 'Informe de Labores: 1 de septiembre de 1966 al 31 de agosto de 1967' (México, D.F., 1967): 40–41.
33. *Voces de Teléfonos de México* 105 (1970).
34. 'San Miguel Canoa, Pue., ya ve, oye, camina y habla', *Voces de Teléfonos de México* 6, 63 (1967).
35. Nick Cullather, 'Development? It's History', *Diplomatic History* 24 (2000): 641–653; Arturo Escobar, *Encountering Development: The Making and Unmaking of the Third World*, paperback reissue, with a new preface by the author (Princeton, NJ: Princeton University Press, 2012).
36. For a complete list from 1962 to 1989, see Saunders, Warford and Wellenius, *Telecommunications*: 417–421.
37. Gwen Urey, 'Infrastructure for Global Financial Integration: The Role of the World Bank', in Bella Mody, Johannes M. Bauer and Joseph D. Straubhaar (eds.), *Telecommunications Politics: Ownership and Control of the Information Highway in Developing Countries* (Mahwah, NJ: Erlbaum, 2012): 113–134; Gerald Sussman, 'Banking on Telecommunications: The World Bank in the Philippines', *Journal of Communication* Spring (1987): 90–105.
38. Patrick Allan Sharma, *Robert McNamara's Other War: The World Bank and International Development, Politics and Culture in Modern America* (Philadelphia: University of Pennsylvania Press, 2017).
39. International Bank for Reconstruction and Development, Appraisal of the Second Telecommunications Project of the Empresa Nacional de Telecomunicaciones, Colombia, 10 March 1971. World Bank, Project Completion Report, Colombia, Fourth Telecommunications Project (Loan 1450-CO), 22 June 1989.

40. Robert F. Gellerman, 'Telecommunication Activities of the Inter-American Development Bank', in *Workshop on Special Aspects of Telecommunications Development in Isolated and Underprivileged Areas of Countries June 26–28, 1978*, Ottawa, Canada: 209–219.
41. 'Technical Co-operation: Seminar on Rural Telecommunications', *Telecommunication Journal* February (1971): 66–68; 'Fifth Session of CITEL', *Telecommunication Journal* November (1970): 735–736.
42. 'Seminar on Rural Telecommunications [English version]': Summary of Discussion and Conclusion Quito, Ecuador 2–13 September 1974' (Geneva: ITU and United Nations Development Programme, 1974): 24.
43. 'Seminar on Rural Telecommunications [English version]': 8.
44. A hop is the connection between two microwave sites through radio waves.
45. 12. Bericht, Misión Alemana para Microondas, 1 October 1967; Dr. Scharrer, Misión Alemana para Microondas to Dreesmann, 5 October 1967; 7. Bericht, Misión Alemana para Microondas, 1 October 1967. Bundesarchiv Koblenz (BArch), BArch B 213/17571.
46. 15. Bericht, Misión Alemana para Microondas, 5 March 1968. BArch B 213/17571.
47. Deutsche Botschaft Bogotá to AA, 21 May 1968, BArch B 213/17571.
48. Ministerio de Comunicaciones, Empresa Nacional de Telecomunicaciones, Telecom, Vicepresidencia Técnica: Programa de Telecomunicaciones para la región oriental del país. Zonas de las intendencias y las comisarias. Bogotá, Enero, 1977. BArch B 213/25430.
49. ITU, World Telecommunication Development Conference, Valletta, Malta, 23 March–1 April 1998, Document 60-E.
50. Walch, Dienstreise nach Paraguay, 28 November 1977. BArch B 213/25476; Dr. Lotz, Entwurf, Inspektion Hauptbericht Teil I: Textband, Fernmeldewesen Paraguay, 30 June 1978, Anlage 57, BArch B 213/25481.
51. Dr. Lotz, Entwurf, Inspektion Hauptbericht Teil I: Textband, Fernmeldewesen Paraguay, 30 June 1978, Anlagen 14 und 16, BArch B 213/25481.
52. Tätigkeitsbericht Schön, Paraguay, 10 January 1975. BArch B 213/25473.
53. Tätigkeitsbericht Schön, 14 April 1976; Vermerk Fernmeldeprojekt Paraguay, 22 June 1977. BArch B 213/25473.
54. Juan Balsevich, *Historia de las telecomunicaciones en el Paraguay (1864–2002)* (Asunción: AGR S.A, 2011): chapter 17.
55. Hugh R. Slotten, 'International Governance, Organizational Standards, and the First Global Satellite Communication System', *Journal of Policy History* 27, 3 (2015): 521–549; Stephen A. Levy, 'INTELSAT: Technology, Politics and the Transformation of a Regime', *International Organization* 29, 3 (1975): 655–680. The European nations were concerned with strong US influence and demanded the replacement of COMSAT as the manager. From 1973 on, Intelsat worked in economic terms as a cooperative. In the late 1970s, US ownership shares decreased to around 30 per cent. In 1973, the official name changed to International Telecommunications Satellite Organization.

56. Larry Martinez, *Communication Satellites: Power Politics in Space* (Dedham, MA: Artech House, 1985): 3–5.
57. Ligia Ma. Fadul G., *Las comunicaciones via-satelite en América Latina,* Cuadernos del Ticom 31 (México, D.F.: UAM Xochimilco, 1984), Investigación realizada en el Centro de Estudios Económicos y sociales del Tercer Mundo AC (CEESTEM).
58. John K. Mayo et al., 'Peru Rural Communication Services Project: Final Evaluation Report' (Center for International Studies, Florida State University, 1987): 29–32.
59. Douglas Goldschmidt, Karen Tietjen and Willard D. Shaw, 'Design and Installation of Rural Telecommunication Networks: Lessons From Three Projects' (Center for International Studies, Florida State University, 1987).
60. Richard L. Horner, 'Communications Satellite Systems for Developing Countries and the Human Rights Policy through the Carter Years: Its Application in Paraguay and El Salvador' (Master's thesis, University of Texas, December 1982); Heather E. Hudson et al., *The Role of Telecommunications in Socio-Economic Development: A Review of the Literature with Guidelines for Further Investigation,* Keewatin Communications (Geneva: ITU, 1978): 82–83; William Pierce and Nicolas Jéquier, 'Telecommunications for Development: Synthesis Report of the ITU-OECD Project on the Contribution of Telecommunications to Economic and Social Development' (Geneva, 1983): 75–80.
61. Horner, 'Communications'; Heather E. Hudson, *Connecting Alaskans: Telecommunications in Alaska from Telegraph to Broadband* (Fairbanks: University of Alaska Press, 2015).
62. Projektprüfungsbericht Telecom/BOG, 4 December 1980. BArch B 213/25444; Detecon to Auswärtiges Amt, 26 May 1983. BArch B 213/25453.
63. Martinez, *Communication Satellites*: 7.
64. Clemente Pérez Correa, 'Treinta años de telecomunicación', *Teledato. Revista de la Dirección General de Telecomunicaciones* 2, 10 (1979): 9–11.
65. 'Obras y realizaciones en materias de telecomunicaciones', *Teledato. Revista de la Dirección General de Telecomunicaciones* 1, 2 (1973): 7–11.
66. Marco A. Fernández Tovar, 'Telecomunicaciones Rurales por corrientes portadoras', *Teledato. Revista de la Dirección General de Telecomunicaciones* 2, 6 (1978): 30–35; 'Tasador telefónico rural', *Teledato. Revista de la Dirección General de Telecomunicaciones* 2, 21 (1982): 3–11.
67. Gabriela Soto Laveaga, 'Searching for Molecules, Fueling Rebellion: Echeverría's "Arriba y Adelante" Populism in Southeastern Mexico', in Amelia M. Kiddle (ed.), *Populism in Twentieth Century Mexico: The Presidencies of Lázaro Cárdenas and Luis Echeverría* (Tucson: University of Arizona Press, 2010): 98–99.
68. Alfredo Bautista Chagoya, 'Antecedentes, estructura y evolución de la CTR', *Teledato. Revista de la Dirección General de Telecomunicaciones* 3, 15 (1976): 32–34.
69. Héctor Arellano Moreno, 'Consideraciones sociales en la planificación del desarrollo de las telecomunicaciones en las regiones distantes y desfavorecidas', in *Workshop on Special Aspects of Telecommunications Development in Isolated and Underprivileged Areas of Countries June 26–28, 1978*, Ottawa, Canada: 33–38.

70. Héctor Arellano Moreno, 'Plan Nacional de Telefonía Rural', *Teledato. Revista de la Dirección General de Telecomunicaciones* 2, 8 (1978): 21–31.
71. 'A todos los rincones de la provincia mexicana sigue llegando el servicio telefónico', *Voces de Teléfonos de México* 10, 120 (1971); *Voces de Teléfonos de México* 150 (June 1974).
72. Dr. Lotz Hauptbericht Teil I: Textband, Fernmeldewesen Paraguay, 30 June 1978. BArch B 213/25481.
73. Dr. Lotz Hauptbericht Teil I: Textband, Fernmeldewesen Paraguay, 30 June 1978. BArch B 213/25481.
74. Saunders, Warford and Wellenius, *Telecommunications*: 401; Antonio Cañas M., 'Development of Rural Telephony in Costa Rica', *Telecommunications for Development: World Communications Year Seminar/Meeting* (Geneva: ITU, 1983), 1–36.
75. Saunders, Warford and Wellenius, *Telecommunications*: 242.
76. Mayo et al., 'Peru Rural Communication Services Project: Final Evaluation Report': 60–67, 74–77.

# Interrelations and Disruptions in the Exchange of Knowledge: Coal, Geology and Industrialization in Mexico

HELGE WENDT

> The greatest problems for the advancement of civilization . . . is how to substitute vegetal combustibles, whose employment brings the most alarming and pernicious effects, with mineral combustibles, whose application by contrast holds the promise of future well-being, of opening new paths for industry, new horizons for work, and new sources for the production of national wealth.[1]

In many contexts across the planet, mineral coal served as the main agent of industrialization. Historiography of industrialization treats this matter concentrating on the history of coal-using machines, on related social questions or – more recently – on environmental effects.[2] In addition to this research, it can be stated that without basic knowledge about the mining, geology and chemistry of coal, industrialization would not have happened – at least, not based on coal as it is normally represented in literature. The present article will discuss the history of knowledge about coal in Mexico during the nineteenth century. It will present a typical account of globalizing knowledge that consists in the interrelation of economies and exchange of knowledge. The second part will discuss the importance to consider disruptions in the transmission of coal-related knowledge, which can be stated in the history of coal in Mexico. For this purpose, the article takes a perspective of regional industrialization in Coahuila, Nuevo León and Puebla inside the Mexican national framework, instead of considering the entire national space.

Many different concepts have intended to define industrialization. A. Rupert Hall, for instance, underlined the importance of technological change, mechanization and the introduction of non-human production modes.[3] This perspective has been challenged by Christopher A. Bayly, Jan de Vries and others, who back a concept called 'industrious revolution'. They consider the industrialization or mechanization of production in some (mostly European) countries as one part of a broader global process that involved the employment of huge numbers of unskilled workers from various regions of the world.[4] The French historian Fernand Braudel also considered technological innovation and change, but emphasized in particular the role of capitalism in this process.[5] Building on Braudel's capitalism-oriented model, Immanuel Wallerstein, in the third volume of *The Modern World-System*, pairs the rise of industry with that of the bourgeoisie or middle class in order to explain the revolutionary potential of the period around 1800 in Europe.[6] Wallerstein challenges economic historiography that portrays England as the role model for the development of industrialization at large, though he still recognizes England as an industrial forerunner.[7] Wallerstein followed the argumentation line of David Landes, Eric J. Hobsbawm, André Gunder Frank and others.[8] Those authors emphasized two changes in industrial production modes: one related to cotton and the other to iron. New production modes for cotton textiles, he states, relied mostly on new machines. And in a complementary process, chemical innovations transformed the iron industry.

While these two changes took place in certain European countries, Mexico and many other former colonies in the Americas were struggling for independence or fighting civil wars. In general, they had quite different economic structures than those existing in Europe.[9] Edward Beatty recently stated that while 'the economies of Britain, the United States, and the North Atlantic became global centres for the invention and use of mechanized technologies in transportation, mining, agriculture, and manufacturing', Mexico was trapped in internal quarrels, not to mention full-scale wars with the northern neighbour and France. In this situation, Beatty underlines constraints to industrialize the country, as for instance high costs to acquire and install steam technology and to procure coal.[10] Industries, which in other countries used coal and operated steam engines, in Mexico employed relatively 'low-tech' technologies and used alternative energy devices such as water power.[11] In Mexico, fossil fuel costs remained high – the petroleum industry would not offer cheaper energy until the 1920s. Nevertheless, the geologist Luis Torón (Villega) reminds us that for some uses coal was more advantageous than petroleum.[12]

Beatty and other historians of Mexican economic history underline the colonial legacy in trade, production, social relations and topo-political penetration, which considerably modified its entry into the era of industrial-capitalism.[13] Yet, this modification meant – among other aspects – a relatively slow and regionally differentiated increase in coal-using industrial production sites and enterprises. Perhaps many of these enterprises would have been more important to the national economy and as competitive as their international rivals, if they could procure cheaper machines and fossil fuel.[14] Despite these disadvantages, it can be emphasized that coal mining and coal-using industries did in fact develop in certain Mexican regions. According to technological demands, some of these industries used imported coal.[15] Other branches could make use of Mexican coal, and some parts of Mexican coal production were exported to the US, where a greater variety of coal-using industries existed.

Thus, the Mexican history of knowledge about coal should be placed into a frame of regional and sectorial differentiation. Jürgen Osterhammel has helpfully pointed out that even European industrialization was a regional phenomenon. Like other historians, he underlines the general importance of regional development to histories of industrialization.[16] Tibor Iván Berend, an economic historian of Europe, formulates the difference between a national and a regional approach to economic statistics in economic history. Additionally, he underlines that national average economic growth rates do not evaluate growth rates of sectors, which at a time were only economic niches. Being still relatively small, those sectors challenged 'old technologies' in the long run.[17]

Developing sectors or regions were never autonomous but rather bound to trans-regional chains of goods and commerce, and entangled in transfer processes for technologies, knowledge, finances and investments, know-how and machines. This also holds true for regions inside Mexico, defined as peripheries of the world economy in world-systems language:

> The impact the Industrial Revolution had in Europe; the presence of the steam engine, fed by mineral coal, and treated mineral coal (coke), used in iron and steel industries, was not a passage of technological history carried out by metropolitan minds alone. It can be observed that countries at the periphery and the entire world changed and were preoccupied with those changes. A scientific study of the history of these issues should show that those irrecusable testimonies are part of a critical tool for reevaluating a history that has until today remained fragmented.[18]

In this quotation, Sánchez Flores points out that even if some regions were considered economically backward, they were nonetheless integrated in a global transition to new forms of production. They also experienced the transition from vegetal to fossil combustibles as the basis for their energy systems. This history of transitions in energy regimes and technologies consists of interrelations and interconnections, but also of disruptions and breaks, between countries, scientific institutions and economies of knowledge.

## INTERRELATIONS OF KNOWLEDGE ON COAL DEPOSITS

Interrelation tends to be the centre of attention in the current historiography of globalization: transfers, circulations, exchange, etc. are the terms that define this part of history. In the case of Mexican coal mining, geological knowledge travelled with European experts or in books imported from Europe. Particularly, many US-Americans and US companies developed activities in Mexico. Very often, they made solid returns on the ideas, goods and sums of money they poured into their Mexican investments. In academia, knowledge gained in Mexico was inserted into scientific works. Alexander von Humboldt, for example, used the knowledge he gained during his travels through late-colonial Mexico to build his reputation as a scholar. Furthermore, circulating knowledge that was less locally defined became useful for foreign investments: mining sciences, geology and national economics were often treated on an abstract level.

To these various histories of continuous interrelations and diffusions of mining knowledge, we can add the transfer of technology. Most machines and technological devices needed, in order to exploit coal deposits in Mexico, were introduced from the US. At Las Esperanzas, a coal mining site exploited by a US-American company, engines of different US-American companies were employed to haul the coal from the depths – among others, two steam engines made by Litchfield Car and Machine Co, a company that built both stationary and mobile steam engines.[19] From its local beginnings in Litchfield (Illinois), the company began to sell its machines within the United States and eventually expanded its activities to equip flour mills worldwide. Furthermore, it gained some initial experience equipping local coal and gas mines before becoming a global provider of steam engines and pumps for mining coal, oil and gas. Nine Ingersoll-Sargeant air-compressor rock drill machines were also employed at Las Esperanzas. This company, whose founder Simon Ingersoll (1818–1894) patented his rock-drilling machine in 1871, was based in the State of New York. The rails for coal-consuming Mexican railroads were also US-American,[20] as well as the built blast furnaces.

From the perspective of global history, a comparison between the development of energy use in Mexico and other nations is in order. Whereas coal became the dominant source of energy for many socio-economic domains in regions that industrialized early, a short overview of the transformation of Mexican energy systems at the end of the nineteenth and the beginning of the twentieth centuries reveals important contrasts. At that time, the growing Mexican energy market depended on diverse international investments. In the early twentieth century, Siemens and AEG started to sell their products in Mexico, just after Canadian and British companies had installed turbines to electrify and illuminate Mexican cities.[21] Furthermore, other foreign companies had established telephone lines (see Berth in this volume). Much more than in Europe and most US-American towns, electricity in Mexico was generated by hydroelectric power plants instead of coal-driven turbines. Public lighting was more often powered by electricity and not by the coal-gas lamps used in European and US-American cities. This difference in the history of public lighting exemplifies the specificity of coal's importance to the industrialization of Mexico: only in certain sectors and regions was coal as dominant as it was for other countries.

Like in most Latin American countries, coal use in Mexico started rather late in the nineteenth century. As in Mexico, mostly hydro power plants provided Latin American cities with electricity for public lighting.[22] Coal, nevertheless was part of geological surveying, and was eventually mined and used in several regions of the continent. The earliest known and exploited coal deposits might have been those of Columbian Zipaquirá, although the extent and production rates still have to be investigated.[23] In Cuba, the government and some private investors were heavily involved in the search for coal since the late 1820s, but were disappointed several times.[24] In Chile, the history of coal was related to the copper boom of the 1830s and heavy British investments. Although, in cities as Santiago, electricity and public lighting was produced by water power, Chilean coal fuelled the copper smelting industry.[25] In Argentina coal deposits were – after initial rumours started in the 1860s – firstly mined in the late 1880s. In contrast to Mexico, the Argentinian national government took immediate advantage of these findings and used this coal to fire steam ships and railways.

In Mexico an extensive economic market developed at the end of the nineteenth century, which suited the interests of the US market. Several US companies, a few European and even fewer Mexican companies extracted subterranean and agricultural resources. In the mining sector copper in Chihuahua and Durango, or iron and grahamite in the state of Veracruz should be mentioned.[26] Asphalt too, as well as agricultural crops like sugar, cotton and sisal were crucial for attracting foreign investments into Mexican rural industrialization.

In the evolution of the energy sector at the end of the nineteenth century, the development of a coal mining industry was a sort of intermediate step before petroleum became the new fundamental fuel.[27] Around 1900, coal mining in Coahuila attracted North American foreign investments. The historiography of coal discovery in the region of the Rio Bravo and Rio Grande started around 100 years earlier with the *Memory about coal deposits* of 1794, written by the physician and naturalist José Antonio Alzate y Ramírez (1737–1799). In its very last paragraph the author mentions having heard about coal reserves in 'San Juan del Río en Nuevo Mexico'.[28] It is difficult to determine exactly which San Juan he might have meant, but his remark does indicate the northern region of Mexico. Alexander von Humboldt (1769–1859) also mentioned possible coal deposits in the regions north of Rio Bravo in the early 1820s. This topographic uncertainty points to a more general problem in the history of Mexican coal mining. The Spanish colonial territories that were organized under the vice-regal crown of New Spain expanded farther north than the actual US–Mexican frontier along the Rio Bravo. As a result of the US–Mexican War and the peace treaty of 1847, vast regions with their immense natural resources came within the borders of the US. Until then, in Alzate's lifetime or during Humboldt's voyage to the Americas, the territories of New Mexico and Texas stood under Spanish colonial (and later Mexican) dominion.

But it was French naturalist Jean Louis Berlandier (1805–1851), who first found lignite and other coal deposits near Laredo. The migrant from France worked with the Mexican Border Commission in the years of 1827 and 1828, which included a survey of the terrains of Texas.[29] Berlandier is part of a history of coal discoveries in the Nuevo Santander region that John A. Adams (2008) studied in *Conflict and Commerce on the Rio Grande: Laredo, 1755–1955*. Jean Louis Berlandier claimed that 'coal, of pitchy quality and equal to Liverpool coals, [had] been discovered upon the grants'.[30] This knowledge was possibly used both during and after the US–Mexican War. But no direct profit was gained from it, as coal mining in that region only began forty years later.

In the years that followed the war, several surveying expeditions in the central states of Mexico were rather unsuccessful. The US-American Mexican Pacific Coal and Iron Mining Company sent Edward Lee Plumb (1825/6–1903) to explore the land for coal and metal deposits in 1854. But three years later, without any considerable findings, Plumb changed employers.[31] Later on in his life, Plumb could earn some money with the development of railroad building and coal mining industries in the northern regions of Mexico, because he was a stakeholder in the Mexican International Railroad Company and acquainted with President Díaz' Secretary of *Haciendas*, José Ives Limantour.[32]

The history of coal exploitation on an industrial scale only starts in the 1860s. In this period, the economic investments of US companies expanded southwards, mostly active in mining copper and iron as well as in building railroads on the northern shores of the Rio Bravo.[33] The German-born surveyor Jacob Kuechler (1840–1907), who having served in the Confederate Army during the US Civil War, had fled to Mexico between 1861 and 1867. During his years in Mexico, he discovered coal in the northern Mexican State of Coahuila.[34] Beginning in 1884, companies with the majority of US capital accessed the deposits at San Felipe, Las Esperanzas and other places. The mined mineral served the *Ferrocarril Internacional Mexicano de Piedras Negras*, who transported it further north to the Southern Pacific Railroad[35] as well as to the nearby iron smelting industry.[36]

Some Mexican–US joint ventures were founded with headquarters in Monterrey, the capital of Nuevo León. Monterrey had a lively economy of sugar refinement, textiles, breweries and iron works, among other industries. A frontier city,[37] Monterrey was home to many of Mexico's leading industrialists and was connected to some of the most important Mexican industrialist families from elsewhere, such as the Creel family, who were bankers and industrialists from Chihuahua. In this region of northern Nuevo León and Coahuila, the big landowners were the same people who participated simultaneously in the incipient exploitation of iron and coal deposits, as well as in the industries in Monterrey. They were additionally engaged in the construction of railways between Mexico City, Monterrey, Laredo, and other northern Mexican towns[38] and in geological surveying expeditions of the early twentieth century.[39]

The inchoate, coal-friendly industrial sphere of this northern region increased in stature and became diversified, but industrial production was not as powerful as it could have been. The rather low domestic demand for Mexican steel forced the Monterrey Fundidora to run its blast furnace at less than 50 per cent of its capacity.[40] The consequence was high prices for its products, which then had difficulties competing with imported steel from the US.[41] Only later, in the 1920s, would production rates increase.

Nevertheless, these northern Mexican coal exploitations had a singular status in Latin America and helped to industrialize the region. In 1891, US Consuls from every region of Latin America turned in an official report to the US-American government on the state of commerce, mining and industry related to coal. The US Consul in Piedras Negras (in the State of Coahuila) was Eugene Féchet (1846–1925),[42] who after serving in the Union Army during the Civil War became a mining entrepreneur in northern Mexico. He mentioned in his report that coal, well-suited for producing coke, was mined by two US stakeholder companies.[43] In 1890 he stated:

> The port of Piedras Negras is the only port in Mexico, and, I believe, in all Spanish America, that exports coal instead of importing it. . . . The coal consumed in this consular district is entirely produced from the Sabinas coal mines, situated in the district. Mines are owned and operated by the Coahuila Coal Company and Alamo Coal Company. Annual output about 10,000 tons. These companies have erected 30 coke ovens, with a daily capacity of 1 ton of coke per oven each day. This coal yields by weight 62 per cent of coke.[44]

The quality of coal from the coal field of San Felipe was analysed in 1891. Those mines were owned by the US-American railroad entrepreneur Collis P. Huntington and in that year produced around 250,000 tons of coal, which were exclusively exported for use by US-American locomotives.[45]

The US-financed coal industries in Mexico extracted this black matter either to use it in self-owned furnaces or to feed self-owned railways. They, additionally, increased knowledge about geology in the northern region. Edwin Ludlow (1858–1924) was another US-American mining engineer and general manager of the Mexican Coal and Coke Company. He came to Mexico in 1889 and was in charge of industrializing the mining site at Coahuila as well as organizing the logistics and sales. He had a good geological knowledge of the Coahuila coal basin, as he described in a contribution to the International Congress of Geology in 1906. In this article he described the geological situation as well as the different types and qualities of coal and their uses.[46] This paper was a follow-up to a contribution that Ludlow had published with the *Transactions of the American Institute of Mining Engineers* in 1902, concentrating on only one site, Las Esperanzas.[47]

The Coahuila coal-market developed greatly as a US-dominated business that included the direct export of this coal and coke to be used in the US. This circumstance might have contributed to the elevated prices of fossil fuel, even in the northern parts of Mexico. Extraction and importation were crucial to the investors as the prolongation of a tax exemption for coal producers of 1910 by José Y. Limantour shows, in order to protect the Coahuila coal industry from losing the US market.[48]

## DISRUPTIONS OF KNOWLEDGE TRANSMISSION

The first focus of this paper was to outline the centrality of global interrelations for the Mexican coal industry: knowledge flowed in, some knowledge and much of the extracted material left Mexico. These interrelations between Mexico, Europe and the US were important, in particular through the influx of knowledge, persons, technologies and funds. Yet, there is another aspect to the history of coal mining in Mexico that is often undervalued. Aside from international interconnections, histories also show moments of interruption and disruption in the sense that some evolutions broke off and did not develop as they could have. A disruptive moment initially concerning only a smaller part of societal development may bear broad consequences for the entire society. For example: the writings of Louis Berlandier were partly known because of his report to the Mexican government in 1828. But with his death in 1851 and Darius N. Couch's obscure acquisition of his legacy (papers and mineralogical collection) for the Smithsonian Institute,[49] any subsequent transfer of knowledge from Berlandier's studies on Mexican geology to the Mexican government or any other Mexican scientific institution was interrupted. Instead, papers and collections passed into the hands of US-American scientists who exploited them in the following years.[50]

The US–Mexican War is another clear interruption to the influx of knowledge. The Mexican system of geological surveying (also practised by some Europeans in Mexico, including Berlandier) was based on the Humboldtian-Freibergian scientific ideal. The aim of geologists of the Mexican mining school until the 1880s was to establish a national geological chart, following the examples of France, Prussia and Spain. This kind of geology was fairly different from the more commercial-oriented geology that was practised in other European countries, and which played a particularly vital role in the US. Whatever the motivations for war and annexation were, the loss of direct territorial access to northern coal fields was a setback for Mexican industrial development. The major losses of territory meant that mostly US-American mining companies were able to carry out survey campaigns in the (formerly) Mexican north, taking advantage of the knowledge about the geological structure detected in Texas, Arizona and New Mexico. Furthermore, the war – leading to incorporation of coal-rich Texas and New Mexico into the United States – also put almost an end to the European influx of men, knowledge and material. Henceforth, geological explorations were carried out by North Americans. Only in the 1860s, Mexicans restarted to survey the central and southern states, but they were still mainly focused on finding silver. Coal was detected in the 1880s, when Santiago Ramírez and some US-American expeditors (such as Plumb, Bancroft and Birkinbine) reported on deposits.

It is unclear how and when knowledge of coal deposits emerged and spread into the Anglophone and Hispanophone parts of North America. Immediately after the outset of the Mexican War, numerous coal sightings were reported by officials of the US Army. Lieutenant Tilden, scouting the Rio Grande north of Laredo, gave a detailed report about coal 'of superior quality' that he and scout Alpheus Rackliff found in November 1846.[51] Bryant P. Tilden wrote in a report (published the following year) that a steam boat was required to navigate the Rio Grande. After acquiring a vessel, the 'Major Brown', he tried to use hardwood for fuel, which proved inefficient. Gladly, after running across a deposit of quality bituminous coal, he was able to switch his fuel source. Tilden repeatedly refers to coal deposits in the vicinity of Laredo, but it remains unclear whether they were already mined. In any case, he states, coal was one of the reasons why Laredo prospered.

Still during the war with Mexico, James William Abert (1820–1897) was sent by his father (the chief topographical engineer of the US Army) to explore New Mexico with General Stephen W. Kearny. Abert wrote in 1848 to the US-American Senate: 'A few miles beyond the scene . . . we notice a high bluff ban, where there were evident signs of coal. I rode over and collected some specimens of fine bituminous coal.'[52] This turned out to be a false alarm, Abert noted. But some days later, as the expedition was travelling along the Rio de Galisteo on its way to Santa Fe, the Army's geologist again spotted 'evidences of coal, and the bed of the main creek was white with saline efflorescences'.[53] He later came back to this area with a coal mining expert from Missouri named Laing. The expert denied that the fossils found could be coal. Undeterred, Abert sent all his findings to Professor Jacob Whitman Bailey at West Point Academy, who agreed to examine the fossils.[54] In contrast to Laing's intuition, this professor of chemistry, mineralogy and geology stated that it was indeed carboniferous material, which came from the region north of Santa Fe, New Mexico, near Raton, and exhibited clear geological differences from other coal formations known in the United States. Bailey could confirm that this material was of 'decidedly a far more recent formation'.[55] The initial confusion about the findings makes it clear that Abert's expedition was commissioned primarily to research the topology and geography of New Mexico, at the expense of geology. Abert did not include the kind of information about rock layers and strata that would figure in a report by experts more acquainted with the field.

During the decades following the war, geologists came from the northern shores of the Rio Grande to explore deposits of industrially and commercially useful materials. Coal and iron were their top priority because US-American firms had started to invest in constructing Mexican railways. This economic expansion was related to earlier military actions and many US-Americans who afterwards started working in Mexican geological explorations were often militaries: Bryant P. Tilden, James William Abert, Jacob Kuechler and Eugene Féchet. Edwin Ludlow himself was a civil engineer, but his brothers were both military officials, one of them becoming governor of Cuba and the Philippines after the Spanish–American Wars.

The division of Mexico into a southern territory, which mostly Mexican geologists explored and surveyed, and a northern part, where predominantly US-American geologists worked, was challenged shortly thereafter by the French invasion. During the period of united Austrian–French government, a Scientific Commission (*Commission Scientifique*) was founded, and some of its members devoted their efforts to mining activities and geological surveying in North America. They published a report in 1867 that mentions some coal deposits and mining, but forwards no suggestions for future industrial advancement. In this report, coal is said to be found near Jalapa (Veracruz),[56] but the only detailed descriptions relate to deposits that were found five years before in California, where the head of a metallurgical survey detected coal deposits near Mount Diablo.[57] Other findings of carboniferous deposits are also mentioned for the states of Chihuahua and Sonora, near the Pacific coast states of Guerrero and Oaxaca.[58] These geological reports by the *Commission Scientifique* ultimately contributed little to expanding the industrial exploitation of coal in Mexico.[59]

Relying on administrative records (as opposed to proper research), Luis Robles Pezuela (†1882), engineer, son of the former director of the Mexican Mining School and presently the secretary of finance, composed a report in 1866 to the Emperor Maximilian on all industrial activity related to mining that had been reported to the ministry. This register contains mostly unconfirmed locations of subterranean resources, reported by developers who were eager to obtain governmental concessions and legal protection. Coal is reported near Jalapa, Reynoso (Tamaulipas), and along the rivers Bravo and Balsas.[60] Others claimed to have found coal or some bituminous matter on the Pacific coast, near Puerto Angel in the South, Tepic in central Mexico and Culiacán in the north, as well as in northern Veracruz, to take some examples.[61] He did not mention any coal discoveries in Coahuila, where mining in general was in a deplorable state.[62] Seemingly, Robles Pezuela's document, written in times of political turbulence, had no consequences for later histories of geological research or Mexican industrialization. With the end of the French–Austrian interlude, neither Robles Pezuela's writings nor the knowledge gathered and published by the Scientific Commission seems to have found any readership in the newly reestablished Mexican Republic.

The pursuit of knowledge on coal deposits only picked up again in the early 1880s. But the writings of that period, such as those of Santiago Ramírez, a former student of the Mexican College of Mines, contain little information about already known deposits in the north or former studies on central Mexico. They underline the apparent split in developments and the disruptions in the flow of information between the northern and the central-southern parts of the country.

In the context of more stable political and scientific conditions, as well as economic necessity during the first presidency of Porfirio Díaz, Santiago Ramírez (1836–1922) carried out one survey expedition in the southern parts of the State of Puebla in 1881. We can only guess why he was commissioned to survey this region.[63] In a special issue of the Mexican *Boletín Minero* of 1921, Luís G. Jiménez states that in 1853 lignite was found in the western parts of the State of Puebla.[64] Jiménez refers to a study by Joaquín Velázquez de León and Felipe Saldívar, published in the *Boletín de la Sociedad Geológica*.[65] Another indication of ongoing coal mining activities in the Matamoros area that Ramírez was in is found in Secretary Robles Pezuela's account of 1864.[66] But Ramírez does not refer to this study, nor to any other previous geological report on the frontier region of Puebla, Guerrero and Oaxaca that he surveyed.[67]

Ramírez mentioned two possible reasons for his geographical choice. In that region, iron-rich strata were already mined in those years, which seemingly also contained carboniferous substances. Knowledge about the geological characteristics from local iron mines may have induced him to search for coal – the fossil combustible that could substitute the increasingly scarce wood used by the local smelting industry. Additionally, even if he had not known about them beforehand, he may have gained knowledge of small coal exploitations in the region during his expedition. What he does not mention in his reports are other industrial usages for coal in the Puebla region, for instance in the flourishing textile industries which, as Carlos Marichal[68] points out, began using modern machines in the 1850s.[69] During his trip along the Tizaac, Balsas and Nexapa rivers, Ramírez found different deposits of coal in the topographical triangle between Izúcar de Matamoros in the north, Chiautla de Tapia at the south-western edge and Acatlán de Osorio in the eastern extremity, whose economies traditionally were based on sugarcane.[70]

Aside from geological and geo-chemical analysis, Ramírez' writings also considered the legal aspects of coal mining. Furthermore, he employed the geological analogy method and factored in the limits of growth on industries relying on vegetal combustibles. Moreover, he considered the multiple industrial uses of coal, its ability to bolster transportation infrastructure, and the economic advantages it offered to national industrial production against imported industrial goods.[71] From these allusions it can be deduced that Santiago Ramírez was familiar with processes of industrialization in the iron industries of both England and the US.

From his own statements it can be deduced that some of the Puebla deposits were already under primitive exploitation (for instance in Acatlán), although in this period the price for coal was higher than that of charcoal. Some other textual testimonies also point to that region. In addition to the generally scarce information about early coal mining activities in that region, no data of quantities of mined coal is available. Jiménez states in his report of 1921 that a *Compañía Carbonífera Mexicana* worked mines in the district of Acatlán. This US–Mexican stakeholder company had acquired the rights for exploiting the site back in 1887.[72] After a period of inactivity, it was bought by a mining company from Virginia.[73] In a report on Mexico from 1891, the Washington-based Bureau of American Republics, an international organization preceding the Pan-American Organization, mentioned these and other Mexican coal deposits.[74] Ten years later, the same organization reported on the year of 1890 that fifty-nine coal mines were exploited within the State of Puebla.[75] The US-American anthropologist Hubert Howe Bancroft included these coal findings in his report on resources from 1893.[76] In this period, a new diffusion of knowledge to the US began on hitherto unknown coal deposits in southern Mexico ready for exploitation.

During the first decade of the twentieth century, rights to exploit iron and coal mines were conceded to the Oaxaca Iron & Coal Company, another US-American company.[77] Visiting the mining area in 1909, John Birkinbine[78] described coal as increasingly replacing charcoal in the iron smelting industries. In contrast to this rather optimistic statement, an anonymous report to José Limantour of the same year claimed that the Oaxaca Iron & Coal Company had not explored the coal field to ensure a continuous production.[79]

## CONCLUSION: ACCELERATING THE MEXICAN ENERGY TRANSITION

> At a great distance from the population centers, it is not very difficult to find inexpensive wood; but in the capital, the scarcity of this fuel that is used in big quantities in the manufacturing business is already alarming. . . . Once they start to exploit coal mines in Matamoros, Chiautla, and Acatlan, there will be cheap fuel to advantageously plan smelts and other mining establishments.[80]

Antonio Pérez Marín, a senior consultant to the Pueblan government stated that by the end of the nineteenth century, knowledge about coal deposits and forms of using this combustible had gained the potential to transform the regional economy. Earlier in the nineteenth century, efforts to induce a transition in energy systems were only regionally and partially successful. Later, coal mining became an established branch of industrial activities, in the northern regions of Coahuila and Nuevo León. The situation was undoubtedly difficult, considering the competitors from north of the Rio Grande frontier coal producers in Mexico had to struggle with.

A regional approach to the history of coal mining in Mexico makes it possible to understand the regional variability of coal's importance. At the end of the nineteenth century, a limited energy transition had partially succeeded in some Mexican states. Rather unsurprisingly, the main transfers of knowledge relating to the coal business in Mexico involved the northern neighbour: either individuals like Kuechler came to Mexico and worked several years as mining geologists, or a mining entrepreneur like Ludlow developed mining sites. US companies made important investments into coal mining and coal-consuming industries in Coahuila and Nuevo León. They improved the extraction economy by building railroads or financing iron works. These investments were accompanied by several Mexican investments into the same industrial branches, as well as into some others.

Generally speaking, the transition in energy systems was a fundamentally global process – and in Mexico, despite its weak growth rates in coal production, the story was very similar. Where coal was most easily found, as in Coahuila and Nuevo León, and where considerable investments into mining and coal-using production sectors were made, a more wide-ranging transition took place. In the central-southern parts of Mexico, the transition happened at a slower pace and was more diverse. There, the usage of coal developed alongside wood, as well as electricity gained from water power. The reasons for these varying developments were differences from the northern regions in terms of topological position, geological conditions and the characteristics of the Puebla coal basin. This explains the delayed influx of international capital into the coal mining industries of Puebla and Oaxaca, as well the late-developing push of the Mexican government to start constructing a railroad connecting the Pacific to the central Highland and the Atlantic coast around Veracruz. There was a clear interruption of information and capital flows between the Coahuila and the Puebla coal fields, which were not comprehensively surveyed.

To portray the role of mineral coal in Mexican industrialization as a history of interrelation and disruption emphasizes more complex global historical narratives than would be allowed by a strict adherence to diffusion models. Knowledge about coal, including its exploitation and usage, came to Mexico from different sources – the United States, France and Germany, among others. Political and economic developments in Mexico meant that these separate knowledge transfers did not merge with one another. Particularly, in the northern coalfields knowledge and technological devices remained in an imperialist context of extraction capitalism. Early on, the status of coal exploitation and usage in the State of Puebla seems to have evolved differently, as more local initiatives were able to set that industry in motion. Once knowledge about the geology of Puebla spread, more internationally oriented companies started to invest into these mining enterprises.

## NOTES AND REFERENCES

1. Santiago Ramírez, *Estudios sobre el carbón mineral* (México, D.F.: Francisco Díaz de León, 1882): 5. Translation by H. Wendt. In Spanish: 'El gran problema que los adelantos de la civilizacion(!) y las necesidades que constituyen su inmediata y natural consecuencia, han venido á plantear en el terreno de la industria(!), y cuya solucion(!) debe buscarse en las investigaciones de la ciencia, consiste en la sustitucion(!) del combustible vegetal, cuyo empleo trae consigo los más alarmantes y perniciosos efectos, por el combustible mineral, cuyas aplicaciones, por el contrario, enuelven una promesa de bienestar futuro, abriendo nuevos caminos á la industria, nuevos horizontes al trabajo y nuevas fuentes de produccion(!) á la riqueza nacional.'
2. There is no place to cite the many studies that could be mentioned. John U. Nef, *The Rise of the British Coal Industry* (London, Hamden: Archon Books, 1966) to E. Anthony Wrigley, *Energy and the English Industrial Revolution* (Cambridge et al.: Cambridge University Press, 2010). Relations of coal and development are analysed by Kenneth Pomeranz, *The Great Divergence: China, Europe, and the Making of the Modern World Economy* (Princeton: Princeton University Press, 2000).
3. Alfred R. Hall, 'Scientific Method and the Progress of Techniques', in E. E. Rich and C. H. Wilson (eds.), *The Cambridge Economic History of Europe. Vol. IV: Economy*

*of Expanding Europe in the Sixteenth and Seventeenth Centuries* (Cambridge: Cambridge University Press, 1967): 96–154.

4. Jan de Vries, 'The Industrial Revolution and the Industrious Revolution', *The Journal of Economic History* 54, 2 (1994): 249–270; Christopher A. Bayly, *The Birth of the Modern World 1780–1914* (Oxford: Blackwell, 2004); Jan de Vries, *The Industrious Revolution. Consumer Behavior and the Household Economy, 1650 to the Present* (Cambridge: Cambridge University Press, 2008); Sheilagh Ogilvie, 'Consumption, Social Capital, and the "Industrious Revolution" in Early Modern Germany', *The Journal of Economic History* 70, 2 (2010): 287–325; R. C. Allen and J. L. Weisdorf, 'Was there an "Industrious Revolution" before the Industrial Revolution? An Empirical Exercise for England, c. 1300–1830', *The Economic History Review* 64, 3 (2011): 715–729.
5. Fernand Braudel, *La Dynamique du capitalism* (Paris: Arthaud, 1985).
6. Immanuel Wallerstein, *The Modern World-System III. The Second Era of Great Expansion of the Capitalist World-Economy, 1730–1840s* (San Diego: Academic Press, 1989).
7. Wallerstein, *The Modern World-System III*: 33.
8. Eric J. Hobsbawm, 'The Seventeenth Century in the Development of Capitalism', *Science and Society* 24, 2 (1960): 97–112; David S. Landes, *The Unbound Prometheus. Technological Change and Industrial Development in Western Europe from 1750 to the Present* (Cambridge: Cambridge University Press, 1969); Eric J. Hobsbawm, *The Age of Capital. 1848–1875* (London: Weidenfeld & Nicholson, 1975); André Gunder Frank, *Dependent Accumulation and Underdevelopment* (New York and London: Monthly Review Press, 1979).
9. Roberto Cortés Conde and Stanley J. Stein (eds.), *Latin America. A Guide to Economic History, 1830–1930* (Berkeley: The University of California Press, 1977).
10. Edward Beatty, *Technology and the Search for Progress in Modern Mexico* (Berkeley: The University of California Press, 2015): 45.
11. Aurora Gómez Galvarriato, *Myth and Reality of Company Stores during the Porfiriato. The 'tiendas' de raya of Orizaba's Textile Mills* (México, D.F.: Centro de Investigación y Docencia Económicas, 2005): 9.
12. Luís Torón, 'El uso racional de los combustibles mexicanos', *El trimestre Económico* 12, 47 (1945): 454–465.
13. Beatty, *Technology and the Search for Progress*.
14. Beatty, *Technology and the Search for Progress*: 164–166.
15. Beatty, *Technology and the Search for Progress*: 212–213.
16. Jürgen Osterhammel, *The Transformation of the World. A Global History of the Nineteenth Century* (Princeton: Princeton University Press, 2014): 638–640.
17. Iván T. Berend, *An Economic History of Nineteenth-Century Europe. Diversity and Industrialization* (Cambridge: Cambridge University Press): 74.
18. Ramón Sánchez Flores, '"Nota preliminar" de José Antonio Alzate y Ramírez', in José Antonio Alzate y Ramírez, *Descubrimientos del carbón mineral y petroleo en México.*

*Documento inédito* (México, D.F.: Sociedad Latinoamericana de Historia de las Ciencias y la Tecnología, 1988): 14. Translated by H. Wendt: 'El impacto que causó la Revolución Industrial de Europa; la presencia de la máquina de vapor, alimentada por carbón mineral, y de un carbón mineral tratado (coque) útil para la industria siderúrgica, no es sólo un pasaje de la historia tecnológica efectuada por las mentalidades metropolitanistas. Como se observa, los países periféricos y el mundo todo se transformaba y se inquietaba por estos cambios. A los estudiosos de la disciplina científica de la historia de estos rubros corresponde demostrar que estos testimonios irrecusables son parte de la función crítica con que debe revalorarse una historia hasta ayer parcelada.'

19. Walter R. Sanders, *The Centennial History of Litchfield, Illinois* (Litchfield: Litchfield, 1953).
20. *El Paso Herald*, Wednesday 18 May 1904: 5.
21. Reinhard Liehr and Mariano E. Torres Bautista, 'Formas y estrategias de expansión de las empresas multinacionales eléctricas alemanas en México, 1894–1942', in Reinhard Liehr and Mariano E. Torres Bautista (eds.), *Compañías eléctricas extranjeras en México (1880–1960)* (Frankfurt am Main, Madrid and México, D.F.: Iberoamericana Vervuert, 2010): 191–220.
22. William J. Hausman, Peter Hertner and Mira Wilkins, *Global Electrification: Multinational Enterprise and International Finance in the History of Light and Power, 1878–2007* (Cambridge: Cambridge University Press, 2008): 99–100.
23. Alexander von Humboldt, *Memoria razonada de las salinas de Zipaquirá* (org. 1801) (Bogotá: Fundación Editorial Epígrafe, 2003). Helge Wendt, 'Coal and Social Transformation in the Works of Alexander von Humboldt' (forthcoming).
24. Helge Wendt, 'Coal Mining in Cuba: Knowledge Formation in a Transcolonial Perspective', in H. Wendt (ed.), *The Globalization of Knowledge in the Iberian Colonial World* (Berlin: Edition Open Access, 2016): 261–296.
25. Luís Ortega, 'The First Four Decades of the Chilean Coal Mining Industry, 1840–1879', *Journal of Latin American Studies* 14, 1 (1982): 1–32. Chris Evans and Olivia Saunders, 'A World of Copper: Globalizing the Industrial Revolution, 1830–70', *Journal of Global History* 10, 1 (2015): 3–26.
26. María del Carmen Collado, *La burguesía mexicana, el emporio Braniff y su participación política 1865–1920* (México, D.F.: Siglo Veintiuno Editores, 1987).
27. See for example the review of mining activities by José G. Aguilera. 'Reseña del desarrollo de la geología en México', *Boletín de la Sociedad Geológica Mexicana* 1, 1 (1905), 35–117. Recent studies on petroleum and economic development: María del Mar Rubio, *Oil and Economy in Mexico, 1900–1930s*, Working paper 690, Barcelona: Universitat Pompeu Fabra / Departament d'Economia i Empresa. http://hdl.handle.net/10230/808. Francesco Girali and Paolo Riguzzzi, 'Entender la naturaleza para crear una industria. El petróleo en la exploración de John McLleod Murphy en el istmo de Tehuantepec, 1865', *Asclepio* 67, 2 (2015). doi:http://dx.doi.org/10.3989/asclepio.2015.20
28. José Antonio Alzate y Ramírez, *Proyecto del Dr. José Alzate Ramírez sobre el descubrimiento y uso del carbón mineral*, ed. Ramón Sánchez (Mexico: Sociedad

Latinoamericana de Historia de las Ciencias y la Tecnología, Cuadernos de Quipu, 198/1794): 78.

29. Russell M. Lawson, *Frontier Naturalist. Jean Louis Berlandier and the Exploration of Northern Mexico and Texas* (Albuquerque: University of New Mexico Press, 2012): 125: 'Berlandier and companions bivouacked at Guerrero for a day, during which time Berlandier explored the town and an abandoned coal mine'.
30. John A. Adams Jr., *Conflict and Commerce on the Rio Grande: Laredo, 1775–1955* (College Station: Texas A&M University Press, 2008): 133.
31. Frank A. Knapp, 'Edward Lee Plumb, amigo de México', *Historia Mexicana* 6, 1 (1956): 9–23.
32. B. W. Aston, *The Public Career of Don José Ives Limantour* (PhD diss.: Texas Tech University, 1972).
33. Cf. Luis Robles Pezuela, *Memoria presentada á S.M. el Emperador* (México, D.F.: J.M. Andrade y F. Escalante, 1866): 28 on coal findings near Reynosa.
34. Camilo Contreras Delgado, 'La explotación del carbón en la cuenca carbonífera de Coahuila (1866–1900). La división espacial del trabajo', *Relaciones. Estudios de la historia y sociedad* 22, 87 (2001): 117–203. McGuire, James Patrick 'Kuechler, Jacob' *Handbook of Texas Online* (http://www.tshaonline.org/handbook/online/articles/fku01), accessed 4 February 2016. Uploaded on 15 June 2010. Published by the Texas State Historical Association.
35. Juan Luis Sariego Rodríguez, *Enclaves y minerales en el norte de México. Historia social de los mineros de cananea y nueva rosita 1900–1970* (México, D.F.: Ediciones de la Casa Chata, 1988): 58.
36. Rodolfo Corona-Esquivel et al., 'Geología, estructura y composición de los principales yacimientos de carbón mineral en México', *Boletín de la Sociedad Geológica Mexicana* 58, 1 (2006): 141–160; 145. Contreras Delgado, 'La explotación del carbón': 188–191. An overview of the mines is given in Luis G. Jiménez, *Boletín Minero* 11 (1921): 628–655.
37. Jorge Balán et al., *Men in a Developing Society. Geographic and Social Mobility in Monterrey, Mexico* (Austin: University of Texas Press, 1973).
38. Mario Cerutti, *Propietarios, empresarios y empresa en el norte de México. Monterrey: de 1848 a la globalización* (Mexico: Siglo Veintinuno Editores, 2000): 86–90.
39. Juan Mora-Torres, *The Making of the Mexican Border. The State, Capitalism, and Society in Nuevo León, 1848–1910* (Austin: University of Texas Press, 2001): 263.
40. Beatty, *Technology and the Search for Progress*: 224.
41. Stephen H. Haber, *Industry and Underdevelopment. The Industrialization of Mexico, 1890–1940* (Stanford: Stanford University Press, 1989).
42. According to Wikipedia, he became an army official in the Philippines 1901–1902.
43. Eugene Féchet, in: *Special-Consular-Reports. Coal and Coal Consumption in Spanish America* (Washington: Government Printing Office, 1891): 12.
44. Féchet, *Special-Consular-Reports:* 10–11.

45. Bureau of the American Republics, *Bulletin of the Bureau of the American Republics 9. Mexico*. Washington (Washington, DC: Government Printing Office, 1891): 79.
46. Edwin Ludlow, 'Les gisements carbonifères de Coahuila', *Guide des excursions du X.e Congrès géologique internationale* 28 (1906).
47. Edwin Ludlow, 'The Coal-Fields of Las Esperanzas, Coahuila, Mexico', *Transactions of the American Institute of Mining Engineers* 32 (1902): 140–156.
48. José Y. Limantour, Letter to Governor of Coahuila Jesús de Valle, 5 April 1910. Fondo CDLIV, Segunda Serie, Año 1910, Carpeta 31, Documento 33 (http://aleph.academica.mx/jspui/handle/56789/95592). In fact, Limantour's decision was rather to leave the tax-exemption laws of 1887 untouched.
49. Joseph Henry, 'Report of the Secretary for 1858', *Annual Report of the Board of Regents of the Smithsonian Institution* (1859): 13–43.
50. Smithsonian Institution Archives, *Jean Louis Berlandier Papers, 1826–1851, and related papers to 1886* (Washington, DC: Smithsonian Institution Archives, n.d.).
51. Adams, *Conflict and Commerce on the Rio Grande: Laredo*: 131.
52. James William Abert, *Report of His Examination of New Mexico, in the Years 1846–1847* (Washington: n.p., 1848): 22.
53. Abert, *Report of His Examination of New Mexico*: 37.
54. Abert, *Report of His Examination of New Mexico*: 107.
55. Abert, *Report of His Examination of New Mexico*: 131.
56. A. Dollfus et al., 'Mémoires et notes géologiques', *Archives de la Commission Scientifique du Méxique II* (1867): 363–403; 389.
57. Edmond Guillemin-Tarayre, 'Rapport sur l'exploration minéralogique des régions mexicaines', *Archives de la Commission Scientifique du Mexique* 3 (1867): 173–470; 197–198.
58. Guillemin-Tarayre, 'Rapport sur l'exploration minéralogique des régions mexicaines': 311. In his introduction to the volume, Edmond Guillermin-Tarayre refers to Humboldt's *Essai politique*, remarking that Humboldt became a second discoverer of Mexico. Guillermin-Tarayre was the head of metallurgical surveying for the *Commission Scientifique*. See Guillemin-Tarayre, 'Rapport sur l'exploration minéralogique des régions mexicaines': 25.
59. Dollfus et al., 'Mémoires et notes géologiques'.
60. Robles Pezuela, *Memoria presentada á S.M. el Emperador*: 28.
61. Robles Pezuela, *Memoria presentada á S.M. el Emperador*: 356. He repeats this list in the work composed while exiled in France: *Apuntes sobre las mejoras materiales aplicables a la América Latina* (Paris, 1869).
62. Robles Pezuela, *Memoria presentada á S.M. el Emperador*: 372.
63. See Lucero Morelos Rodríguez, 'Brief History of Geological and Mining Exploration in Nineteenth Century Mexico', in W. Mayer et al. (eds.), *History of Geoscience: Celebrating 50 Years of INHIGEO* (London, 2017): 303–313. María Teresa Sánchez Salazar, 'La minería del carbón y su impacto geográfico-económico en el

centro-oriente y noreste de Coahuila, México', *Investigaciones Geográficas* 31 (1995): 93–112.

64. Jiménez, *Boletín Minero*.
65. *El carbón mineral en Mexico* was a special issue of the *Boletín Minero* 11 (1921): 698. This issue gives an account of the locations and chemical features of carboniferous material in every state of Mexico. The reports do not necessarily refer to exploitation rates, nor to the history of how the coal was found.
66. Robles Pezuela, *Memoria presentada á S.M. el Emperador*: 401.
67. Enrique Canudas Sandoval, *Las venas de plata en la historia de México. Síntesis de Historia Económica*, Siglo XIX (Villahermosa: Editorial Utopía, 2005).
68. Carlos Marichal, 'La economía de la época borbónica al México independiente, 1760–1850', in Sandra Kuntz Ficker (ed.), *Historia económica general de México. De la Colonia a nuestros días* (México, D.F.: Colegio de México, 2010): 173–209; 205.
69. Manuel Miño Grijalva, La protoindustria colonial hispanoamericana (México, D.F.: El Colegio de México, 1993); Marichal, 'La economía de la época borbónica'; Ernest Sánchez Santiró, 'El desempeño de la economía mexicana, 1810–1860. De la colonia al estado-nación', in Sandra Kuntz Ficker (ed.), *Historia económica general de México. De la Colonia a nuestros días* (México, D.F.: Colegio de México, 2010): 275–301.
70. On modern knowledge of geology and constraints to exploit the coal deposits of Oaxaca, see Sánchez Salazar, 'La minería del carbón'.
71. This understanding of Ramírez' works is also expressed in José Alfredo Uribe Salas and Maria Teresa Cortés Zavala, 'Andrés del Río, Antonio del Castillo y José G. Aguilera en el desarrollo de la ciencia mexicana del siglo XIX', *Revista de Indias* 66 (237): 491–518.
72. Antonio Pérez Marin, Minería, 'Informes y documentos relativos á comercio interior y exterior'. *Agricultura, minería é industrias* 21 (1887): 153.
73. Jiménez, *Boletín Minero*: 639.
74. Bureau of the American Republics, *Bulletin of the Bureau of the American Republics*: 80.
75. Bureau of the American Republics, *Mexico. A Geographical Sketch* (Washington, DC: Government Printing Office, 1900): 195.
76. Hubert H. Bancroft, *Resources and Development of Mexico* (San Francisco: The Bancroft Company, 1893): 153.
77. John Birkinbine, 'The Mixteca Country in the State of Oaxaca, Mexico', *Journal of the Franklin Institute* 168, 3 (1909): 201–216.
78. His research is reviewed in the 1910 *The Engineering and Mining Journal* 9 (1910).
79. Anonymous, Report to Mr. José Y. Limantour, 29 October 1909. Fondo CDLIV, Segunda Serie, Año 1909, Carpeta 29, Legajo 13 (http://aleph.academica.mx/jspui/handle/56789/81704).
80. Pérez Marin, 'Minería': 159. Translated by H. Wendt. The Spanish original text says: 'A mucha distancia de los centros de poblacion no es tan dificil encontrar leña á buen

precio; pero en la capital es ya alarmante la escasez de ese combustible, que se necesita en grandes cantidades para las negociaciones fabriles . . . En Matamoros, Chiautla y Acatlan, una vez que exploten las minas de carbon de piedra, habrá combustible barato para plantear con ventaja, fundiciones y otros establecimientos mineros.'

# Machines and Texts: Writing the History of Educational Technology in Latin America

JOSEP SIMON

Today, Information and Communication Technologies (ICT) are portrayed as the solution to underdevelopment, for their capacity to improve education, business and good governance.[1] A significant amount of pioneering ICT initiatives for education designed in the US or Europe have used Latin America as a privileged ground to become global. A new literature that critically analyses projects such as One Laptop per Child or Computers for Schools in several Latin American countries is currently burgeoning in the field of Science and Technology Studies. Concomitantly, work produced within a range of disciplines interested in educational research is contributing to dismantle the restrictively presentist understanding of educational ICT as mere computing solutions, in favour of more comprehensive and nuanced definitions of educational technologies.

In this novel perspective, educational ICT would be a contemporary case study of a *longue durée* history of technologies in the classroom, including items such as blackboards, slide rules, textbooks, student and teacher notebooks, slide and film projectors, television and film, laboratory guides, examinations and tests, programmed instruction manuals, museum and laboratory collections, teaching machines and a number of utopian projects conceived to improve education. Research analysing these technologies and the practices associated with them is still scarce and scattered in several academic fields that do not often interact – among others, science and technology studies, history of education, history of science, history of technology, media studies, museum studies and contemporary science education. While the literature is large, its analytical quality is unequal, and a focus from the history of technology perspective is rare.

In this paper, I will provide an exploratory analytical review of this literature and my own research experience. I will start by providing a historically contextualized characterization of educational technology. Subsequently, through case studies on different Latin American national contexts, I will discuss the potential of investigating nineteenth-century patents of invention of pedagogical tools and teaching machines for primary education. Finally I will analyse the rise of audio-visual and mass communication in the context of US–Latin American cultural diplomacy during the Cold War. By necessity, the approach in this paper is more empirical than theoretical, and it presents a series of sketches that offer a partial coverage of a comprehensive view of this history in the Latin American region. Nonetheless, by adopting a broad perspective, combined with exemplary case studies and an interdisciplinary literature review, the main aim of this paper is to suggest a new historiographical and methodological agenda for research on education in Latin America within the history of technology.

## TECHNOLOGIES OF EDUCATION

In 1970, the first Brazilian symposium on the teaching of physics was held in São Paulo under the auspices of the Sociedade Brasileira de Física (Brazilian Physics Society). A major panel was devoted to the teaching of physics in secondary education with the participation of university physicists, school teachers and science education researchers across the country. Presentations dealt with multiple aspects ranging from pedagogical philosophy to syllabi, examinations, teacher training, careers and salaries. A critical point was the discussion of the role of pedagogical packages imported from the United States – such as those of the Physical Science Study Committee (PSSC) and Harvard Project Physics – in changing how physics was taught in Brazil. Participants considered that these foreign initiatives had had great impact, especially through their textbooks. However, Brazilian schools required not just mere translations of foreign textbooks, but the national production of pedagogical packages adapted to the specific needs of national schools, teachers and students.[2]

Textbooks had been a traditional way of teaching physics in Brazil since at least the nineteenth century, and the problems of contemporary science teaching were often attributed to its bookish nature.[3] However, in this context, a novel and broader definition of these traditional teaching tools emerged. In discussion with his colleagues, Antônio S. Teixeira Jr., professor of physics at the Universidade de São Paulo and the author of a standard secondary school physics textbook, contended that it was not just a question of *texts* but of *technologies*:

> I consider this as technology too. Texts come together with a teaching material, an orientation, a work; it is an importation of a whole educational technology, linked to the text. The PSSC is not only a text but a set of materials.[4]

In fact, the PSSC pedagogical package, released in 1960 in the US and translated in Brazil a few years later, comprised not only a main textbook, but also other teaching materials.[5] However, Teixeira referred not only to these elements, but to the design features of a textbook, as a technology with an in-built pedagogical philosophy, technique and knowledge structure. Thus, the appropriation of a textbook into teaching practice commonly required specific training for teachers on how to use it, and its particular design could contribute to drive particular actions involving teacher and students in the classroom. Technology and technique went hand in hand with the production and use of an object (a textbook) that many science and technology researchers oblivious to the epistemological relevance of education would qualify as 'the last existential act' in the making of knowledge.[6]

The PSSC project had originated in a proposal for the production of 'Movie Aids for Teaching Physics in High Schools' presented at the Massachusetts Institute of Technology (MIT) in 1956. Between 1956 and 1960, through a military-industrial inspired organization and grants from the National Science Foundation, the Sloan Foundation and Ford Foundation, the PSSC, constituted by a team of several hundred physicists, high school teachers, instrument makers, filmmakers, photographers, editors, typists and educational test designers, developed a new physics course which comprised a main textbook, teacher's guides, laboratory handbooks, scientific instrument kits, films and a series of popular science books. Its development included experimental trials at pilot schools prior to the commercialization of the packages, and training workshops for acculturating teachers in the use of PSSC materials. A non-profit corporation was established to administer the PSSC products.[7]

The project aimed at breaking with the ways in which physics had previously been taught – including the inadequacy of available textbooks and the excessive textbook-dependence of teaching – and to align the new applications of cinema and television with education. It grew from a proposal for the production of educational films into a more complex pedagogical package which also included scientific instruments, a textbook and several other printed materials. In line with contemporary educational psychology, these pedagogical materials emphasized depth over breadth by focusing contents and pedagogical rationale on the conceptual *structure* of subjects and on teaching through *inquiry*.[8]

The PSSC project was a success in the US and internationally. The earliest foreign editions of its main course were translations into Spanish and Portuguese produced and used in Latin America during the early 1960s.[9] Latin America was a privileged region for the production and trial of science teaching materials originally designed in the US or on site by local educationists and international agencies such as UNESCO. In the US, the PSSC project opened the way to subsequent projects of pedagogical innovation in physics (such as the Harvard Project Physics), other sciences (biology, chemistry, geology) and the social sciences. It contributed to establish a permanent base at MIT for innovative educational projects with international projection which – although not a cumulative history – has built a tradition running until our days. The type of projects launched by the PSSC at MIT would move during the 1960s and 1970s to educational projects on other subjects (e.g. technological and environmental studies, community education and medicine), the design of programmed instruction texts inspired in B. F. Skinner's teaching machines and Fred S. Keller's Personalized System of Instruction, and the production of computer generated films, and computer educational programmes and tutorials (e.g. ELIZA and PLATO).[10] It was this context that shaped the views of MIT entrepreneurs such as Seymour Papert and Nicholas Negroponte on the potential of computers for education, through concepts such as the 'Knowledge Machine' and 'constructionism' that contributed decades later to the establishment of MIT Media Lab and the development of projects such as One Laptop per Child (OLPC).[11]

Antônio Teixeira's definition of textbooks as technologies was received at first with some puzzlement at the 1970 Brazilian physics teaching symposium. But in the 1970s, in practice this conception became standard in the development of Brazilian projects for the teaching of physics such as the PEF (Projeto de Ensino de Física), the PBEF (Projeto Brasileiro para o Ensino de Física), and the FAI (Física Auto-Instrutivo).[12] Brazilian educational researchers were perfectly familiar since at least the early 1960s with the new overarching concept of educational tools, teaching methods and pedagogical innovation projects as technologies. Moreover, they were part of the international development of systems of programmed instruction that blurred the boundaries between texts and machines. In this period, Brazil was also the centre of operations for UNESCO's pilot project of experimental science teaching which developed a programmed instruction course in physics combining texts, instrument kits and films.[13] Analogously, the Keller Plan developed by American behavioural psychologist Fred S. Keller (based on his Personalized System of Instruction), originated from conversations between Keller, a colleague at Arizona State University, and two psychology professors at the newly created Universidade de Brasília, where the method was first tried before being successfully implemented in the US.[14]

By the late 1960s Brazilian physicists were also experimenting with the implementation of computers in education. Sérgio Mascarenhas, director of the Institute of Physics at the Universidade Federal de São Paulo, considered that the great revolution of the twentieth century would not be satellites or nuclear physics, but the making of a whole new educational technology using computers, television, magnetic tapes and films, among others. According to him, in contrast to other type of industries limited by market saturation, this was a vast and ever-expanding market due to population growth and the increase of free time brought by machine civilization. In this context the rise of the electronic computer industry for education was an enormously lucrative business as developments in the US were demonstrating.[15] Efforts in this direction were also being developed in Brazilian universities. Mascarenhas argued against objections to the use of computers in teaching, that:

> It is not an inhuman technique for the fact that the student is learning with a machine. On the contrary: there are many teachers in Brazil that are machines much colder than computers.[16]

Mascarenhas' enthusiasm could not hide that the utopian forecast of the new machine-based pedagogy intended in some ways the substitution of teachers, and concealed that the design of each educational technology involves the implementation of a particular pedagogical philosophy.[17]

The close relationship between Latin America and the US in the making and definition of educational technology would not be accurately characterized by a history of continuities over half a century. Nonetheless, today, a number of projects originally designed in North America such as OLPC and Computers for Schools, exemplify again the intensity of these interactions. The impact of these type of projects in Latin America is currently receiving a great deal of attention within educational research, producing a large body of literature which however is often biased by a literal replication of the triumphalist statements of its developers.[18]

However, there is also a growing body of critical literature on this topic within the field of Science and Technology Studies which deserves mention.

Research on OLPC has shown the relevance of Latin America for contemporary computer in education projects. In 2012, more than 80 per cent of OLPC's international reach (affecting globally forty-six countries) was focused on ten Latin American countries. An ethnographic investigation of this programme in Peruvian rural primary schools revealed the sharp contrast between OLPC leaders' fundamental belief in educational transformation by expansion of technological access, and the serious failures of this programme. Designed at MIT as a global solution, the programme did not take into account the educational and cultural diversity of the numerous countries in which it was implemented. In contrast, engineers and social anthropologists working locally identified the need for developing collaboration with local teachers and students to produce a fruitful pedagogical appropriation of this technological package.[19] Thus, the traditional conception of centre and periphery mediated by diffusion – held by the project designers – was reconfigured as a productive process of negotiation in which a diversity of agents in rural areas located in so-called peripheral countries would be able to reshape the design of educational technologies in order to fulfil their intended mission.[20] An analogous study of the introduction of OLPC laptops in primary and secondary education institutions in Paraguay displayed the stark difference between OLPC designers' pedagogical ideals and the pedagogical philosophy required for these schools. Engaging with the culture of students, teachers and parents in Paraguayan schools was deemed a fundamental requirement to develop this educational programme, which paradoxically the MIT ideologues had not contemplated. In this perspective, OLPC leaders can be aligned with a view of educational technology shaped by a material and conceptual *fetishism* of machines. Moreover, this ethnographic investigation pinpointed the interest of applying the concept of *charisma* to research on the users and uses of educational technologies, as Paraguayan students, teachers and parents often thought that they were responsible for the programme's failure due to their inability to use its computers properly. This object charisma was ingrained in their more general expectation of joining the information society promoted by the Paraguayan government.[21] Similar conclusions and approaches have arisen in other Latin American countries for other computer in education programmes such as *Computadores para Educar* in Ecuador.[22]

The fact that understanding educational technologies requires dealing not only with technology but also with education, and that both involve particular social, political and economic rationalities which are contingent and changing in time and place, might seem a truism. However, as we have seen, it defies notions about new educational technologies commonly held by designers and the public.[23] In 1989, the second edition of a review volume on the computer revolution recalled that for more than two decades the claim that computers were about to revolutionize education had been periodically announced without actually happening.[24]

Although a particular definition of educational technology crystallized between the 1960s and 1970s in Latin America and beyond, it was far more diverse than what is usually meant today as ICT for education. Furthermore, a quick retrospective glimpse at our recent history is sufficient to evince the limitations of holding a single definition of technologies of education. The presence of technology in classrooms and its use in teaching is much older and diverse. Accordingly, in the next sections I will go back in time and present briefly two case studies that exemplify an interdisciplinary but historically nuanced approach to educational technology.[25] The first section deals with patents for primary school teaching machines and techniques, the second with audio-visual and mass communication media for education.

## PATENTS OF PEDAGOGICAL INNOVATION

The provision of teaching materials for schools was a high priority for the newly independent republics in Latin America since the early decades of the nineteenth century. This involved the supply of textbooks, which initially were often translations of European works or books especially produced for Latin America in cosmopolitan book trade capitals such as London and Paris.[26] Subsequently, this production was typically diversified and strengthened with the concurrence of national authors, and by the mid-twentieth century governments were making large investments in a national textbook production for primary and secondary education.[27] Textbooks as research objects for Latin American history have caught the interest of historians of education and educational researchers.[28] In contrast, research on other types of educational technologies is more limited. This is in fact an international problem for the history of education which is still finding its way towards the production of more analytical investigations on its material culture, through dialogue with other disciplines such as museum and media studies, science and technology studies and the history of science.[29]

Thus we know for instance that, in the first decades of the nineteenth century in Colombia, schools based on the Bell-Lancaster monitorial system of instruction were catered with writing samples, blackboards, carbon pencils, pens, inkwells, paper and charts dealing with topics such as spelling, morality, national constitution, grammar, arithmetic and geography. Although there was a plan to provide schools and universities with cabinets of physics, chemistry and natural history, it took several decades to introduce this type of educational technology. Towards the end of the century, collections of machine models, scientific instruments and geological samples were purchased through agreements with European countries.[30] Similar characteristics are found in the supply of primary school materials in Mexico during the same period.[31] In the last decades of the nineteenth century, in Brazil, school materials imported for object-teaching included for instance: Parker letters, Prang models, Deyrolle and Saffray collection cabinets, lathes, anatomical models, physics and chemistry apparatus, and geographical, natural history and weights and measures system charts. In the first half of the twentieth century new items were added, for instance audio-visual means such as projectors and turntables.[32] In Argentina comparable collections were created and arranged in school museums through the acquisition of pedagogical materials produced in France and Germany. Towards the end of the nineteenth century some local teachers and manufacturers were producing encyclopaedic boxes designed for object lessons in the national perspective.[33] By the second half of the century the secondary schools in the major cities of Argentina, Brazil or Mexico had major collections for the teaching of physics, chemistry and natural history.[34] The account provided here is eminently descriptive, as is the sample secondary literature in which it is based. In the following, I exemplify a more analytical approach that can show the potential to get deeper into our historical understanding of nineteenth-century educational technologies in Latin America. I have chosen a case study on patents for its ability to integrate several relevant lines of inquiry in this context. Dealing with patents, advertisements or extant teaching collections is a mandatory prerequisite to document the Latin American contribution to the production and use of educational technologies. A selection of patents preserved in Mexico can serve as an illustrative example of this approach.

Extant patents or privileges for the exploitation of teaching inventions in nineteenth-century Mexico are in fact a very small part of all patents preserved in Mexican archives. They range from 1836 to 1890 (although primarily concentrated in the second half of the century) and represent less than one per cent of a corpus of documents privileging other fields of technological invention such as agriculture, transport, construction, mining or metallurgy. Pedagogical inventions were included in the class of 'stationery, teaching and popularization articles' subdivided in two further categories: 'office articles, letter press printing machines, bookbinding, teaching articles, and typewriters' and 'book jackets, publicity, mailing, communication by carrier pigeons and signals'. Documents on pedagogical invention preserved at the Mexico City and Mexican national archives are of two types: applications for permission to introduce or commercialize teaching inventions in the city's schools, evaluated by a municipal school commission (called *privileges*) and patents per se, which were subject to examination by a government commission. In practice, the process of patent evaluation rarely discussed the usefulness of inventions, but only their compliance with legislation and public safety. The loose practice of patent registration and control justifies a combined analysis of these two archival sets.[35]

Due to the marginal and fragmentary character of this documentation,[36] it would be difficult to produce a quantitative analysis able to explain the role of educational technology inventions in the transformation and improvement of educational practice, as it is often customary in economic history and the history of technology. In spite of the benefits of such an approach, it is well known that patents are not an exact measure of all technological inventions in a national context, that inventing is different from patenting, and it is still a matter of discussion whether patents have promoted or obstructed technological change.[37] The approach used here is qualitative, but it still serves the aim of suggesting ways in which the study of educational technology inventions could contribute to the study of educational practice and change.

If we discard the privileges awarded for the introduction of textbooks, primers, treatises or encyclopaedias in schools, there was a relevant number of privileges and patents that presented inventions of techniques and technologies aimed at improving teaching practice. These included reading, writing and counting techniques, and technologies integrating a number of different school subjects in the form of pedagogical systems, encyclopaedic boxes and teaching machines. Their authors are in general little known, they were Mexican,[38] and were teachers (in public or private schools), mechanics, traders or independent entrepreneurs. Among these inventions, writing techniques were in general implemented through the design of notebooks that included letter patterns that the pupil would have to fill with ink following carefully the path formed by the pattern, or different types of horizontal, vertical and diagonal lines constituting a scaffold allowing the young writer to get an exact calligraphy through repetition. Most of these inventions insisted on the importance of hand and body posture, thus being techniques and technologies intended to discipline students' bodies.[39] There were also machines devised to print these types of writing guide techniques, a blackboard made of darkened glass allowing to copy different sets of calligraphy placed in the background, or a mechanical spelling machine formed by a turning disc with four concentric rows of letters (from simple letters to three and four letter combinations) with windows displaying letter combinations that changed when turning the disc, and a machine aimed at 'solving any arithmetic problem'.[40]

Some of these inventions targeted several school subjects and advocated for their integrated learning through objects. In 1887, Pomposo Becerril, an artist and teacher who directed the School of Arts and Trades in Toluca, the Mexico state capital, presented an application for a patent on a method for the simultaneous teaching of geography and arithmetic, designed as a board game.[41] Another teacher in the state of Mexico, Clemente Antonio Neve, proposed a school instruction system designed to teach reading, writing, arithmetic, grammar, cosmography, political and physical geography and history through a series of chests of drawers containing three-dimensional coloured letters, numbers and balls, dolls, writing samples, pictures, diagrams, blackboards, square and circular panels and an arrangement of circular tables. Analogously, there were patent applications for encyclopaedic boxes as small chests of drawers containing hundreds of objects (minerals, vegetables, preserved animals or parts of them, woods and industrial products) conceived for object lessons.[42]

Neve is known for his firm advocacy for the teaching method called 'intuitive teaching' or 'objective teaching' which organized teaching through the manipulation of everyday simple objects and the use of the senses by school pupils. For instance, he used rosaries to teach mathematics as a way to materialize arithmetic.[43] He was also the inventor of an 'Intuitive Machine for the teaching in schools', a cube (occupying a volume of one cubic metre) to whose sides were attached different movable figures. Each movable piece (discs or squares) had printed information on its surface and was attached to the cube by poles that allowed their spinning. Some of these discs also had radial metal rods where sets of coloured balls could be placed. According to Neve, working with his machine two hours in the morning and two hours in the afternoon would be enough for adequate school performance. With this mechanical system he planned to teach reading, writing, arithmetic, grammar, geography and cosmography by establishing arithmetic, geometric and chromatic homologies between syllables, numbers and planets with the arrangement of the discs and balls in his machine. Not in vain, in his patent application Neve referred to the logic machine invented by Ramon Lull in the thirteenth century, and it is likely he was inspired by it. He also accused a contemporary competitor of plagiarism: according to him, a machine presented the same month by Antonio P. Castilla with similar aims to his, under the name of 'Caleideslojico' (a combination of 'kaleidoscope' and 'logical') was in fact a copy of an invention well known in the schools of Germany and the US and described for instance in a *Manual of School Material* published in New York in 1874, which he cited. In contrast, his invention was the result of a thirty-year experience as a teacher, and it had deserved awards in the Agricultural and Mechanical Annual Fair held in St. Louis, Missouri in 1879, and a similar competition in Puebla.[44] Castilla was also a teacher in Mexico City (where Neve had moved and opened a private school) and an active author of patent applications for teaching inventions.

Accusations of plagiarism or lack of novelty were usual in the process of these Mexican patent applications, and in disputes over their originality one of the most usual arguments of competing applicants or patent official reviewers was that the pretended invention had already been invented abroad and copied or adapted by the Mexican applicant. Some patent applications included quite detailed drawings; they were in general figurative (not technical) representations. It appears that the Mexican authorities did not require inventors to deposit at least their invention prototype. As a result, and taking into account that – as far as I know – there are no preserved collections of nineteenth-century educational technologies (including teaching machines) for primary education in Mexico, it is difficult to know if these inventions were actually produced, if they were manufactured in a relevant number, or if they were introduced and used in schools at all.

Historians of the nineteenth-century scientific instrument trade have noted that instrument makers often included in their catalogues reference to instruments that had never been produced, as a way to attract customers by displaying an encyclopaedic repertoire of available products.[45] Historians of technology have demonstrated that patent applications were in several ways literary constructions aimed at protecting for as long as possible inventor's rights, discarding competitors, recovering investments and claiming profit over future inventions based on similar scientific or technological mechanisms.[46] The educational technologies presented in this section were technically rudimentary, both in design, materials and production, and possibly they were produced in small numbers. In the same period, in Europe, countries such as France and Germany were producing scientific instruments for primary and secondary education in large workshops which not only catered for national demand, but were massively exported abroad to other European countries, the Latin American republics and North America, among others.[47] In spite of what appears to be a much more modest production, both qualitatively and quantitatively,[48] the Mexican patent documents presented here are extremely valuable since they can show how Mexican teachers, mechanics and traders saw pedagogical innovation for the improvement of national school education, what their horizons were, and how they conceived teaching and learning. Moreover, through these patents we can investigate the expectations of the technification and technologization of education and the ideas on what a teaching machine should be, in a Latin American country such as Mexico, during the second half of the nineteenth century.

## AUDIO-VISUAL AND MASS COMMUNICATION MEDIA FOR EDUCATION

In the course of the nineteenth century, new technologies progressively found their way into Latin American classrooms through the initiatives of teachers, inventors, traders and governments. The machines and techniques that they imagined and designed were grounded in particular pedagogical philosophies (such as object lessons and experimental physics demonstrations). Thus, they contributed to definitions of educational technology related but different to those arising in the twentieth century that I presented in the first section of this paper.

By the mid-twentieth century the use of audio-visual and mass communication media for teaching were on the rise through a more massive distribution of educational technologies in schools, and efforts to standardize their use that contributed to the diversification and specialization of teaching practices. In Mexico, Roberto Moreno García and María de la Luz López Ortiz, two teachers and professors at the national teacher training school and the faculty of Philosophy and Letters of the Universidad Nacional Autónoma de México (UNAM), published an important manual on audio-visual education synthesizing their experiences.[49] Its preface opened with a controversial claim:

> Film projection, applied with anarchy and lack of pedagogical systematization is one of the greatest evils that are currently threatening our national educational system.

The authors compared the risks of contemporary enthusiasm for film projection in schools with the advent of the printing press in the sixteenth century that initially had turned teaching into a bookish practice. The problem was defined as follows: films and slides had arrived in large amounts to schools because of their affordable prices, promoting the belief that all school subjects could be taught exclusively through their projection, and discarding other relevant teaching tools. Retrospectively films had occupied the privileged space that once books had, and the idea that teaching equalled technical mastery of projection machines was becoming established.[50] We can see here the charismatic status that new educational technologies often take.[51] The authors reminded that no more than 20 per cent of the more than 20,000 schools in Mexico had electricity supply. Only a very limited number of schools would be able to use these new educational technologies. Moreover, it was a mistake to think that mechanical apparatus could substitute teachers' action. Without denying their interest, teaching involved a wider range of techniques and technologies within a long history of educational experience, including models, blackboards, charts, school field trips (to nature, the city, or museums), graphic materials, drama representations, clubs, a long repertoire of audio-visual projection techniques, scientific instruments, radio and television.

Their manual provided a description of more than one hundred such materials and stressed their inscription in the framework of several pedagogical philosophies and techniques that should guide teaching practice, organized in five classes that they entitled 'Mexico Classification'. Furthermore, it displayed empirical data on educational technologies and teaching practice surveys carried forward in Mexico, and a good knowledge of international developments.[52] In 1951, Moreno had established a Service of Audiovisual Education at UNAM, and through different executive positions in the national school system he developed a sustained programme for the training of teachers as a way to produce a rational and efficient introduction of educational technologies in Mexican schools. For this purpose a good knowledge of pedagogical foundations and techniques was mandatory, in order to avoid improvisation and to subdue the economic interests of the educational technology marketplace to school needs.[53]

Although a good amount of the new educational technologies (e.g. films and their projection apparatus) were still imported, during the second half of the twentieth century there was a rising Latin American production of audio-visual and mass communication media for education. Countries such as Mexico, Brazil, Argentina and Chile had started to produce educational films since the 1920s.[54] Moreover, since the 1930s, Colombia, Argentina, Uruguay and Venezuela had national educational radio networks.[55] The first national radio networks in Latin America were created with a cultural and educational goal, and they originally broadcasted specially designed educational programmes. These pioneer radio stations were called in some of these countries 'schools of the air'.[56]

The experience of these educational networks inspired the establishment of television broadcasting between the 1950s and 1960s in all Latin American countries.[57] Countries such as Colombia were offering in the 1960s teaching through television on selected primary school subjects. Colombian broadcasters received technical support from US agencies. In the same period the Organization of American States agreed the creation of a Latin American Television Centre to train educational television professionals, with headquarters in Colombia, Mexico, Chile and Argentina. The development of educational television over large territories also involved technical challenges such as the establishment of terrestrial or satellite networks.[58] Educational television and radio have survived until our days especially in support of education in rural areas in Latin American countries.[59] However, their history displays more ruptures than continuities, since many projects were ephemeral and ended after several years for financial, technical or political problems.[60]

In spite of the difficulties to establish stable platforms for the production of educational technologies in Latin America, there were countries such as Mexico, Chile and Brazil that developed relevant experiences in the training of teachers from across the continent and the production of pedagogical packages. Through agreements with UNESCO a Latin American institute of educational cinematography was established in Mexico (1956),[61] a Latin American mission on audio-visual aids in Chile (1960s),[62] and a project producing science teaching educational technologies in Brazil (1950s to 1970s). UNESCO's Latin American experience was subsequently exported to other parts of the world. The Brazilian project was an exemplary experience, originally starting in university laboratories where science kits were produced for some schools in the São Paulo state, subsequently moving to factory premises in order to expand production to the nation's schools, and finally broadening its production to the medical technology market.[63] These projects were not only national but constituted poles of educational production and training across the Latin American region.

## CONCLUSIONS

Computers, tablets and online platforms are invading the way we think and practise teaching and learning, as inevitable references in government policies for educational improvement, and international discourses of socio-economic development. The degree to which the idea of education has been technified and technologized, and educational technology features in the public arena is perhaps unprecedented. Nonetheless, electronic machines are just one type of educational technology among a large set of technological objects designed for educational action since at least the nineteenth century. With hindsight, we can see that most new educational technologies were endowed at the time of their introduction with a charisma that contributes to obscure the substance of their contribution to teaching and learning. In spite of the escalating technologization of our society and the undeniable power of educational technologies, a historical and sociological analysis demonstrates that education is still in the hands of human teachers whose practice is shaped by socio-cultural features connected to geopolitical location. Technologies per se would be unable to perform education and to produce educational change. While this might be obvious for historians of technology and Science and Technology Studies scholars, it is still a relevant point to raise towards education scholars and policy makers, and overall to claim the need to historicize educational technology and its practices.

In the twenty-first century Latin America is a privileged field for the trial and commercialization of new educational technologies, and as we have seen, this has been the case since at least the nineteenth century. The importation of teaching instruments, machines and techniques was common in the nineteenth century and it went on during the twentieth century. Moreover, as this paper suggests since the nineteenth century Latin America has also produced educational technology inventions, transformative appropriations of imported technologies, and significant international experiences in pedagogical innovation. While we move towards the inflexible implementation of 'global' educational quality indicators, the techno-pedagogical inventiveness of teachers such as Antonio Neve, the participation of Brazilian researchers in the development of programmed instruction and school science kit design, or the experience of Mexican, Brazilian, Argentinian and Colombian teams in the development of educational radio, cinema and television, are relevant landmarks to pinpoint.

In contrast to the restrictive definition that characterizes common understandings of ICT for education today, there were particular definitions of educational technology in different historical periods. In nineteenth-century Mexico, printed materials, writing, reading and counting techniques and tools, and encyclopaedic boxes, could be arranged together with teaching machines which developed pedagogical sequences in the classroom. Between the 1920s and 1960s in Latin America, the rise of cinema, radio and television, and the growth and diversification of technologies of image projection opened the way to new terms and meanings such as audio-visual aids and mass communication media. Between the 1960s and 1970s, the production of pedagogical packages for science teaching in Brazil contributed to the conception of educational technology as a kit integrating many – if not all – of the elements previously mentioned in a different pedagogical framework.

In parallel, the production of educational technologies from the nineteenth century to our days has seen major changes such as the move from the solitary inventor to large and interdisciplinary teams, the technification and specialization in educational technology design, commercialization and teaching manipulation, and its role in the institutionalization and professionalization of educational research.

Our discussion of textbooks as technologies and programmed instruction in the context of Brazilian educational research in the 1960s to 1970s, and on the textual qualities of nineteenth-century Mexican teaching machines, are illustrative examples of these trends. As I have argued in this paper, the research objects that I have empirically defined as educational technologies (from textbooks to computers) and the historical phenomena in which they are inscribed, require an integrated approach considering their historically nuanced technological and pedagogical diversity, but dealing with them as a whole. In this context, Latin America features prominently for the historical and contemporary relevance of its educational technology experiences and the great potential of case studies in this geopolitical region of the world. This research agenda requires collaboration between a wide range of disciplines interested in educational practice as a research field, and thus it constitutes a great opportunity for future research in the history of technology.

## NOTES AND REFERENCES

1. See for instance Ministerio de Educación Nacional, *Revolución educativa: Plan sectorial 2006–2010. Documento nº8* (Bogotá: República de Colombia, 2008); Gobierno de la República, *Plan nacional de desarrollo 2013–2018* (México: Gobierno de la República. Estados Unidos Mexicanos, 2013); Barack Obama, '47 – Address Before a Joint Session of the Congress on the State of the Union', *The American Presidency Project* (25 January 2011), http://www.presidency.ucsb.edu/ws/?pid=88928 (accessed 26 March 2018).
2. Sociedade Brasileira de Física, *Atas do Simpósio Nacional de Ensino da Física-janeiro 1970. Edição preliminar* (São Paulo: Sociedad Brasileira de Física, 1970).
3. Beatriz Alvarenga in *Atas do Simpósio Nacional de Ensino da Física*: I-9.
4. Antônio Teixeira in *Atas do Simpósio Nacional de Ensino da Física*: I-22–23. All translations from Portuguese and Spanish into English are by the author of this paper.
5. Josep Simon, 'The Transnational Physical Science Study Committee: The Evolving Nation in the World of Science and Education (1945–1975)', in John Krige (ed.), *How Knowledge Moves: Writing the Transnational History of Science and Technology* (Chicago: University of Chicago Press, 2019): 308–342.
6. John Hedley Brooke, 'Introduction: The Study of Chemical Textbooks', in Anders Lundgren and Bernadette Bensaude-Vincent (eds.), *Communicating Chemistry: Textbooks and Their Audiences, 1789–1939* (Canton, MA: Science History Publications, 2000): 1–18; Josep Simon, 'Textbooks', in Bernard Lightman (ed.), *A Companion to the History of Science* (Chichester: Wiley-Blackwell, 2016): 400–413.
7. Uri Haber-Schaim, 'PSSC PHYSICS: A Personal Perspective', in *PSSC: 50 Years Later* (American Association of Physics Teachers, 2006) http://www.compadre.org/portal/pssc/pssc.cfm?view=title (accessed 26 March 2018).
8. John L. Rudolph, *Scientists in the Classroom: the Cold War Reconstruction of American Science Education* (New York: Palgrave, 2002); Simon, 'The Transnational Physical Science Study Committee'.

9. These were produced in Brazil (1962–1964) and Colombia (1964). An earlier translation into Spanish of the main PSSC textbook was published in Spain (1962) by a publisher with branches in many Latin American capitals. This edition circulated widely in all Latin American countries, except Colombia.
10. PLATO (Programmed Logic for Automatic Teaching Operations) was actually designed and run at the University of Illinois, but also used at MIT. Educational Services Inc, 'Records, 1956-1970', MC 79, Box 11; Folder 'Eliza', Records of the Physical Science Study Committee, MC.626, Box 7; Folders 'NSF Comp. in Educ. 12/7/67 Md.' and 'Computer-Aided Displays Proposal', and Roger L. Johnson, 'The Use of Programmed Learning and Computer-Based Instruction Techniques to Teach Electrical Engineering Network Analysis', Report R-297, July, 1966 (Urbana, IL: Coordinated Science Laboratory, University of Illinois, 1968), and D. L. Bitzer, E. R. Lyman and J. R. Suchman, 'REPLAB. A Study in Scientific Inquiry using the PLATO System', Report R-260, December, 1965 (Urbana, IL: Coordinated Science Laboratory, University of Illinois, 1965), Records of the Physical Science Study Committee, MC.626, Box 10, MIT, Institute Archives and Special Collections.
11. MIT Media Lab was created in the 1980s, based on the Architecture Machine Group created by Negroponte in 1967. The OLPC OX laptop was released in the twenty-first century, but it synthesized the ideas on machines and education developed by Papert since at least the 1960s. Morgan G. Ames and Daniela K. Rosner, 'From Drills to Laptops: Designing Modern Childhood Imaginaries', *Information, Communication & Society* 17, 3 (2014): 357–370.
12. Fuad D. Saad, *Análise do projeto FAI – Uma proposta de um curso de física auto-instrutivo para o 2° grau* (São Paulo: Universidade de São Paulo, 1977); Idely Garcia Rodrigues and Ernst W. Hamburger, *O 'Grupo de Ensino' do IFUSP: histórico e atividades. Publicações. IFUSP/P-1035, Março/1993* (São Paulo: Instituto de Física – Universidade de São Paulo, 1993); Roberto Nardi, 'Memórias da educação em ciências no Brasil: a pesquisa em ensino de física', *Investigações em Ensino de Ciências* 10, 1 (2005): 63–101.
13. Simon, 'The Transnational Physical Science Study Committee'.
14. Fred S. Keller, 'Engineering Personalized Instruction in the Classroom', *Revista Interamericana de Psicología* 1, 3 (1967): 189–197.
15. Historians of computing in Latin America have remarked on Brazilian leadership in the development of a national computer industry between the 1960s and early 1990s. But they hardly make any mention of the market of computers for formal education. Ivan da Costa Marques (ed.), 'History of Computing in Latin America', *IEEE Annals of the History of Computing* 37, 4 (2015): 10–86.
16. Dirceu Soares, 'Êles não estão brincando', *Realidade* 33, December (1968): 202–216.
17. Edward Hamilton and Andrew Feenberg, 'Alternative Rationalisations and Ambivalent Futures: A Critical History of Online Education', in Andrew Feenberg and Norm Friesen (eds.), *(Re)Inventing the Internet: Critical Case Studies* (Rotterdam: Sense Publishers, 2012): 43–70.

18. A large amount of the educational papers on this topic reproduce the propagandist statements of One Laptop per Child's leader, or the Computers for Schools website official statements, instead of analysing the real impact of these programmes.
19. Anita S. Chan, 'Balancing Design: OLPC Engineers and ICT Translations at the Periphery', in Eden Medina, Ivan da Costa Marques and Christina Holmes (eds.), *Beyond Imported Magic: Essays on Science, Technology and Society in Latin America* (Cambridge, MA: MIT Press, 2014): 181–206.
20. In concordance with this view the general argument of David Edgerton, *The Shock of the Old: Technology and Global History since 1900* (London: Profile Books, 2006), and Nelly Oudshoorn and Trevor Pinch (eds.), *How Users Matter: The Co-Construction of Users and Technologies* (Cambridge, MA: MIT Press, 2003).
21. Morgan G. Ames, 'Translating Magic: The Charisma of One Laptop per Child's XO Laptop in Paraguay', in Medina et al., *Beyond Imported Magic*: 207–224.
22. María Belén Albornoz, Mónica Bustamante Salamanca and Javier Jiménez Becerra, *Computadores y cajas negras* (Quito: Flacso Ecuador, 2012). This work in addition includes a concise review of contemporary projects in other countries such as Costa Rica, Chile, Argentina, Uruguay, Brazil, Mexico and Colombia.
23. Hamilton and Feenberg, 'Alternative Rationalisations and Ambivalent Futures'.
24. James A. Levin and Yaakov Kareev, 'Personal Computers and Education 1980, 1984', in Zenon W. Pylyshyn and Liam Bannon (eds.), *Perspectives on the Computer Revolution* (Norwood: Ablex Publishing Corporation, 1989, 2nd edn): 369–393.
25. This approach is partially supported by proposals such as R. Murray Thomas, 'The Nature of Educational Technology', in R. Murray Thomas and Victor Kobayashi, *Educational Technology – Its Creation, Development and Cross-Cultural Transfer* (Oxford: Pergamon Press, 1987): 1–23; Neil Selwyn, *Education and Technology: Key Issues and Debates* (London: Continuum, 2011); Keri Facer, *Learning Futures: Education, Technology and Social Change* (London: Routledge, 2011); Bill Ferster, *Teaching Machines: Learning from the Intersection of Education and Technology* (Batimore: Johns Hopkins University Press, 2014).
26. Eugenia Roldán Vera, *The British Book Trade and Spanish American Independence. Education and Knowledge Transmission in Transcontinental Perspective* (Aldershot: Ashgate, 2003).
27. Lorenza Villa Lever, *Cincuenta años de la Comisión Nacional de los Libros de Texto Gratuitos: cambios y permanencias en la educación mexicana* (México: Comisión Nacional de los Libros de Texto Gratuitos, 2009); Aníbal Bragança and Márcia Abreu (eds.), *Impresso no Brasil: Dois séculos de livros brasileiros* (São Paulo: UNESP, 2008); Laurence Hallewell, *O livro no Brasil: sua história* (São Paulo: EdUSP, 2005).
28. Gabriela Ossenbach Sauter and José Miguel Somoza Rodríguez (eds.), *Los manuales escolares como fuente para la historia de la educación en América Latina* (Madrid: UNED, 2001). For textbooks as a general object of research, see Simon, 'Textbooks' and 'Physics Textbooks and Textbook Physics in the Nineteenth and Twentieth Centuries', in Jed Z. Buchwald and Robert Fox (eds.), *The Oxford Handbook of the History of Physics* (Oxford: Oxford University Press, 2013): 651–678; Josep Simon

and Antonio García-Belmar, 'Education and Textbooks', *Technology and Culture* 57, 4 (2016): 940–950.

29. Martin Lawn and Ian Grosvenor (eds.), *Materialities of Schooling: Design, Technology, Objects, Routines* (Oxford: Symposium Books, 2005); Paulí Dávila and José M. Anaya (eds.), *Espacios y patrimonio histórico-educativo* (Donostia: Erein, 2016). For illustrative examples of history of science approaches to this topic see José R. Bertomeu Sánchez and Antonio García Belmar, *Abriendo las Cajas Negras: Instrumentos Científicos de la Universidad de Valencia* (València: Universitat de València, 2002); Peter Heering and Roland Witje, *Learning by Doing: Experiments and Instruments in the History of Science Teaching* (Stuttgart: Franz Steiner Verlag, 2011).
30. Luis Rubén Pérez Pinzón, *Historia del Empresarismo en el nororiente de Colombia. Tomo 3* (Colombia: The author, 2015): 79, 84–87, 130, 148–149, 196–197.
31. Josefina Granja Castro, *Métodos, aparatos máquinas para la enseñanza en México en el siglo XIX: Imaginarios y saberes populares* (Barcelona – México: Ediciones Pomares, 2004): 31–34.
32. Rosa F. de Souza, 'Objects of Learning: The Pedagogic and Material Renovation of Elementary School in Brazil, in the 20th Century', *Educar Em Revista* 49, jul./set. (2013): 103–120.
33. Susana V. García, 'Museos escolares, colecciones y la enseñanza elemental de las ciencias naturales en la Argentina de fines del siglo XIX', *História, Ciências, Saúde – Manguinhos* 14, 1 (2007): 173–196.
34. In 2013 I catalogued the extant science collections of the oldest secondary school in Mexico City, affiliated to the Universidad Nacional Autónoma de México, with the collaboration of Felipe León. They comprised more than 500 physics instruments and some chemical products, mathematical instruments and experimental physiology and psychology instruments, produced between the nineteenth and early twentieth centuries, mostly by French, British and German makers but also some local artisans. The catalogue was not published and the project could not continue for lack of interest of the Secretaría de Difusión Cultural de la Escuela Nacional Preparatoria. In Mexico there are preserved physics and chemistry teaching collections at the universities of Zacatecas, Puebla and the state of Mexico. For Argentina and Brazil, see María C. von Reichenbach, 'Historic Instruments for the Teaching of Physics: A Chronology of the Situation in Argentina', *Museologia E Patrimônio* 8, 2 (2015): 123–142; Marcus Granato and Marta Lourenço, *Coleções científicas de instituições luso-brasileiras: Patrimônio a ser Descoberto* (Rio de Janeiro: MAST/MCT, 2010).
35. Granja Castro, *Métodos, aparatos máquinas para la enseñanza*: 27–39. On the difference between privileges and patents see Mario Biagioli, 'Patent Republic: Representing Inventions, Constructing Rights and Authors', *Social Research* 73, 4 (2006): 1129–1172.
36. The two archival series used here are arguably incomplete, but in addition, there is a larger amount of primary sources that would allow building a larger set of patent records. Edward Beatty, 'Patents and Technological Change in Late Industrialization: Nineteenth-century Mexico in Comparative Context', *History of Technology* 24, 2 (2002): 121–150.

37. See for instance, Juan Ignacio Campa Navarro, 'Patentes y desenvolvimiento tecnológico en México: Un estudio comparativo entre la época de industrialización proteccionista y el régimen de apertura', *América Latina En La Historia Económica* (2017), DOI: 10.18232/alhe.789; Edward Beatty and Patrício Sáiz, 'Propiedad industrial, patentes e inversión en tecnología en España y México (1820 1914)', in Rafael Dobado, Aurora Gómez Galvarriato and Graciela Márquez (eds.), *México y España ¿historias económicas paralelas?* (México: Fondo de Cultura Económica, 2007): 425–467; Carolyn C. Cooper (ed.), 'Patents and Invention', *Technology and Culture* 32, 4 (1991): 837–1093; Robert Fox, *Technological Change: Methods and Themes in the History of Technology* (Abingdon: Routledge, 2004).

38. In contrast with the patenting of industrial technologies in the same period, largely due to foreign entrepreneurs. Beatty and Sáiz, 'Propiedad industrial, patentes e inversión en tecnología'.

39. 'Método de escribir de Castillo' (1858), 'Método de escribir Murguía' (1858), 'Nuevo Método para enseñar a escribir letra inglesa' (1858), 'El copiador popular de Antonio P. Castilla' (1870), in Granja Castro, *Métodos, aparatos máquinas para la enseñanza*: 53–65. On technique, training and body discipline see Peter Dear, 'A Mechanical Microcosm: Bodily Passions, Good Manners, and Cartesian Mechanism', in Christopher Lawrence and Steven Shapin (eds.), *Science Incarnate. Historical Embodiments of Natural Knowledge* (Chicago: University of Chicago Press, 1998): 5–82; Andrew Warwick, *Masters of Theory: Cambridge and the Rise of Mathematical Physics* (Chicago: Chicago University Press, 2003).

40. 'Máquina para estacigencias de escritura' (1869), 'Pizarras calcantes' (1877), 'Aparato para enseñar a leer, llamado "Silabario Mecánico". Autor: Valeriano Lara' (1881), 'Máquina para resolver cualquier problema aritmético, llamada "Contador infalible"' (1857), in Granja Castro, *Métodos, aparatos máquinas para la enseñanza*: 66–73.

41. 'Método práctico recreativo para aprender simultáneamente la geografía y la artimética' (1887), in Granja Castro, *Métodos, aparatos máquinas para la enseñanza*: 74–83.

42. 'Sistema de enseñanza en las escuelas. Autor: Clemente A. Neve' (1875), 'Caja enciclopédica para la enseñanza intuitiva. Autor: Ildefonso Estrada y Zenea' (1878), in Granja Castro, *Métodos, aparatos máquinas para la enseñanza*: 88–99, 116–121.

43. Mílada Bazant, 'La mística del trabajo y el progreso en las aulas escolares, 1874-1911', in Alícia Civera Cerezedo (ed.), *Experiencias educativas en el Estado de México. Un recorrido histórico* (México: FOEM – El Colegio Mexiquense, 2013): 131–150.

44. 'Sistema de enseñanza en las escuelas. Autor: Clemente A. Neve' (1875), 'Caja enciclopédica para la enseñanza intuitiva. Autor: Ildefonso Estrada y Zenea' (1878), in Granja Castro, *Métodos, aparatos máquinas para la enseñanza*: 100–107.

45. Paolo Brenni, 'Nineteenth Century Scientific Instrument Advertising', *Nuncius* 17 (2002): 497–514.

46. Charles Bazerman, *The Languages of Edison's Light* (Cambridge, MA: MIT Press, 1999); Greg Myers, 'From Discovery to Invention: The Writing and Rewriting of Two Patents', *Social Studies of Science* 25, 1 (1995): 57–105.

47. Paolo Brenni, 'The Evolution of Teaching Instruments and Their Use Between 1800 and 1930', *Science & Education* 21, 2 (2012): 191–226.
48. This appears from the analysis of educational technology patents that I have performed here. However, the historical study of the production, circulation and use of educational technologies in Latin America is still a much understudied research area.
49. Roberto Moreno y García and María de la Luz López Ortiz, *La enseñanza audiovisual* (México: Editorial Patria S. A., 1952).
50. Moreno y García and López Ortiz, 'Prólogo de la primera edición', in *La enseñanza audiovisual* (2nd edn, 1960): 7–9.
51. Ames and Rosner, 'From Drills to Laptops'.
52. Moreno y García and López Ortiz, *La enseñanza audiovisual* (2nd ed., 1960).
53. Libertad Menéndez Menéndez, 'Roberto Moreno y García', in Facultad de Filosofía y Letras (ed.), *Setenta años de la Facultad de Filosofía y Letras* (México D.F.: Universidad Nacional Autónoma de México, 1994): 443–444.
54. Rosa M. Gudiño Cejudo, 'Un recorrido filmográfico por la Secretaría de Educación Pública: México (1920–1940)', *Revista Tempos E Espaços Em Educação* 11, 26 (2018): 91–112; Sheila Schvarzman, *Humberto Mauro e as imagens do Brasil* (Sao Paulo: UNESP, 2004); Alicia Alted Vigil and Susana Sel (eds.), *Cine educativo y científico en España, Argentina y Uruguay* (Madrid: Editorial Universitaria Ramon Areces, 2016); Armando Rojas Castro (ed.), *Cine Educativo. Boletín Quincenal del Instituto de Cinematografía Educativa de la Universidad de Chile* 2, 15 de junio (1932): 1–4.
55. Moreno y García and López Ortiz, *La enseñanza audiovisual* (2nd edn, 1960): 218–219.
56. Milcíades Vizcaíno, *Estado y medios masivos para la educación en Colombia (1929–2004)* (Bogotá: Universidad Cooperativa de Colombia, 2014); Christina Ehrick, *Radio and the Gendered Soundscape: Women and Broadcasting in Argentina and Uruguay, 1930–1950* (Cambridge: Cambridge University Press, 2015); Ayder Berrio et al., *Radio Sutatenza: una revolución cultural en el campo colombiano (1947–1994)* (Bogotá: Banco de la República, 2017); Eugenia Roldán Vera, 'Los orígenes de la radio educativa en México y Alemania: 1924–1935', *Revista Mexicana de Investigación Educativa* 14, 40 (2008): 13–41.
57. John M. Culkin, 'Televisión educativa en América Latina', *La Educación en América* 11, enero-marzo (1963): 57–65.
58. William Bollay, 'Advanced Technological Approaches to Education', in *Report of the Colombia-U.S. Workshop on Science and Technology in Development, Fusagasugá, Colombia, February 26–March 1, 1968* (Washington, DC: National Academy of Sciences in Cooperation with the Colombian Ministry of Education, 1969): 30–32; Vizcaíno, *Estado y medios masivos*: 272–291.
59. Dorothy Tanck de Estrada (ed.), *Historia mínima de la educación en México* (México D.F.: El Colegio de México, 2010).
60. Vizcaíno, *Estado y medios masivos*.

61. UNESCO, *Regional Seminar on the Use of Visual Aids in Adult and School Education in Latin America. Mexico, 28 September–17 October 1959. Final Report* (Paris: UNESCO, 1960).

62. J. D. Kimball, *Audiovisual Aids. Improving Productivity of the First Nine Years of Schooling. 1964, 1965. End of Mission, January, 1966. Chile* (Paris: UNESCO, 1966).

63. Antônio C. Souza de Abrantes, *Ciência, Educaçao e Sociedade: O caso do Instituto Brasileiro de Educaçao, Ciência e Cultura (IBECC) e da Fundaçao Brasileira de Ensino de Ciências (FUNBEC)* (Rio de Janeiro: Casa de Oswaldo Cruz – Fiocruz, 2008); Simon, 'The Transnational Physical Science Study Committee'.

# Technology and the Fates of Three Caribbean Commodities

DAVID PRETEL

## INTRODUCTION

The transformative power of new technologies has become something of a cliché. Technological change has served to explain everything from industrial revolutions to political regimes, from cultural values to social structures. It has been noted that technology has been a major historical driver of environmental transformations through, for example, natural resources exploitation. After all, new tools, machines and practices have determined many of the patterns of commodity production, as the well-known histories of silver, sugar, oil and rubber demonstrate. At the same time, the processes of technological-driven industrialization in some parts of the world have often been connected to the commodity frontier-making process in far-flung locations. Of course, the interconnections between technological change and expanded commodity production have been diverse and uneven throughout history. It seems evident that the interplay between technological change and commodity production has been specific to material and historical conditions, including such factors as state and imperial politics, geography and the environment.

This article considers the evolution of natural resource frontiers in the Caribbean from the perspective of the history of technology – specifically the changing relations between technological change and the life cycles of commodities between the mid-nineteenth century and the Second World War. Drawing on the cases of three basic resources produced in the Maya-populated Yucatán peninsula – logwood, henequen and chicle – it offers an account of the rise and decline of global commodities. The article contends that the shifting trends of production, trade and consumption of these raw materials were primarily determined by the historical interplay of global technologies, local knowledge and practices and environmental conditions of production.

In adopting such a technological-historical vision over a long time frame, this article seeks to appraise the technological imperatives for tropical commodity production through a comparative study. The purpose is to understand how technological forces shaped commodity production in the Yucatán peninsula, illuminating common patterns and contrasting certain features and distinctions. Rather than attempting a comprehensive coverage of all the dimensions of logwood, chicle and henequen commodity chains, this article pursues a deeper understanding of the links and co-evolution of global industries and shifting commodity frontiers in the context of the consolidation of industrial capitalism. Such a perspective also exposes industrial imperatives, mutual impacts, political confrontations, transnational networks and unintended consequences.

The economic and socio-cultural history of henequen, chicle and logwood have received considerable scholarly attention. Despite this literature – often confined to area studies – the consideration of these commodities' technological dimensions within a comparative framework has not been given the attention it deserves. Rural communities in Central America and the Caribbean, particularly Maya people in the remote peninsula of Yucatán, have produced and used these natural resources for centuries. It was during the second part of the nineteenth century – the heart of the so-called Second Industrial Revolution – that the exploitation and manufacturing of these basic natural resources peaked as they became basic supplies for the growing chemical, agro-industrial, textile and food industries in Western Europe and the United States. At this point, Maya peoples and global industries became inextricably linked and their histories entangled until well into the twentieth century.

Frontier capitalism and consumer demand during this period were closely related, but such a link cannot be understood without considering how they were mediated by technological and environmental imperatives. In the three cases under consideration, technological changes in industrial and agricultural production significantly transformed the patterns of commodity production throughout the nineteenth century. Technological transformations also played a relevant part, though ambivalent, in their eventual decline in the twentieth century, through the development of artificial substitutes, such as chemical colourants, artificial fibres and gum bases made of synthetic polymers such as resins and waxes. Today, the production of henequen, chicle and logwood are no longer the main economic activities in the Yucatán peninsula, even though cordage, chewing gum and dyestuffs are still traded in large quantities in international markets.

After this introduction, this article consists of three additional sections. The first section discusses the historical impact of technology on the dynamics of tropical commodities extraction, building on conceptual categories such as 'technological frame', 'frontier capitalism' and 'commodity chain'. It considers not only the commodification of nature and the production of raw materials for world markets but also the impacts on social and labour relations at commodity frontiers. The second section traces the global rise, and eventual decline, of the henequen, chicle and logwood commodity chains. It reconstructs and analyses the long-term technological patterns of their extraction at different scales, identifying the specific technologies and expert knowledge involved in their production, including local agricultural and forestry practices. Additionally, it considers the later development of chemical innovations for the industrial production of foodstuffs, dyes and plastics in Europe and North America, particularly the development of mechanical innovations for processing and chemical substitutes. The final section of the article reflects on the interrelated trajectories of technological frontiers and the conjunctures of commodity production from a broader comparative perspective, looking beyond national historiographies and political events to illuminate long-term patterns.

## TECHNOLOGY, COMMODITIES AND FRONTIER CAPITALISM

There is compelling historical evidence to argue that technological innovations have been a source of profound socio-cultural changes at distinct scales, from the local to the global. As sociologist Stephen Hill has noted, the establishment of a new 'technological frame' throughout a given society pushes the realignment of its culture. In that sense, technological change is a double-edged sword: serving as a source of opportunities and economic change but also bringing about the relocation – even the subordination or destruction – of prevailing cultures.[1] The cultural impacts of technological change are apparent when we consider the extraction of raw materials at the peripheries of the world economy. The capitalist penetration into tropical lands often results in the transformation of indigenous cultures, the weakening of local economic systems and the deterioration of social structures, introducing previously unknown dependencies on the international economy.

A fertile point of departure for understanding the relationship between technology, culture and commodities is the notion of 'frontier'. The apprehension of the dynamics at the commodity frontier as a specific place of transition requires attention to its flexible nature and its interrelation with capitalist spatial expansion. The process of frontier expansion in tropical areas is often driven by commercial and industrial demands in far-flung industrial and urban localities. The dynamics of the incorporation of previously uncommodified land occurs alongside technological changes that increase the demand for raw materials such as agricultural and forestry products. At the same time, the possibilities presented by the extraction of specific natural resources are constrained by geographic, geological and environmental conditions that in turn influence the patterns of technological innovation in commodity extraction and production.[2] The appropriation and commodification of nature relied, in the words of Jason Moore, on 'an ingenious combination of technology and frontier-making'.[3] For example, Dale Tomich has shown that technological and environmental forces restructured the labour regimes that were prevailing at Caribbean frontiers of sugarcane production during the nineteenth century. From this perspective, nature is recreated through technologies and expertise, which in turn are bound to socio-economic conditions.[4]

Commodity frontiers are more than boundaries marked by nature. Historically, capitalist expansion to previously uncommodified lands has created zones of socio-political and cultural encounters.[5] Therefore, the study of extractive frontiers is also useful for unpacking the historical process of incorporating new peoples to empires, nations and the world economy. A clear historical example is the commercial exploitation of tropical rainforests, which became borders of contact between colonizers and indigenous people.[6] Exogenous and indigenous technological cultures and epistemologies also met at shifting commodity frontiers, which were not physically fixed but fluid. Commodity frontiers are historically constructed spaces of interaction where socio-political and economic structures of tropical rural societies are reconfigured, particularly labour relations and land tenure. Concentrating on the process at work at commodity frontiers, historical anthropologists such as Scott Cook and Eric Wolf have shown that the commodification of Central American and Caribbean natural resources radically transformed indigenous cultures, economies and epistemologies, often involving exploitations and dependencies.[7] This process was not an exception of this region, but a general historical trend of global capitalist development.

The political ecology of tropical commodity frontiers is often related to exogenous industrial and technological changes, such as developments in botany and chemistry.[8] For instance, historian Ian Inkster has shown how global chemical advancements and the development of the celluloid industry in England directly impacted colonial dynamics and the politics of warfare in the tropical forests of Taiwan between the 1860s and the First World War.[9] According to Inkster, the exploitation of Taiwanese camphor trees was a pervasive unintended source of marginalization and conflict of Tayal indigenous communities. Another example would be the resistance from indigenous people in the Spanish Philippines to the expansion of mechanized sugar plantations during the second half of the nineteenth century. As environmental historian Richard Tucker shows, land-grabbing and commercial agriculture politicized the indigenous frontier at the Philippine island of Negros to the point that warfare broke out on the island and its tropical forests were decimated.[10] For the case of Latin America, there are many other exemplary cases, most notably the expansion of rubber and palm oil frontiers in the Amazon.

Also central to this article is the analytical category of 'commodity chain', understood as an underlying socio-economic structure of modern global capitalism. The category of commodity chain – often also called 'value chain' or 'production network' – is widely used among political economists and economic historians to shed light on the entangled processes of extraction, transportation, production, commercialization and consumption of commodities. Commodity chains are fluid material and knowledge-based circuits, where some links can concentrate the power of the whole chain through, for example, the control of prices and production quotas. Using this category of commodity chain, economic historians such as Carlos Marichal, Steven Topik and William Clarence-Smith, among others, have shown how specific agricultural raw materials – from coffee to cotton, from rubber to sugar – were basic inputs for European and US industries and foodstuffs in their mass consumption markets during the nineteenth and twentieth centuries.[11] It seems clear that the ever-increasing widespread global consumption of Latin American raw materials that emerged in the nineteenth century was the driving force behind frontier extractivism in the region. What remains a disputed question among economic historians is whether commodity monoculture and the unequal terms of the exchange of commodities have been prime factors behind Latin America's relative underdevelopment.[12]

A technological history of frontier capitalism needs to concentrate on local–global articulations. During the second half of the nineteenth century, the penetration of commodity frontiers went hand in hand with the consolidation of manifold global structures, systems and institutions, which together transformed the scale and scope of tropical agricultural production.[13] These global structures were not only political (e.g. empires) but also socio-environmental, technological and economic. A good example is patent systems, but there are many others, from transnational transport networks to international capital markets. Again, these global structures not only impacted local frontiers but were themselves shaped by local conditions. There is, of course, a temptation to narrate a unified history of global commodities vis-à-vis global technological changes. A more rewarding approach is to rewrite the local, national and regional histories of commodities with an emphasis on their technological dimension, including world connections, exchanges and dependencies.

One valuable lesson to be drawn from recent works examining the history of global commodities through the lens of the history of science and technology is that any global history of commodities first requires a close examination of local technologies-in-use, indigenous knowledge systems and hybrid or creole practices, including the impact of chemical and mechanical transformations on the production of cash crops.[14] This approach requires an understanding of technology in its broadest sense, including its more informal forms, such as practical, tacit, hybrid and creole. It not only requires consideration of technology as an object (e.g. an artefact), but to related issues such as the practices, actors, spaces and institutions of knowledge production, circulation, appropriation and contestation. Another important reason to adopt a broad definition of technology is to transcend the artificial division between technology and the environment and instead look at the interplay between the two.[15] Such an approach to the history of technology may help open up the study to include actors and societies that have been less visible or even marginal in the history of technology, such as indigenous communities and local peasantry in tropical agricultural and forest frontiers.

Several factors have been the historical drivers of commodity booms and busts. The patterns of the production, trade and consumption of Latin American commodities can be correlated with fiscal, financial, consumption and trade-related matters. Technological transformations have also been a major force of the life cycle of commodities. Intersecting circuits of technologies and expertise developed in tandem with the making of these global commodity chains, thereby facilitating their production, trade and consumption. This is especially clear for the years of the so-called Latin American export boom (*c.* 1880–1929), which were characterized by accelerated international trade, new imperialism, knowledge globalization and cultural encounters. These decades saw the expansion and diversification of Latin American commodity chains of foodstuffs and basic raw materials. The Great Depression of the 1930s diminished the volume of exports and altered the prices of Latin American commodities, although the international demand for tropical raw materials continued to be remarkable.

The appetite for tropical raw materials followed the industrialization processes and concomitant technological transformations in Europe and the United States during the years of the so-called Second Industrial Revolution. That was the era associated with technological breakthroughs in mechanics, organic chemistry and telecommunications. The development of international transport networks made of railroads and steamships facilitated global commodity production and trade. It was also a time of practical scientific research that resulted in, among other things, important advancements in agricultural technologies and processes, which were unplanned spin-offs from other economic sectors (refrigeration, barbed wire, steamship) or new equipment and inputs deliberately conceived for agriculture.[16] Along with botanical and agricultural sciences, an array of mechanical and chemical innovations served as tools for the commodification and standardization of tropical landscapes. This was not, however, a period characterized solely by global technological breakthroughs but also by the ubiquity of local technologies. Global technological innovations coexisted with a large number of indigenous artefacts and traditional practices, many of which had a central economic and cultural relevance.

One cannot complain that the history of Caribbean commodities is a neglected topic.[17] As many historians have shown, this region has long been an area of domestication and production of a large variety of agricultural products, many of which – like sugar, coffee and cocoa – have been objects of global exchange. The plantation system was grounded in the Caribbean in a context of imperial clashes. In their expansion, plantations met local ecosystems and agricultural fields at the margins of the world economy, creating new areas of contact and conflict. Therefore, a central concern in the study of such Caribbean plantations is the impact on local farming communities, smallholding agricultural practices and crops oriented principally towards subsistence and local markets. The exploitation of Caribbean tropical forests was also linked with new imperialist endeavours and the making of industrial capitalism. Forest frontiers sometimes cut across traditional national and imperial borders, resulting in the politicization of the sites of resource extraction. Of course, crops oriented towards subsistence and local markets – sometimes called anti-commodities – continued to be valuable for Caribbean peoples during the years of the export boom.[18]

## A TECHNOLOGICAL HISTORY OF MAYA COMMODITY FRONTIERS

The peninsula of Yucatán was one of the last frontiers in the Greater Caribbean.[19] The situation of this remote and forgotten land of Maya peoples changed during the second half of the nineteenth century, when the production of raw materials for international markets became widespread in the area. Comprised of three Mexican states (Campeche, Quintana Roo and Yucatán), the colony of British Honduras (today's Belize) and the northern part of Guatemala (Petén), this region was sparsely populated and politically uncontrolled. The Yucatán peninsula was a contested frontier, where colonization and commodification encountered indigenous communities that resisted political expansion and capitalist extraction. Although it would soon become an area of production of large quantities of commodities for distant markets, local cultivation of maize and other staples – vital for the survival of the Maya population – continued to occur at the margins of large-scale commercial estates and rainforests. The Maya way of farming and relationship with nature were very different from those of large monoculture plantations, whose production was limited to a single crop, or massive exploitation of forests. For instance, indigenous communities used the swidden farming system – involving the slashing and burning of vegetation – which yielded only enough for self-sufficiency or at most for sale at local markets.[20] This subsistence-style system of agriculture was achieved by means of traditional technology.

During the second part of the nineteenth century and the first decades of the twentieth century, the Yucatán peninsula became the main world-producer of henequen, logwood and chicle. Foreign capital investments and international trade drove the exploitation of these resources. In the words of the historian Hernan W. Konrad, 'available natural resources were exploited in a fashion regulated by international demand and the price for primary products'.[21] These raw materials were not only the object of capitalist development but were native natural resources that had had a central relevance in the Maya economy and society for centuries. Resource exploitation happened at the expense of local 'campesinos' and their agricultural lands and forests, often involving conflicts with indigenous socio-economic and knowledge structures. Maya workers came to engage in the process of commodification, but not without resistance. It was a response not only to proletarianization and land dispossession but to marginalization, violence, displacement, acculturation, the destruction of nature and the expansion of infrastructures. The boom of Yucatecan commodities took place in the midst of the Caste War (1847–1901), a major conflict between the Mexican state and indigenous populations.[22] It was a political conflict with an international dimension. In addition to its economic interests in the region, Britain controlled the southern part of the peninsula. Nationals from other countries such as Spain and the United States also became involved in the conflict.

The rise of henequen, also known as sisal, is the foremost example of Yucatán's transition to agrarian capitalism.[23] During the mid-nineteenth century, sugarcane became the major cash crop of the area but was soon replaced by henequen. A hard cellulosic fibre stripped from a plant native to the Yucatán peninsula, henequen had been used since ancient times by Mayans to make ropes, clothes and mats. Known as *kih* in Maya language, the henequen fibre is obtained by scraping the sword-like leaves of different varieties of agave that were first domesticated in this region for cultivation on commercial plantations. It was during the 1880s, coinciding with the Porfiriato regime, that henequen became the leading export commercial crop of Mexico.[24] From that moment on, labour-intensive large-scale henequen estates were built in Yucatán. Mayan communities were pushed to work as wage labourers in henequen estates built on the arid lands of the northern part of the peninsula. They became peasant-peons who harvested this fibre crop – a simple task – while continuing to maintain their subsistence farming. The golden age of Yucatecan henequen lasted until the First World War.

Technological change set the stage for the boom of henequen. Agricultural technologies were improved during the final decades of the century, with the objective of saving both labour and time and thereby accelerating grain harvesting on a massive scale. Such agricultural innovations were only made possible by a combination of practical knowledge about both machinery construction and agriculture. For instance, the invention of the reaper-binding machine in the United States in 1872 reduced labour intensity in grain harvesting but demanded large quantities of binder twine. For farmers in the Midwestern United States, henequen fibre offered several advantages. Metal wire was expensive, bulky and a risk for animals. By contrast, henequen was cheap, resistant and could be obtained in large quantities from the Yucatán peninsula. Other commercial uses of henequen were as rope, cordage and rigging cable for a range of manufacturing and commercial activities, such as fishing and shipping.

The perfected twine-knotting machine soon became the most widespread harvesting technology in the vast grain fields of the United States. Several Chicago-based companies, such as McCormick Machinery Harvesting Company, Deering Harvester Company and Plano Manufacturing Company, manufactured reaper-binding machines during the 1880s and 1890s. American commercial houses, such as Thebaud Brothers of New York, also sold machinery and tools to Yucatecan planters. This company happened to be one of the largest buyers of raw Yucatecan henequen. In 1902, the International Harvester Company was established through a merger of the era's leading manufacturers of agricultural machinery; this company monopolized the American market of mechanical reaper-binders from that point on.

Equally important for the rise of henequen was the mechanization of its production. As early as in the 1840s, US inventors had tried to mechanize the decortication of henequen in Yucatán, without much success.[25] It was Yucatecan locals who first successfully invented machinery for scraping and processing henequen starting in the 1850s. Some sixty patents were requested in Mexico for this purpose between 1857 and 1889 as well as other various public prizes awarded to local inventors.[26] Among them, the mechanical rasping called 'Solis wheel' was especially important.[27] During the 1880s and 1890s, successive innovations were devised and manufactured following Yucatecan inventions. Although the defibration machine was a Mexican invention, US and British manufacturers supplied most of the machines used in Yucatán. Other technological equipment, such as henequen presses, was likewise imported.[28]

Each agave leaf only yields a small portion of henequen fibre. On the early henequen haciendas, the separation of the fibre from the pulping matter was done by hand, yielding insufficient cord for exportation. Until the development of mechanized devices, rasping was done using indigenous hand tools (*Tonkós* and *pashké*). Productivity was low, large quantities of fibre were wasted, and the henequen cord produced was of inconsistent quality, durability, length and weight. Moreover, it was impossible to meet the increasing foreign demand with such a labour-intensive and time-consuming process. As the inventor and carpenter José Esteban Solis himself declared, before the rasping machine there were not enough workers to make the mass production of henequen profitable.[29]

The improvement and use of decorticating technologies fostered the introduction of new sources of energy. To extract the fibre, planters started using steam-powered spinning rasp machines in the 1860s. In 1892, over 1,300 imported steam-powered rasping machines were employed in Yucatán.[30] Steam engines moving the wheels of the rasping machines required a constant supply of firewood and water until gas and oil were introduced in the 1910s and 1920s.[31] The northern part of the peninsula had hundreds of windmills and animal-operated pumps raising water as well as an extensive network of transport infrastructures moving supplies.

By the turn of the century, fibre-processing technologies were present on most sizeable haciendas – still family-owned – which had machine houses stocking presses, rasping trains, mills, mechanical cleaners, boilers and steam engines. The most modern automatic rasping machines could process around 20,000 leaves per hour.[32] Biological conditions – especially the perishability of agave – gave rise to the development of agro-industrial complexes in henequen fields. Agave leaves dry fast, and therefore the pulp had to be removed within a day after harvesting. Processing had to be done on the plantation or at a nearby location. An endless conveyor separated the henequen fields and the processing factory. As in the modern sugarcane factories established throughout the Caribbean region, work on the plantation fields and the machine house had to be accurately synchronized in time and space. It seems, however, that the coordination between field and factory was not as efficient as in the most advanced sugarcane complexes.[33] The rise of mechanized henequen production also fostered the establishment of a local machine and tools industry in Yucatán that, although small and intensive in labour, had the capacity to maintain and repair the equipment and provide spare parts.[34]

In contrast to processing techniques, innovations in henequen cultivation and harvesting were very limited and continued to be labour-intensive activities. The dry Yucatecan plains and the climate of the north of the peninsula were perfect for henequen cultivation. There was no need to plough the land or use pesticides or fertilizers. Henequen plants lived between fifteen to twenty years, at which point they had to be replaced. From the time the plant was six or seven years old, the large agave leaves were cut using machetes during the dry season – that is, between November to May. The leaves were transported by carts and then brought to the factory through a mechanical belt. After the fibre was mechanically extracted by decortication, it was washed and hung upon lines for drying under the sun. Large henequen haciendas had warehouses to store the baled fibres of dried henequen.[35]

The henequen complex induced the expansion of a modern transportation and communication infrastructure in the area.[36] In the 1880s and 1890s, this included railways, roads, steamships, telegraph lines, telephones and wharves. Infrastructure networks were needed for the export of raw henequen fibre. The railway infrastructure connected the main cities in the henequen region (such as Merida) with the harbours of Sisal and Progreso. In addition, large commercial haciendas had portable narrow-gauge tramways to ship henequen from the estates to the railway stations. Portable Decauville tracks drawn by mules extended through the fields to load the bulk agave leaves and wood necessary for steam engines.

Yucatán specialized in the export of raw fibre. The final manufacturing of henequen products was rare in Yucatán; for the most part, local estates only accomplished the primary processing. Most of the final manufacturing of henequen into commercial twine and cordage was done in American factories, similar to those of the cotton industry. Cordage manufacturing companies in the United States – such as Plymouth Cordage Company in Massachusetts – had automatic twine spinning machines.[37] Some cordage plants were built in Yucatán during the late nineteenth and early twentieth centuries but were ultimately unsuccessful. Later, in the mid-1920s, cordage factories producing rope and binding twine were established in the city of Merida, supplying national and international markets. However, they only represented a small share of the market. The cordage-making technology was relatively simple, which meant that some of the machinery and equipment could be designed and produced locally.

By the First World War, Yucatán was a world leader in the international market for hard fibres, supplying more than 85 per cent of the US demand for binder twine.[38] This explains why sisal was, at the international level, one of the few leaf fibres whose production was automatized.[39] However, during the interwar period, the henequen industry entered a depression, and Yucatán lost its world primacy. The main competition came from alternative sources of twine and rope, such as Manila hemp from the Philippines and jute produced in India.[40] The competition also came from other parts of the world such as Java, Cuba, Kenya, Dutch Indies and Brazil, where sisal was naturalized. In some of these countries, hybrid sisal plants yielded a higher quantity and more resistant fibres after fewer years of cultivation.

The gradual transition in US grain fields, during the 1920s, from the binder machine to the combined harvester also reduced the demand for binder twine.[41] Although the invention of the hay baler in 1937 would give a second life to henequen during the 1940s, given an increased need for baler twine, the replacement of henequen was inevitable. The gradual development of cheaper and more resistant synthetic fibres after the Second World War – made of nylon, Dacron and polythene – only solidified the decline of henequen.[42] An important factor behind the decline of henequen was that it could not be supplied on demand, as the plant needed to be cultivated for several years before it yielded fibre. Another aspect triggering the crisis of the industry (which needs further investigation) was the increasing inefficiency and low quality of production. It seems that the mechanical obsolescence of Yucatecan rasping machines was not unusual, with regular breakdowns due to lack of maintenance. This resulted in foreign complaints about the quality and thickness of Mexican henequen. There were also criticisms about the traces of other materials in Mexican henequen in comparison with the hard fibres produced elsewhere.[43]

In a parallel history to henequen, the dynamics at the rainforest frontiers of the Yucatán peninsula were likewise defined mainly by technological changes on various scales. However, traditional forestall forms of commodity production and technological imperatives were very different from those of the henequen-producing estates.[44] Forest endowments were given and fixed, yet technology transforms them into commodities that were traded in international markets. Gum- and colourant-producing trees were found in significant quantities in the southern lowlands of the Yucatán peninsula. This tropical rainforest area contains a wide range of other trees such as cedars, mahoganies and strangler figs. The specific extraction and processing techniques that could be used were determined by the biological characteristics of the trees themselves.

The extraction of forest resources – particularly chicle, logwood and exotic hardwoods – drove infrastructures to access southern parts of the peninsula of Yucatán. The expansion of the frontier of forestry extraction required further penetration into the heart of Maya territory, which was still controlled by indigenous peoples. Railroads, roads and ports became the key means of exporting chicle to the United States. With the boom of the American chewing gum industry during the first decades of the twentieth century, tractors and carts arrived in the forest and isolated indigenous villages of the southern part of Yucatán peninsula. Chicle stations in the tropical forest were thus connected with Yucatecan cities and ports (and later on with airstrips financed by chewing gum companies) from where large quantities of raw chicle were exported to the United States. The American anthropologist Cyrus L. Lundell explained this situation on an ethnobotanic trip to the southern Mexican state of Campeche in 1931: 'The exploitation of the sapodilla forest has made the remote interior accessible. Road and trails have been opened to bring out chicle by truck and mule-trains.'[45] Lundell also worked for the Tropical Plant Research Foundation, a US institution that carried out botanical research on chicle production in the Caribbean and Central America between 1927 and 1931.

Between the 1870s and 1940s, Yucatecan chicle became the world's main source of chewing gum production.[46] The ever-increasing expansion of chicle extraction in Caribbean forests followed the boom of chewing gum consumption in the United States during those years. Mechanical and chemical improvements were the backbone of the growing chewing gum industry. In 1871, the Brooklyn inventor Thomas Adams was granted a US patent for a gum manufacturing machine that required natural chicle as an essential ingredient. Before Adams' patent, chewing gum was made largely of sugary paraffin wax – invented in the 1840s – and was already a relative commercial success in the United States. However, Adams realized that natural Maya chicle was far superior, thus driving a paradoxical transition from a chemical to a natural product source. Adams' gum-making machine was further improved during the following decades, and different flavours were added to the final product. Chewing gum manufacturing would become a business success from the late nineteenth century, as the cases of the large American companies Wrigley and Adams make clear.

Chicle (known as *sicte* by Maya people and as *tziktli* in Nahuatl) was the basis for the rise of the American chewing gum industry. A type of natural latex, similar to rubber, chicle was obtained from the sapodilla evergreen tree (*Manilkara zapota* or *zapotilla*), abundant in the forests of the Yucatán peninsula.[47] Maya techniques and practices of chicle extraction in Yucatecan sapodilla forests did not change during the chewing gum boom. Tapping incisions to the living tree were made using machetes during the rainy season (June to February), allowing the sapodilla tree to exude its liquid latex. This was a labour-intensive process that relied on traditional practical knowledge and expertise. The necessary tools used by tappers were often provided by intermediary contractors who supplied chicle to foreign chewing gum manufacturing companies.[48] Extraction was not only constrained by the age and distribution of the trees – often irregular and spread over a vast territory – but also by ecological conditions. Each tree could only be tapped every three or four years through diagonal cuts on the bark of the tree. The exuded latex was boiled at forest camps, reducing its moisture and thickening it. Then the crude sticky latex was poured into wooden moulds to form chicle blocks.

Although chicle was extracted from other parts of Central America, such as the Miskito coast of Nicaragua, the US chewing gum industry mostly relied upon the local communities of the Yucatán peninsula, which almost had a world monopoly on the supply of chicle. In contrast with the case of henequen, indigenous communities and other migrant workers were ultimately independent seasonal labourers, although they were often tied to contractors, landowners, cooperatives or concessionary companies.[49] Companies that received forest concessions in the area during the late nineteenth and earlier twentieth centuries included the Mexican Exploration Company and the East Coast of Yucatán Colonization Company.

American chewing gum companies bought crude chicle in blocks of varying consistency, moisture and quality, with prices varying accordingly. Final standardized chewing gum was mass-manufactured in the United States, where chicle was washed, dried, centrifuged and heated at high temperatures. Sheeting and wrapping machines were used to cut and pack the final product. Finished chewing gum sticks consisted of usually no more than 10 per cent of pure natural chicle; the remainder consisted of various ingredients, including synthetic gums, flavours and above all sweeteners such as sugar or syrups.[50] US chewing gum manufacturers increasingly relied on modern laboratory science and industrial research to produce a consistent and sterile product. For instance, the chemist Frederic Dannerth, lead researcher of the Rubber Trade Laboratory in Newark (New Jersey), noted at an April 1917 meeting of The American Chemical Society that, due to the growing importance of the US chewing gum industry, it had 'become necessary to establish standard methods for the purchase of the crude block chicle'.[51] To meet these standards – after cleaning the chicle of dirt, leaves and bark – manufacturers had to assess four factors: moisture, colour, quality and volume. Similarly to henequen primary processing at the plantation site, latex-gatherers (*chicleros*) accomplished a basic *in situ* refining through which they obtained an intermediate product (chicle) that was somewhere between exuded latex and chewing gum.[52] Mexico would not manufacture commercial chewing gum until the 1920s.[53]

After a decline during the Great Depression, chewing gum consumption peaked in the United States during the early 1940s. In 1940, 80 per cent of the chicle consumed in the United States came from the Yucatán peninsula, with the state of Campeche contributing over 50 per cent of total production.[54] Chicle was in short supply after the Second World War, a circumstance that hindered its further commercial expansion. The spatial distribution and ecological conditions of sapodilla trees not only shaped the production process but constrained the size of the industry. These trees could not yet be cultivated on plantations and had not been naturalized to other regions. Yucatecan sources were becoming less abundant and more difficult to access.[55] Also, chicle extraction was seasonal; trees' latex vessels lay dry for long periods after tapping; and supply was heavily dependent on local labour, which retained the task of tapping the trees and preparing the raw blocks. Such environmental and labour limits, together with the exhaustion of the sapodilla trees, drove the development of synthetic substitutes during the post-war era. In the end, the fortune of the local chicle economy was decided by international forces. Again a chemical innovation was behind the patterns of commodity production, but the entanglement between technological changes and the dynamics at chicle frontiers were neither linear nor inevitable. The transition from a natural gum base to bases derived from synthetics was increasingly a reality by the late 1940s. This resulted in a sharp decline in the demand for natural chicle. Industrial chemistry opened and closed the cycle of chicle extraction in the forests of Campeche, Quintana Roo, Belize and Petén, directly impacting the socio-economic and labour structures of indigenous people and triggering the politicization of Maya frontiers and decades of local conflict.

The third commodity this article considers is logwood (*Haematoxylum campechianum*), a tree known as *ek'* by the Maya – which has properties similar to those of brazilwood – and grows near rivers and bays. From pre-Hispanic times, Mayas used this strong wood for construction and its extract, obtained from the heartwood, as a dye and medicine. Beginning in the late sixteenth century, logwood was exported to Europe in large quantities along with other natural colourants produced in the Caribbean and Central America such as indigo, madder and cochineal. Used with mordants such as copper, logwood yields black, purple, yellow and blue dyes valuable for colouring and printing textiles. Historically the most significant areas for the commercial exploitation of logwood trees in the Yucatán peninsula have been the banks of the River Hondo, in the southern border between Mexico and Belize, and the New River in Belize (then the colony of British Honduras) as well as the forests of Laguna de Términos in the southern Mexican state of Campeche.[56] Logwood cutting was a central factor behind the colonization of British Honduras, sparkling political clashes between the Spanish and British empires to control its regular supply. During the nineteenth century, logwood trees were naturalized to other parts of the world, including Asia and the United States but particularly other places in the Greater Caribbean such as Jamaica, Cuba and Haiti, where it was exploited on plantations.[57] Although logwood cutting and trade declined somewhat in the British colonial outpost of Belize during the late nineteenth century, it increased in other places, including Jamaica and the Mexican state of Campeche, during the second part of the nineteenth century.[58] For instance, in 1892 The West Indies Chemical Works Ltd., a manufacturing company that went on to supply logwood extracts and crystals to international markets for over half a century, was established in Spanish Town, Jamaica.[59]

Nineteenth-century industrialization processes – particularly the mechanization of the textile industries – in Britain, France, Germany and the United States intensified the Atlantic trade of logwood extracted from rainforests in the Maya region. As in the cases of henequen and chicle, technological changes in Europe and the United States explain the quest for logwood, with long-term implications for the exploitation of peoples and environments in the remote forests of the Yucatán peninsula. During the Caste War, Maya refugees were pushed into tropical forests, where they worked in logwood and chicle extraction but also on sugar estates in newly cleared lands.[60] The deep red extract obtained from logwood had various additional uses in European and American industries, such as the papermaking, printing, chemical and pharmaceutical sectors. Its primary use was as a colourant to dye cotton, wool, paper and crystals, but it was also a basic raw material in the large-scale manufacturing of ink. Hematein, which results from the oxidation, by either natural or artificial means, of logwood extract, also had important uses. By the turn of the century, hematein, in its purified form, started to be used as a routine histological stain for microscopy in laboratories. Although the extract of logwood (haematoxylin) was first successfully used as a biological staining formulation in 1865, it was not until the 1920s that researchers began to realize its full value.[61]

As in the cases of henequen and chicle, new transport and communication infrastructures penetrating indigenous lands proved a critical factor in the colonization of the forest and the expansion of logwood extraction, although limited by the sabotage of Maya rebels. Logwood cutting was done manually using hand tools, mainly axes and saws. Then, the bark was removed *in situ*, and the logs cleaned and squared into smaller pieces. To reduce costs, from the mid-nineteenth century, it became common to isolate the logwood extract by mechanical means rather than exporting the raw log, thereby reducing the shipping expenses. Cutting logwood was more expensive than tapping chicle, which may explain the shift to chicle extraction in the area during the first decades of the twentieth century. As in the case of chicle, forest concessions for the exploitation of logwood were granted by the Mexican federal government to both Mexican and foreign companies.

A good example is the Mexican company Cuyo & Anexas, established in 1876 in the north-eastern part of the Yucatán peninsula with the backing of German capital investment. This firm set up a large company town that exploited forestry resources including logwood and chicle. It had a narrow-gauge railroad, a telephone network, large warehouses and a company pier.[62] In 1895, however, German investors withdrew as artificial dyes were becoming cheaper and more reliable than the exploitation and transport of logwood extracts to Europe.[63] As in the case of chicle, most probably overproduction was also a constraining factor, as logwood was less accessible than other tropical trees. A few years earlier, in 1892, a company called the Mexican Exploration Company, backed by British capital, had been created to exploit Yucatecan forest products. Another example was the Compañia Colonizadora de la Costa Oriental de Yucatán, which produced large quantities of both chicle and logwood.

Synthetic dye chemistry opened a new technological paradigm in the textile industry. The first synthetic dye had been invented in England back in 1856 by chemist and businessman William Henry Perkin. From then on, and throughout the later nineteenth century, artificial dyes would be invented in the context of routine industrial research, predominantly in the German chemical sector. German dyestuff companies would, from the 1880s, employ large numbers of trained chemists and engineers. Chemical companies protected their invention through both patents and trade secrets.[64] In 1869 alizarin was synthesized and in the 1880s synthetic indigo. The first commercial fast synthetic black dye was invented in 1862 in Manchester by the German chemist Heinrich Caro, who worked for the chemical manufacturer Roberts, Dale & Co.[65] Following this invention, additional synthetic black dyes obtained from coal-tared hydrocarbons were made commercially available, such as John Lightfoot's black process, which was applied to cotton fabrics.

The invention of aniline dyes did not immediately displace natural dyes from the international market of colourants. This was certainly true in the case of logwood. As the New York periodical *American Dyestuff Reporter* pointed out in a 1918 article 'despite unwearyingly attempts by armies of chemists during the recent decades, no perfect substitute for logwood has yet been synthesized'.[66] The transition from natural colourants to synthetic dyestuffs was a long process that took over half a century. In a context of growing but volatile international markets for colourants, logwood coexisted with artificial dyes. During their early years, the costs of producing the new synthetic dyes continued to be high. The artificial dyes did not allow for the same range of colours and shades as natural logwood. Natural dyes were sometimes used together with synthetic ones.[67] Dyeing processes in textile manufacturing remained complex, requiring the collaboration of experts in industrial chemistry and colourists with practical experience in natural colourants. Mordants still needed to be used with precision to obtain particular shades. Similarly, as occurred with indigo, logwood extracts kept the capacity to compete for international markets, as they were exported in large quantities to Europe – notably France and Britain – during the late nineteenth and early twentieth centuries.[68] Innovations in synthetic dye manufacturing and testing also had direct applications to natural colourants and mordants, reducing costs and improving the consistency of natural dyes.[69] The synthetic dye industry had emerged in the late nineteenth century, the same period during which logwood began a second life, which lasted through the First World War. For example, the port of El Carmen, in Campeche, had enjoyed a golden period of logwood trade during the decade of the 1890s, when it exported to the United States, United Kingdom, Germany and France. The decline of logwood extraction in the Yucatán peninsula had to wait until the decade of the 1910s.[70] Logwood exports stopped to be of commercial and industrial importance in the textile industry during the 1920s. By then, aniline dyes were already cheaper, readily available in large quantities and easy to use.

## FROM TECHNOLOGICAL TRAJECTORIES TO COMMODIFIED NATURE AND CULTURE

The history of technology explains major changes in the geographies of capitalist development. This article highlights how the interplay of global technological changes and local technologies and practices allows for large-scale extractive forms of production of tropical resources. This technological view of commodity extraction provides a number of insights for a connected history of global industries and resource frontiers. Indeed, a comparative analysis of commodity chains could help us rethink the ways in which global technological changes, indigenous knowledge and local practices were intertwined with a rise in the production of export commodities between the mid-nineteenth century and the Second World War. Technological trajectories affecting commodity production were the result of several political-institutional factors such as the appropriation of knowledge through patents, the development of standardized technological cultures and local responses to the exploitation of these natural resources. Factors limiting the extraction of these resources also affected global innovation in the context of a rise in the professionalization of research at industries and multinational companies.

Looking at just the cases of logwood, henequen and chicle, it can reasonably be argued that, from a long-term perspective, mechanical and chemical innovations served as the primary driving force behind rural transformations in the Yucatán peninsula. These transformations were not only the result of the introduction of new processing devices but also of new infrastructures and logistical capacities along the commodity chain. Yet it was not only distant trajectories of technological change that drove the cycles of exploitation of Yucatecan forests and agricultural lands. In defiance of technological determinism, local machinery and practices of commodity extraction in the Yucatán peninsula simultaneously pushed global technological innovations and a quest for synthetic substitutes. This is especially clear for the case of the henequen–wheat complex, which was characterized by a mutual dependence and synchrony between the technologies of grain harvesting in the United States and the mechanized henequen estates in Yucatán.

What is paradoxical is that technological changes were bound to ecological conditions. Overexploitation and the increasing difficulty of accessing forestry products drove a quest for alternative sources of supply. The development of synthetic substitutes eventually led global industries to lose their interest in Yucatecan materials. For many years natural and synthetic substitutes competed and coexisted in international markets, although the dependency on natural commodities was in the long run broken by the development of a global programme of chemical innovations for the production of foodstuffs, fibres, dyes, stains and plastics. The abandonment of these resource frontiers brought about a crisis of extractive economies in the Yucatán peninsula, which were highly dependent on foreign markets and industries.

The impact of the 'technological frame' was, however, not immutable, nor did it inevitably determine the fate of natural, political or human frontiers. Technological change may explain major discontinuities at frontiers of extraction, but it did so differently in each case, with distinctive paths, variable chronologies and unintended consequences. The ways in which new chemical, refining and processing techniques influenced the exploitation of natural resources in the Yucatán peninsula followed multiple paths. The interplay between technology and commodities was neither linear nor straightaway but was instead bound to geographical and environmental conditions. Henequen, chicle and logwood possessed very different intrinsic physical and biological properties. Extractive and cultivation practices, refining techniques and manufacturing processes varied from one commodity to the other. Labour systems also varied, from the hacienda regimes of henequen plantations to seasonal indigenous work in tropical forests.

During the years of the so-called Second Industrial Revolution, commodity frontiers in the Yucatán peninsula became deeply entwined with industries and mass markets in Western Europe and the United States. It was a 'dissonant connectivity', to use Ian Inkster's eloquent words.[71] The frontiers of extraction, manufacturing and consumption of these commodities were separated, that is, nature, industries and the market had different socio-economic conjunctures. The commodification of Yucatecan lands and forests also reorganized indigenous agrarian socio-economic systems and the social relations of rural communities. This is especially clear if we look at the international division of labour in the production of these commodities. Maya peoples were necessary workers for the production of these commodities and, although they were not slaves, they suffered exploitative conditions. However, indigenous communities were not passive actors. The pressure to produce vast quantities for mass markets triggered the politicization of commodity frontiers and indigenous resistance to the introduction of new technologies such as railroads, telegraphs and steam-machines.[72] In fighting the new mode of production, the Maya people were ultimately resisting the extinction of their culture.

## NOTES AND REFERENCES

1. Stephen Hill, *The Tragedy of Technology: Human Liberation versus Domination in the Late Twentieth Century* (London: Pluto Press, 1988).

2. Dale Tomich, 'Commodity Frontiers, Spatial Economy, and Technological Innovation in the Caribbean Sugar Industry, 1783–1878', in Adrian B. Leonard and David Pretel (eds.), *The Caribbean and the Atlantic World Economy: Circuits of Trade Money and Knowledge, 1650–1914* (London: Cambridge Imperial and Post-Colonial Studies Series, Palgrave Macmillan, 2015): 184–216.
3. Jason W. Moore, *Capitalism in the Web of Life: Ecology and the Accumulation of Capital* (London and New York: Verso, 2015): 129. On the impact of technological changes on natural resource exploitation see also pages 155–161.
4. Tomich, 'Commodity Frontiers'.
5. Michael R. Redclift: *Frontiers: Histories of Civil Society and Nature* (Cambridge, MA: MIT University Press, 2006): Chapter 2.
6. See, for example, Christopher R. Boyer, *Political Landscapes: Forests, Conservation, and Community in Mexico* (Durham: Duke University Press, 2015); Eric R. Wolf, *Europe and the People Without History* (Berkeley and Los Angeles: University of California Press, 1982); Catherine LeGrand, *Frontier Expansion and Peasant Protest in Colombia, 1850–1936* (Albuquerque: University of New Mexico Press, 1986); Martin Daunton and Rick Halperin (eds.), *Empire and Others: British Encounters with Indigenous Peoples, 1600–1850* (Philadelphia: University of Pennsylvania Press, 1999); Claudia Leal, *Landscapes of Freedom* (Tucson: The University of Arizona Press, 2018).
7. Scott Cook, *Understanding Commodity Cultures: Exploration in Economic Anthropology with Case Studies of Mexico* (Maryland: Rowman & Littlefield Publishers, 2004); Eric R. Wolf, 'Types of Latin American Peasantry: A Preliminary Discussion', *American Anthropologist* 57, 3 (1955): 452–471.
8. Daniel R. Headrick, 'Botany, Chemistry, and Tropical Development', *Journal of World History* 7, 1 (1996): 1–20.
9. Ian Inkster, 'Indigenous Resistance and the Technological Imperative: From Chemistry in Birmingham to Camphor Wars in Formosa, 1860s–1914', in David Pretel and Lino Camprubí (eds.), *Technology and Globalisation: Networks of Experts in World History* (London: Palgrave-Studies in Economic History, 2018): 41–74.
10. Richard P. Tucker, *Insatiable Appetite: The United States and the Ecological Degradation of the Tropical World* (Berkeley: University of California Press, 2002): 101–103.
11. Steven C. Topik and Allen Wells, *The Second Conquest of Latin America, 1850–1930: Coffee, Henequen and Oil* (Austin: University of Texas Press, 1997); Steven Topik, Carlos Marichal and Zephyr Frank (eds.), *From Silver to Cocaine: Latin American Commodity Chains and the Building of the World Economy, 1500–2000* (Durham and London: Duke University Press, 2006); William Clarence-Smith and Steven Topik (eds.), *The Global Coffee Economy in Africa, Asia and Latin America, 1500–1989* (Cambridge: Cambridge University Press, 2003).
12. Sandra Kuntz-Ficker (ed.), *The First Export Era Revisited Reassessing its Contribution to Latin American Economies* (London: Palgrave Studies in Economic History, 2017).
13. Jonathan Curry-Machado (ed.), *Global Histories, Imperial Commodities, Local Interactions* (Basingstoke and New York: Palgrave Macmillan, Cambridge Imperial and Post-Colonial Studies Series, 2013).

14. Lisa Roberts (ed.), 'Special Issue: Exploring Global History through the Lens of the History of Chemistry', *History of Science* 54, 4 (2016); Daniel Rood, *The Re-invention of Atlantic Slavery: Technology, Labor, Race, and Capitalism in the Greater Caribbean* (Oxford: Oxford University Press, 2016); Francesca Bray, Peter A. Coclanis, Edda L. Fields-Black and Dagmar Schäfer (eds.), *Rice: Global Networks and New Histories* (Cambridge: Cambridge University Press, 2017).
15. Martin Reuss and Stephen H. Cutcliffe (eds.), *The Illusory Boundary: Environment and Technology in History* (Charlottesville: University of Virginia Press, 2010)
16. Ian Inkster, *Science and Technology in History: An Approach to Industrial Development* (Basingstoke and London: Macmillan, 1991).
17. See, for example, Sidney Mintz, *Sweetness and Power: The Place of Sugar in Modern History* (New York: Penguin Books, 1985); Stuart McCook, *States of Nature: Science, Agriculture, and Environment in the Spanish Caribbean, 1760–1940* (Austin, TX: University of Texas Press, 2002); Victor Bulmer-Thomas, *The Economic History of the Caribbean Since the Napoleonic Wars* (Cambridge: Cambridge University Press, 2012); Adrian Leonard and David Pretel (eds.), *The Caribbean and the Atlantic World Economy: Circuits of Trade, Money and Knowledge, 1650–1914* (Basingstoke and London: Palgrave Macmillan, Cambridge Imperial and Post-Colonial Studies Series, 2015).
18. Harro Maat and Sandip Hazareesingh (eds.), *Local Subversions of Colonial Cultures Commodities and Anti-Commodities in Global History* (Palgrave Macmillan, Cambridge Imperial and Post-Colonial Studies Series, 2016).
19. Alfredo Cesar Dachary and Stella Maris Arnaiz Burne, *El Caribe Mexicano: Una frontera olvidada* (Chetumal: Universidad de Quintana Roo, 1998).
20. Arthur Demarest, *Ancient Maya: The Rise and Fall of a Rainforest Civilization* (Cambridge: Cambridge University Press, 2005); Anabel Ford and Ronald Nigh, *The Maya Forest Garden* (Walnut Creek: Left Coast Press, 2015). See as well the Mexican documentary *Maize in time of War* (2016).
21. Hernan W. Konrad, 'Capitalism on the Tropical Forest Frontier: Quintana Roo, 1880s to 1930', in Jeffrey T. Brannon and Gilbert M. Joseph (eds.), *Land, Labor, and Capital in Modern Yucatan: Essays in Regional History and Political Economy* (Tuscaloosa: University of Alabama Press, 1991): 143–171.
22. Nelson A. Reed, *The Caste War of Yucatan* (Stanford: Stanford University Press, 2001); Terry Rugeley, *Rebellion Now and Forever: Mayas, Hispanics, and Caste War Violence in 1800–1880* (Stanford: Stanford University Press, 2009).
23. Fernando Benítez, *Ki: el drama de un pueblo y de una planta* (FCE, 1956); Alan Wells, *Yucatán's Gilded Age: Haciendas, Henequen, and International Harvester, 1860–1915* (Albuquerque: University of New Mexico Press, 1985); Sterling Evans, *Bound in Twine: The History and Ecology of the Henequen–Wheat Complex for Mexico and the American and Canadian Plains, 1880–1950* (College Station: Texas A&M University Press, 2007).
24. María Cecilia Zuleta, *De Cultivos y Contribuciones. Agricultura y Hacienda Estatal en México en la 'época de la properidad'*. Morelos y Yucatán 1870–1910 (México, D.F.: Universidad Autónoma Metropolitana, 2006): 232.

25. Howard F. Cline, 'El Episodio del Henequén en Yucatán', *Secuencia* 8 (1987): 186–203.
26. Alfonso Zamora Pérez, *Catalogo Crítico de las Máquinas desfibradoras México (1830–1890)* (México, D.F.: Universidad Autónoma Metropolitana, 1999); Rafael Barba, *El Henequén en Yucatán* (México, Secretaria de Fomento: 1895).
27. Decreto de 13 de enero de 1857: 'Privilegio para José Esteban Solís por una Máquina de Raspar Henequén'.
28. Eric N. Baklanoff and Jeffery T. Brannon: 'Forward and Backward Linkages in a Plantation Economy: Immigrant Entrepreneurship and Industrial Development in Yucatán, Mexico', *The Journal of Developing Areas* 19, 1 (1984): 83–94.
29. Archivo General de la Nación (AGN), Caja 12, Exp. 494 and Caja 12, Exp. 787.
30. Rafael Barba, *El Henequén en Yucatán* (México, Secretaria de Fomento: 1895).
31. Sterling Evans, 'King Henequen: Order, Progress, and Ecological Change in Yucatán, 1850–1950', in Robert Boyer (ed.), *A Land Between Waters: Environmental Histories of Modern Mexico* (Tucson: The University of Arizona Press, 2012): 150–172.
32. Alfonso Fabila, 'Exploracion economico-social del estado de Yucatan', *El Trimestre Económico* 8, 31(3) (1941): 399.
33. Allen Wells, 'From Hacienda to Plantation: the Transformation of Santo Domingo Xyucum', in Jeffrey T. Brannon and Gilbert M. Joseph (eds.), *Land, Labor, and Capital in Modern Yucatan: Essays in Regional History and Political Economy* (Tuscaloosa: University of Alabama Press, 1991): 112–142.
34. Baklanoff and Brannon, 'Forward and Backward Linkages'.
35. 'Hemp Twine: How and of What It Is Manufactured', *Boston Daily Globe* (28 August 1910).
36. Victor M. Suarez Molina, *La evolución económica de Yucatán* (Mérida: Universidad de Yucatán, 1976): Vol. II., 141–227.
37. Plymouth Cordage Company, *The Plymouth Cordage Company: Proceedings At Its Seventy-fifth Anniversary, 1824–1899* (Cambridge, MA: Printed at the University Press, 1900).
38. Wells, 'From Hacienda to Plantation': 115.
39. H. R. Carter, 'The Decortication of Fibrous Plants, with Special Reference to the Belgian Flax Industry', *Journal of the Textile Institute Proceedings and Abstracts* 4, 2 (1913): 231–265.
40. Inés Ortiz Yam, *De Milperos a Henequeros en Yucatán, 1870–1937* (México, D.F.: El Colegio de México, 2013): 140–141.
41. Evans, *Bound in Twine*: 199–210.
42. Antonio Rodríguez, *El Henequén: una planta calumniada* (México: CostaAmic, 1966): 339–342.
43. Rodríguez, *El Henequén*: 327–332 and 375–377.
44. Peter Klepeis, 'Forest Extraction to Theme Parks: The Modern History of Land Change', in B. L. Turner, Jaqueline Geoghegan and David R. Foster (eds.), *Integrated*

*Land-Change Science and Tropical Deforestation in the Southern Yucatán: Final Frontiers* (Oxford: Oxford University Press, 2004): 39–62.

45. Cyrus L. Lundell, 'Archeological Discoveries in the Maya Area', *Proceedings of the American Philosophical Society* 72, 3 (1933): 147.

46. On the history of chicle in the Yucatán peninsula and the chewing gum industry in the United States see Michael R. Redclift, *Chewing Gum: The Fortunes of Taste* (New York and London: Routledge, 2004); Jennifer P. Mathews, *Chicle: The Chewing Gum of the Americas, from the Ancient Maya to William Wrigley* (Tucson: University of Arizona Press, 2009).

47. Luis G. Giménez, *El chicle: su explotación forestal e industrial* (México: Imprenta Manuel Casas, 1951).

48. Redclift, *Frontiers*: 131–159.

49. Redclift, *Frontiers*: 131–159.

50. Frank E. Egler, 'The Role of Botanical Research in the Chicle Industry', *Economic Botany* 1, 2 (1947): 188–209.

51. Frederic Dannerth, 'The Industrial Chemistry of Chicle and Chewing Gum', *Journal of Industrial and Engineering Chemistry* 9, 7 (1917): 679–682.

52. Egler, 'The Role of Botanical Research'.

53. Gimenez, *El chicle*: 92–94.

54. Claudio Vadillo, 'Una historia Regional en tres tiempos. Campeche s. XIII-XX', *Península* 3, 2 (2008): 46–56; M. Ramos Díaz, 'La bonanza del chicle en la frontera caribe de México', *Revista Mexicana del Caribe* 4, 7 (1999): 172–193.

55. Giménez, *El chicle*: 95–97.

56. Alan K. Craig, 'Logwood as a Factor in the Settlement of British Honduras', *Caribbean Studies* 9, 1 (1969): 53–62.

57. Agustí Nieto-Galán, *Colouring Textiles: A History of Natural Dyestuffs in Industrial Europe* (Boston and London: Kluwer Academic Publishers, 2001): 16.

58. Pascale Illegas and Rosa Torras, 'La extracción y exportación del palo de tinte a manos de colonos extranjeros: El caso de la B. Anizan y Cía', *Secuencia* 90 (2014): 79–93.

59. *The Chemical Trade Journal and Chemical Engineer* 57 (1915): 518.

60. Claudio Vadillo, 'Extracción y comercialización de maderas y chicle en la región de Laguna de Términos, Campeche, siglo XIX', in Mario A. Trujillo and José Mario Bolio (eds.), *Formación empresarial, fomento industrial y compañías agrícolas en el México* (México, D.F.: CIESAS): 299–318; Angel E. Cal, 'Capital–Labor Relations on a Colonial Frontier: Nineteenth-Century Northern Belize', in Jeffrey T. Brannon and Gilbert M. Joseph (eds.), *Land, Labor, and Capital in Modern Yucatan: Essays in Regional History and Political Economy* (Tuscaloosa: University of Alabama Press, 1991): 83–107.

61. H. J. Conn, 'The History of Staining Logwood Dyes', *Stain Technology* 4, 2 (1929): 37–48.

62. Víctor Sánchez Molina, *La evolución económica de Yucatán a través del S XIX* (Mérida: Universidad de Yucatán, 1977).
63. Molina, *La evolución económica*: 214–216.
64. Joris Mercelis, 'Corporate Secrecy and Intellectual Property in the Chemical Industry through a Transatlantic Lens, c.1860–1930,' *Entreprises et Histoire* 1 (2016): 32–46.
65. Carsten Reinhardt and Anthony Travis, *Heinrich Caro and the Creation of Modern Chemical Industry* (Dordrecht: Kluwer, 2000): 57–59.
66. 'Logwood: The Historic and Standard Black', *American Dyestuff Reporter* 2, 20 (17 June 1918). This article also pointed out that: 'The chemical superiority of Logwood as black is doubtless accentuated by its superior tinctorial properties.'
67. Berthold Wuth, 'Substitutes for Indigo, Aniline Black, Logwood & c.', *Journal of the Society of Dyers and Colourists* 25, 4 (1909).
68. Michael A. Camille and Rafael Espejo Saavedra, 'Historical Geography of the Belizean Logwood Trade', *Yearbook – Conference of Latin Americanist Geographers* 22 (1996): 77–85; Nieto-Galán, *Colouring Textiles*.
69. Reinhardt and Travis, *Heinrich Caro*.
70. Vadillo, 'Extracción y comercialización de maderas': 299–318 and 307–311.
71. Personal communication with Ian Inkster (4 September 2017).
72. Redclift, *Frontiers*: 136–144.

# Brazil's Mid-twentieth-century 'Techno-class' and the Search for Moderate Reform

EVE BUCKLEY[1]

Drawing on recently renewed interest in the historiography of Latin America's middle class, this essay examines the emergence of a 'techno-class' in twentieth-century Latin America: a group of technically trained middle-class professionals who believed that their scientific expertise offered a rational route to social betterment that would benefit the poor through infrastructural changes palatable to elites. Civil engineers, agronomists and members of related applied scientific professions believed that technological modernization could resolve political debates about landholding inequality and other contentious issues without necessitating violent confrontation. In the words of Mexicanist historian Michael Ervin, they embraced a moderate 'middle politics' through which they hoped to mediate the conflicting interests of elite and popular sectors by offering technological solutions to long-standing challenges. In this paper examples are drawn primarily from Brazilian technocrats' struggle to reduce the suffering caused by drought in their country's semi-arid northeast region, with a focus on the work of agronomists in the 1930s to 1950s. However, the technocrats who focused on drought works (especially roads, reservoirs and irrigation canals) as a way to alleviate poverty and power inequities in the Brazilian *sertão* have parallels throughout Latin America, discussed briefly below.

Despite their decidedly unromantic bureaucratic positions, members of Latin America's mid-twentieth-century techno-class occupied a crucial nexus in modernizing nations as they attempted to realize ambitious visions for social transformation founded on technological change. As members of an expanding middle class, the region's technocratic professionals believed firmly that their level-headed scientific expertise would provide much-needed remedy for entrenched social challenges. Many members of the new techno-class eschewed adherence to any explicit political ideology in favour of depoliticized scientific rationalism. They viewed decadent landowning elites and the untutored masses as obstacles to progress and presented themselves as uniquely able to navigate a middle path that would address the most urgent needs of the poor without provoking the ire of the powerful. These shoals often proved more treacherous than anticipated; veering too far in either direction could provoke leftist revolution or conservative reaction. Brazil's drought technocrats provide an instructive example of this dilemma, as some (notably development economist Celso Furtado) were accused of allying with radical 'peasant leagues' during the late 1950s and exiled following the rise of a right-wing military regime in the mid-1960s.

## TECHNOCRATS AND THE EXPANDING MIDDLE CLASS IN TWENTIETH-CENTURY LATIN AMERICA

The term 'technocrat' has been used by historians of various world regions to describe a range of scientifically minded and technically trained actors during the nineteenth and twentieth centuries. Specific meanings vary over time and place, but typical definitions encompass the embrace of technological systems and expertise as central to national advancement and policymaking. In the European context, technocrats were often middle-class political moderates, pursuing rational, efficient administration as a route to both modernization and social peace. They wished to avoid the depredations of unrestrained capitalism and the bloodshed and upheaval of socialist revolution, through a guided modernization process rooted in science. British technocrats, as analysed by Mike Savage, viewed themselves as industrious and progressive; less self-indulgent than the sons of the ruling classes whose influence they sought to supplant, and more ambitious (for themselves and their nation) than members of the poorly educated labouring class.[2] Their professional experience and prior training released them from the constrictions of a social class structure based solely on birth. French historian Antoine Picon traces 'the roots of modern technocratic ideals' to 'the [early nineteenth-century] Saint-Simonian doctrine advocating a totally organized society in which scientific and technological competence take precedence over traditional social distinctions'.[3] The positivist doctrine of French social theorist August Comte, so influential in late-nineteenth-century Latin America (especially at institutions like the Escola Politécnica in Rio de Janeiro), is a philosophical descendant of this view. Its motto, 'order and progress', remains emblazoned on the Brazilian flag ('*ordem e progresso*').

In the Latin America context, particularly under neoliberal administrations of the late twentieth century, the term technocrat has often been used to reference the apolitical self-conception of social engineers like Pinochet's 'Chicago boys', economists who viewed their own expertise as a politically 'neutral' mechanism for propelling social change. Historians have deftly critiqued this stance for its deeply political motivations and uses; many note that Latin America's technocratic professionals have depended heavily on political patronage for their employment and social status.[4] Yet prior to 1960, Latin American technocrats were most often trained in fields like engineering, public health and the natural sciences, rather than in economics, and they displayed many characteristics similar to those of their British and French counterparts described above: middle class, politically moderate and progressive intellectuals who hoped to mediate between elite self-interest and the ignorance (or recalcitrance) of the poor, to promote national improvement.[5] In her study of Colombian mining engineers, Pamela Murray defines technocrats as 'those who, in the interest of economic modernization, have applied technical criteria—rationality, efficiency—to the realms of government and industry'.[6] Colombia's mining engineers during the late nineteenth and early twentieth centuries, like Brazil's drought works personnel of roughly the same period, believed they had a new and important role to play in formulating public policy for their industrializing and forward-thinking nation.[7]

Some mid-twentieth-century Latin American technocrats adopted a posture of political neutrality early in their careers but began to articulate an overtly political vision for the uses of their expertise as their exposure to entrenched poverty and inequality increased. This is the case for many engineers who oversaw the construction of Brazil's drought works beginning in the 1910s.[8] Others were self-consciously politicized actors who hoped to direct national bureaucracies towards particular ends.[9] For example, historian Mikael Wolfe asserts that the hydraulic technology promoted by Mexico's new National Irrigation Commission in the mid-1920s, following a decade of revolutionary bloodshed, was intended to 'bring social liberation to the agrarian masses without the government radically altering existing land-tenure patterns', and to 'create a prosperous American-style agrarian middle class that would be a source of sociopolitical moderation and advance technical agricultural skills for Mexican campesinos'.[10] Wolfe's book-length analysis demonstrates how central the work of engineers and agronomists was to realizing the agrarian reform goals of Mexico's revolutionary governments, particularly during the administration of President Lázaro Cardenas (1934–1940).

For her study of cybernetic engineers in Salvador Allende's Chile during the early 1970s, Eden Medina coined the term 'technologist' to emphasize the deliberate embedding of political values within the sociotechnical systems that her historical subjects created.[11] This terminological innovation is rooted in the assumption that actors described as technocrats (by historians or in their own writings) generally adopted an overtly apolitical stance, although recent scholarship indicates that there were some exceptions to this in twentieth-century Latin America. But in Chile, according to Patricio Silva, technocrats prior to the Allende administration embraced middle-class values of meritocracy and progress rather than any more specific political ideology. Silva depicts Chile's early-to-mid-twentieth-century technocrats as aiming to 'buffer' the potentially inflammatory opposition between their country's social and political extremes (landed elites and revolutionary Marxists, in the simplest terms).[12] Similarly to the post-revolutionary Mexican agronomists that Ervin and Wolfe have studied, such technocrats tried to articulate a 'middle politics', more progressive than that of landowning and industrial elites but often unintelligible or undesirable, in its statist nationalism and modernizing enthusiasm, to rural *campesinos*. In Ervin's depiction, Mexico's early-to-mid-twentieth-century agronomists were men of often modest middle-class background with considerable scientific training, who saw themselves as ideally positioned to navigate a middle road that would propel their country into a prosperous, industrial future while avoiding a return to the civil strife of the recent revolutionary decade.[13]

Latin America's mid-twentieth-century technocrats were quintessential members of the middle class that grew throughout the region as a result of immigration, urbanization, an expanding state bureaucracy and the establishment of new technical schools and universities. One historian describes the relationship between urbanization and the rising middle class (somewhat exaggeratedly) in these terms: 'as Latin American cities exploded, lost was the face-to-face familiarity that had once characterized town life, and social position came to be less ascribed than acquired. Family antecedents counted for little when people no longer knew exactly who was who.'[14] Social historians ascribe many of the same qualities to the twentieth-century's transnational middle class that their colleagues attribute to technocrats. Members of the middle class saw themselves as democratic, meritocratic, modern, educated and they hoped to exert a moderating force between the elite and the poor. Middle-class professionals in particular played 'an indispensable role in pressing the elite to meet their responsibilities and in teaching the laboring classes proper discipline in the workplace and good hygiene at home'.[15] In mid-twentieth-century Colombia, A. Ricardo López argues, 'a middle-class professional was imagined as the foundational democratic figure who could exercise state rule by educating and preparing both the elites and the laboring classes to coexist harmoniously—and hierarchically—in peace'.[16] In Brazil, particularly during the 1930s under Getúlio Vargas's early administrations, middle-class bureaucrats were invited 'to minimize attachments to conventional politics and instead align themselves with an administrative state'. Vargas's authoritarian Estado Novo (1937–1945) was viewed by many professionals as a triumph of 'experts and expertise' over self-interest, in pursuit of national progress.[17]

In historian Brian Owensby's formulation Latin America's twentieth-century middle-class professionals were 'the social embodiment of progressive modernity, avatar[s] of a supposedly reproducible, universal, and teleological process' of cultural advancement.[18] Eschewing the raw exercise of power indulged in by elites, on the one hand, and the strident militancy of working-class party politics, on the other, middle-class professionals – increasingly with university degrees – believed that they could guide their nations into a peaceful and prosperous future. In the case of engineers, agronomists, and other technical professionals, including those employed by state bureaucracies, this guidance depended on the clear-headed deployment of technical expertise.

## BRAZIL'S DROUGHT TECHNOCRATS AND THE POLITICS OF THE PROFESSIONAL MIDDLE CLASS

In mid-twentieth-century Brazil many technocratic professionals trained in fields like engineering and agronomy were hired by new and growing federal agencies, particularly those focused on expanding infrastructure like roads and sewage systems. One such agency was the national department for works to combat droughts in the country's northeast region, known since 1945 as DNOCS (in Portuguese the *Departamento Nacional de Obras Contra as Sêcas*). Drought was a recurrent plague in the northeast's semi-arid interior *sertão,* and in 1909 Brazil's federal government established an agency tasked with ameliorating drought as a way of mitigating regional poverty. This work was intended to complement the improvements attained via modern science in Brazil's southern cities, notably public health reforms in Rio de Janeiro, and to mirror the regional development efforts of the US Bureau of Reclamation, among other global examples.[19]

Until 1945 the drought agency's managerial staff were stationed in Rio de Janeiro, and even once DNOCS' headquarters were relocated to the northeastern city of Fortaleza, Ceará, its directors remained on the relatively cosmopolitan coast. But the technical staff who oversaw the construction of roads, reservoirs and irrigation canals intended to combat the drought scourge often travelled deep into the interior and remained there for the duration of a drought (from several months to a few years). Their task was to supervise low-tech and largely unskilled construction work by male heads of famished households, in return for which the workers and their families received sustenance rations of rice, beans and brown sugar. At the height of severe drought episodes, the agency's largest *sertão* construction sites might enrol several thousand labouring men, with many thousand family members encamped nearby. Conditions in such places were precarious, with infectious disease and water scarcity constant threats to inhabitants' survival.

The civil engineers and agronomists who supervised DNOCS's work sites and experimental farms were rarely, if ever, from the *sertão* themselves. Most came from coastal cities and wealthier southern states like Rio de Janeiro and Minas Gerais. There they had attended one of a small number of engineering and agricultural schools established during the late nineteenth and early twentieth centuries. These men were exemplars of Brazil's rising middle class; they believed in the potential of modern science and technology to resolve social ills, and they relied on their own training in applied sciences to secure the public sector employment that guaranteed their own families' economic security and social status. Drought agency construction sites introduced these middle-class professionals to compatriots whose aspirations and worldviews did not generally mirror those of coastal urban modernizers. The reports and manuals that DNOCS's site managers produced reveal how Brazil's drought technocrats sought to situate themselves between landowning elites (whom they viewed as self-serving and decadent) on the one hand and a largely illiterate mass of (often landless) farmers on the other. In the eyes of DNOCS staff, rural families required significant moral and civic instruction in order to be transformed into productive citizens.

Such views are starkly evident in the reports and publications of drought agency agronomists stationed in the *sertão* during the 1940s and 1950s. During its first two decades the agency focused primarily on civil engineering projects like road networks and packed-earth dams. In 1932 it established an agricultural service just as agronomy was gaining recognition as a profession in Brazil. The first director of the agency's agronomic service, José Augusto Trinidade, and his successor through the 1950s, José Guimarães Duque, believed that their staff had an important educational role to play in the *sertão*. Both men focused on smallholders as the most likely to adapt to irrigation's demands, since they suffered so much more than large estate owners did during droughts. Trinidade established schools at his agricultural posts where he hoped to 'educate rural boys and girls, children of [landowning farmers] or of poor field workers, to live in a renovated *sertão*'.[20] Instruction, practical and 'adopted to the mentality of the illiterate',[21] would 'cultivate a new attitude that knows how to make use of all the gigantic efforts which the Government of the Union is undertaking throughout the Northeast'.[22] For his part, Duque advocated formal and informal education to remake ordinary *sertanejos* (as the region's inhabitants were known) into active 'citizens of the *sertão*'. At one of his experimental farms the workers' children divided their days between classroom instruction and 'practical education' in the fields, after which they worked on the farm to supplement their families' income.[23] By 1953 the drought agency's agricultural service operated seven schools under Duque's guidance, serving more than four hundred children. Students learned reading, writing and maths along with home economics, modern hygiene and civic virtues. Duque also used radio programmes to publicize new agricultural techniques and methods of soil conservation to far-flung *sertanejos*.

Consistent with the conservative reformism of Brazil's mid-twentieth-century administrations led by Getúlio Vargas, Duque argued that *sertanejos* needed to be taught how to avert calamity through a process of acculturation guided by scientifically trained men. *Sertão* farmers' periodic misery was a result of weak moral character, he asserted, proclaiming in a 1951 publication:

> The hour has arrived for the people [*o povo*] to participate actively in the destiny of their environment, to help resolve, definitively, the questions that will determine the survival of all, and not continue to be mere spectators to the government's initiatives, accomplished through the work of the *técnicos*. It is impossible to overcome the climatic irregularity and the obstacles to agricultural production permanently with a group of people indifferent to the fortune of their environment, inactive and self-centered.[24]

This emphasis on moral character and education overlooked the structural inequities, particularly concentrated landholding, that exacerbated many *sertaenjos*' vulnerability to recurrent drought.

In the early 1950s Duque published multiple editions of a book called *Soil and Water in the Drought Polygon*, which was purportedly about dry-farming techniques (adopted from farmers on the Western US plains) but increasingly centred on the role of education in any effort to modernize the *sertão*. In this the agronomist accused *sertanejo* farmers of engaging in 'a collective, organized movement to destroy the natural wealth that will affect . . . the livability and productivity' of their region, by burning fields to clear them.[25] He argued that it was up to DNOCS's agricultural service to 'prepare the common man to better take advantage of [engineering works to] complete the work of the *técnico* who abruptly introduced irrigation, like a wedge, into the Northeast's social organization'.[26] Promoting habits of cooperation among farmers was as central to irrigation's success as any technical effort, Duque asserted:

> Knowledge of engineering, botany, agronomy and medicine are not sufficient [for the success of drought works]. An elevated level of sacrifice, a deep human understanding of the population's needs, an almost messianic Christian spirit, an absence of egoism are all needed in order to resolve the drought problem . . . The social aspect of droughts is equal to its technical aspects.[27]

For irrigation to succeed, Duque believed, *sertanejos* needed to develop a disciplined work ethic and learn to cooperate with their neighbours. He described irrigated farming as requiring 'Christian virtue' while engaged in

> the joint exploitation of a public reservoir by a group of families financed by a social drought fund; the rental of lots with control over their fertility exercised by an agronomist representing the government; the sale of harvested crops coordinated under the direction of a private association; and the collective purchase of materials and necessary items by the same administrative organization: these are the only ways to avoid the exploitation of man by man and to maintain production throughout generations.[28]

Approaching agronomy as a civilizing mission, Duque portrayed *sertanejo* farmers as suited to modernized production only if they were provided with appropriate scientific tools and moral instruction.

Following a devastating drought in 1953, Duque published a final edition of *Solo e Água no Polígono das Sêcas* in which he made the most forceful case for the need to transform the *sertão* both technologically and socially. Regionally appropriate education was essential, he insisted, to take advantage of the 'robust but latent intelligence in the soul of the masses'.[29] Adopting an ambitious view of his agricultural service's mandate, Duque argued that expropriation of land from estate owners, and rental by the state of irrigated plots on that land to farmers, would enable both soil conservation and sufficient food production. Estate owners typically devoted their best land to commercial crops, which left only less fertile soil for food cultivation and contributed to soil depletion. By settling poor families around publicly funded reservoirs, Duque's Serviço Agro-Industrial would guarantee food sufficient for the growing *sertanejo* population (12.5 million in 1950 and increasing by 2.4 per cent annually). This, he hoped, would minimize outmigration during droughts, which damaged 'the state, society, and the family' by draining young, productive men into urban capitals that did not readily absorb their labour.[30] In this way, Duque described land expropriation – a politically volatile proposal – as having technical and economic motivations as much as social aims. Large cooperatives of smallholding irrigators would combine the economic advantages of *fazenda* estates with the social advantages of *minifundia* (smallholding), he argued. As the *sertão's* population grew, new smallholder colonies could be established along the more humid western periphery of the drought zone, with roads linking their products to northeastern markets. Agronomists would 'establish medical and religious assistance, civic education and hygiene, and technical agricultural instruction' at these settlements, helping farmers learn to relate to each other as plants do in a harmonious ecosystem.[31]

Agronomists who reported to Duque during the 1940s and 1950s reflected his conviction that *sertanejos*' adaptation to new cultivation methods could form the foundation for much broader social transformation. Analysing the economy and population along an extension of the São Francisco River in Pernambuco state, agronomist Trajano Pires da Nobrega proclaimed in 1941 that an irrigated smallholder colony there would 'initiate the formation of new Brazilians, namely these transformed families'.[32] In his report for the drought agency Nobrega described a modest local economy in which most crops and livestock were cultivated for domestic consumption. Many manufactured items and even basic foods had to be imported from coastal capitals, and an unnavigable stretch of river at the Paulo Afonso falls impeded easy transport of goods to or from surrounding communities. A postal bus traversed part of the region weekly; the remainder was served by an unreliable mule post that covered a two hundred kilometre route. Many farmers did not have documented title to the land that they claimed ownership of based on long-standing occupation and cultivation. Unreliable rainfall discouraged agricultural investment except along the riverbank, where *vazante* cultivation in the evaporated riverbed was reliable even during dry seasons. Along the river most farmers did own their land, but farms had been 'pulverized' over generations of partible inheritance into narrow plots, sometimes three by forty metres in dimension.

Despite these challenges to economic and cultural development, Nobrega viewed the inhabitants optimistically, as extremely independent and hardworking farmers and ranchers. He describes them as '*caboclo* (mixed-race native-European) types' who inherited the 'independence and lack of discipline of their indigenous forebears' but for whom the strong desire to own and improve their land had resulted in 'genuine evolution that should form an excellent basis for selecting future irrigators'.[33] Like Duque, Nobrega thought the greatest impediment to the success of irrigated farming was the prevailing individualism and suspiciousness (*desconfiança*) of local farmers; agronomists thus needed to instil a more cooperative spirit within the community. He hoped that the wealth generated by initial irrigators would break other farmers' distrust of novelty and impel them to join the project. Nobrega also recommended 'introducing one or two foreign families, preferably from regions best known for their agricultural capacity, like Germans or Poles', to model profitable irrigated farming. This was a common suggestion among mid-twentieth-century *sertão* developers, reflecting prevailing assumptions about the cultural and racial superiority of Europeans.[34]

Nobrega proposed that an initial trial be conducted at a drought agency agricultural post. Plots could be rented to the post's *sertanejo* workers who would be guided by agronomists in cultivation techniques, use and repair of machinery, methods of crop transport and home economics – the latter so they would not squander their profits, thereby jeopardizing the value of the experiment as a model for other farmers. Undisciplined settlers would be dismissed, but successful ones could move after three years to their own plots in an area of the river valley acquired for this purpose by the federal government. Nobrega estimated that if every family in the new settlement received five hectares of irrigated land and two to three dry hectares for livestock, a three-thousand-hectare stretch of river would accommodate five or six hundred families. Along with the regional economic benefit that such a scheme could provide, it would retain *sertanejos* during droughts, temporarily accommodating up to five thousand people (and thousands more cattle), and reducing their undesirable migration to industrializing cities like São Paulo where they congregated in informal settlements (*favelas*).

Ultimately Nobrega envisioned a city emerging around the model irrigated colony, offering modern amenities like electricity, water and sewage services; a well-appointed school; a cinema, market and health clinic; and clean parks and squares – in short, 'a center for credit, production, and consumption'.[35] Private investors, perceiving the unfolding commercial opportunities, would fund additional irrigation canals, accelerating the region's economic development. 'This will be the process by which the *sertão* of the São Francisco [River] valley is transformed into the promised land', Nobrega concluded grandly. 'This is not a utopian plan', he insisted. Yet, judging by the drought agency's quarterly bulletins, few such settlements were actually established during the 1940s outside of the agricultural commission's own posts. This was due to both the conflicting priorities of civil engineers and agronomists within the drought agency and to regional elites' resistance to the creation of autonomous smallholder communities from among a dependent labouring class.

Despite such challenges, during the 1950s several agronomists published training manuals for new *sertanejo* irrigators that emphasized the technological and social goals of the drought agency's smallholder settlements. An undated, handwritten pamphlet illustrated with simple line drawings, *Irrigante amigo! Seja bem vindo ao Projeto São Gonçalo* (Irrigator friend! Welcome to the Project São Gonçalo [in Paraíba state]) presented settlers with a list of their contractual rights and responsibilities.[36] From DNOCS they would receive technical assistance, year-round water, medical and dental care, schooling for younger children, financial credit and membership in the post's agricultural cooperative; in return, they were to use their lot for farming and ranching and maintain good relations with surrounding families. Irrigators should accept agronomists' guidance about what crops to plant and how to cultivate and market them through the cooperative. They were to work hard, maintain their canals, drains and homes and participate in meetings, classes and social activities sponsored by the management. 'If you follow these instructions, you will be a good *colono*', the pamphlet concluded. A more formal publication along the same lines specified that the irrigator's primary goals were to avoid soil erosion and salinization and to maximize his family's income by following agronomists' advice.[37]

Carlos Bastos Tigre's moralizing *Catecismo do Agricultor Irrigante* (Irrigator's catechism) was published by DNOCS in 1954 as a wide-ranging advice manual for participants in the agency's irrigated settlements. Tigre encouraged families to construct their homes of cement, tile and wood, use 'modern and elegant' kerosene or electric lamps, install piped water and sewage pipes leading to a cistern or septic tank, construct separate housing for livestock, hang attractive curtains and place perfumed plants and fruit trees around their property. 'If a farmer pastures cattle on lands served by irrigation canals, this shows that he is ignorant, has no sense of responsibility, no spirit of cooperation, is an egotist and unpatriotic', Tigre admonished.[38] He encouraged irrigators to seek a range of advice and services from the drought agency's agricultural posts. By cooperating with each other they would foster 'community progress and national prosperity', demonstrating their gratitude to the government for providing them with valuable infrastructure and assistance.[39]

Tigre's catechism extended to nutrition, advising settlers to eat fruits and vegetables as well as fresh eggs, meats and fish of known provenance, and warning that 'a bad diet saps your strength, disposition, and initiative'.[40] He also offered guidance about personal hygiene, pregnancy and sanitary infant care, opining, 'The death of so many children in the *sertão* is due to poor food hygiene and their parents' lack of the most basic understanding of nutrition.'[41] (This is a strikingly depoliticized view relative to the contemporaneous analyses of the economic structures underlying malnutrition, by Pernambucan doctor Josué de Castro and others.[42]) Tigre encouraged moral discipline of adolescents, especially boys, so that 'abuse of their bodies' would not lead to imbecility, insanity or venereal disease; he assured parents that 'they will thank you and God when they reach maturity for having given them such protection'.[43] It is clear from such sources that the *sertão*'s technocratic modernizers understood their educational purview to encompass much more than farming techniques.

## THE MIDDLE POLITICS OF AGRONOMISTS IN BRAZIL'S MID-TWENTIETH-CENTURY *SERTÃO*

DNOCS agronomists' vision for *sertão* transformation entailed restructuring the material basis of political power insofar as regional dynamics allowed – particularly by promoting land redistribution – while simultaneously reforming the most marginal farmers into disciplined, productive workers. To accomplish this they tried to combat both elite corruption and self-interest and the seeming recalcitrance of the poor whom they intended to help. They thus engaged in the 'middle politics' described by Ervin with regard to Mexican agronomists in 1930. During that post-revolutionary period Mexican *técnicos* promoted land reform, which pitted them against elites; yet they had difficulty persuading *campesinos* (peasant farmers) to provide census data – for fear of tax implications – or to embrace nationalist motivations for intensifying agricultural production.[44] *Sertão* development also involved multilevel negotiations, including between technocrats employed by the federal drought agency and the 'targets' of their well-meaning assistance.

Generally, irrigation was desired by *sertão* farmers during droughts but was rarely used or maintained in years when it was not essential for survival. In one instance, the director of the agricultural post around Forquilha Reservoir in Ceará complained to Duque (his superior) about farmers' use of irrigated plots for pasture. This damaged both canals and drainage systems. The 'recalcitrant' livestock owners apologized when agency staff brought this problem to their attention, but did not change their behaviour – in short, the *cooperantes* (as irrigators were called) 'weren't cooperating!' Citing such cases, Duque petitioned Brazil's Ministry of Public Works to promote legislation that would expropriate all land crossed by federal irrigation canals, along with one hundred metres on each side to act as a buffer against roving cattle. This would give the agricultural service a stronger legal basis for punishing those who abused their infrastructure, much of which had been constructed on private land.[45]

Conflict between agronomists and irrigators escalated around Forquilha over the 1940s. At one point, water serving the property of Silvestre Gomes Coêlho, who allowed his cattle to roam across neighbouring irrigators' land, was turned off, and he was fined for damages to the canals. In response, Gomes Coêlho brandished his rifle at agency staff who arrived to halt water flow to his property and threatened to blow up the water meter if his irrigation access was not restored. In Duque's assessment, such 'acts of low education' (meaning low social class) were typical of this fellow, who could rarely be persuaded to follow 'rules of good comportment' (*regras do bom caminho*). The agronomist stationed at Forquilha deemed such behaviour 'unconscionable', especially when engaged in by the very *sertanejos* for whose benefit the nation had constructed and maintained irrigation networks at great effort and expense.[46] Like their colleagues throughout Latin America, Brazil's middle-class technocrats often found the behaviour and worldview of the far less educated compatriots whom they intended to help to be both uncouth and incomprehensible. From the mid-1940s onward canal restoration became a reliable source of employment for drought migrants, due to the significant damage caused to them by free-range livestock.

Eight years after the fraught interaction with Gomes Coêlho, Duque was mired in debate about how much water could be drained from the Forquilha Reservoir by irrigators. The basin also yielded a three-thousand-kilogram annual fish harvest, providing sustenance for nearly forty thousand people in the area.[47] Due to a drought in 1953 and resulting heavy water use by irrigators, reservoir volume had fallen to 5 per cent of capacity. By 1957 only one-fifth (100 hectares) of the land served by irrigation canals could be farmed due to diminished water volume and leakage from canals that were in poor repair.[48] As such examples demonstrate, drought agency staff were frequently embroiled in disputes over which populations would benefit from their management of land and water and which local residents would bear the cost of this restructuring.

Oversight of the Choró Reservoir in Ceará and its irrigation canals, which watered approximately 950 hectares (roughly 2,300 acres), passed from the drought agency's engineering division to its agricultural service in the 1940s. Agronomists' survey of the project at this time revealed various irregularities in land use arrangements. Many irrigators who rented plots from the agency were in flagrant disregard of legislation limiting each family to ten hectares of *terras secas* (dry land) for their home, livestock and less water-demanding crops and four hectares within the irrigated basin for food cultivation. When participants were selected for an irrigated colony, preference was supposed to be given to the poorest qualified applicants. Yet some settlers around Choró had accumulated up to twelve times the legal limit for property rental, acquired in the names of various family members. Such families often sublet parcels to sharecroppers or wage labourers, including to drought migrants who were supposed to be the primary beneficiaries of federal reservoir construction. This corrupt system effectively replicated the exploitative social organization that *sertão* reformers aimed to displace. In 1948 DNOCS's director explained to his superior, the minister of public works, that he was rescinding several contracts of families who had accumulated unjustifiably large parcels in order to 'rescue' 282 indigent families (1,270 people) from the 'inhumane' means of sustenance in which they were presently engaged. This would deliver on the agency's mandate to serve 'innumerable poor workers who genuinely need it' and to increase food production, creating a 'more vibrant outlook' in the *sertão*.[49]

Not surprisingly, many renters of Choró's irrigated plots protested vigorously against the rescinding of their rental contracts. José Delfino de Alencar, who effectively controlled 122 hectares of dry land and fifty-eight *vazante* plots within the reservoir basin (accumulated in the names of his children) successfully sued DNOCS in court. Agronomist Duque and his engineering colleagues in the agency bitterly disputed this ruling (which was appealed multiple times), calling the outcome 'a matter of life or death' for several poor families who would be able to farm the contested property if it were wrested from the Alencar clan. Duque's supporters within DNOCS depicted such conflicts as a struggle over the civic health of the *sertão*, pitting social progress against 'begging, affliction, crime, and communism'.[50]

By 1958, under President Kubitschek's aggressive programme of infrastructural expansion, more than 65,000 people were living in irrigated reservoir basins administered by Brazil's federal drought agency, and one thousand students participated in schools or farming clubs at agricultural posts.[51] Agronomists continued to proselytize rational use of the *sertão's* semi-arid land, ecologically adapted for maximum crop yield, economic benefit and enjoyment. Progress was limited not by natural or financial resources, according to Duque, but by *sertanejos'* 'incapacity to cooperate and elite egoism' combined with a lack of political will to prioritize the economic needs of the poor.[52] Many public works in the *sertão* remained underutilized for these reasons, he asserted. Yet Duque remained confident that with strategic use of seed selection, irrigation, dry farming, fertilizer, pesticides, intercropping and rotation of human food and livestock forage, *sertão* farmers could adequately feed their region's growing population and increase their own per capita income. Nonetheless, he feared cultural impediments to such progress, particularly the reluctance of young *sertanejo* men to subordinate their capricious whims to the discipline of intensive cultivation, and the cultural tendency to waste profits on ostentatious trinkets or alcohol rather than investing in property, education and agricultural tools. Speaking in 1959 to Rio's engineers club, Duque remarked that much of the DNOCS agronomists' work required gaining farmers' trust and persuading them to break with their communities in the interest of regional and national progress.[53]

## CONCLUSION

Brazil's mid-twentieth-century drought technocrats exemplify the moderate reformism pursued by technocratic professionals across Latin America during this period. These new cohorts of professionals, trained in a range of applied sciences, saw technical expertise as an essential engine of social transformation, one that would foster national progress without provoking civil strife. In their view the value of their recommended changes to transportation and hydraulic infrastructure, farming methods and household economies was self-evident, simply a matter of embracing modern science. Agronomists' and engineers' recommendations for alterations to patterns of land use and water distribution were rooted in science and should therefore (in their view) be immune to political critique.

Of course the issues that twentieth-century agronomists, civil engineers and their colleagues grappled with in Latin American nations were often at the heart of contentious political disputes. Questions of land ownership and the organization of agricultural economies were deeply intertwined with regional and national power dynamics, and any promotion of expanded smallholding, in particular, was bound to confront significant resistance from traditional landowning sectors. This was certainly the case in the Brazilian *sertão,* where young men who arrived from coastal cities eager to apply the knowledge they had gained at agricultural and engineering schools soon confronted the social tensions at the heart of any discussion about redistributing access to – and control over – farmland, food and water. Some of these individuals became staunch defenders of impoverished drought migrants, in the course of their work. But many drought technocrats also found the *sertanejos* whom they believed would benefit from their expertise to be frustratingly dismissive of, even antagonistic towards, their prescriptions for greater productivity. These men (in the mid-twentieth-century *sertão* all such individuals were male) were rarely if ever cognizant of the social and political values that they brought to the region and its inhabitants, based on their own class location, political views and educational experience.

Members of Latin America's emerging techno-class were significant actors in the twentieth century and deserving of much greater scholarly attention than they have received thus far. These professionals are of interest for their attempt to navigate a middle road between the competing interests of traditional elites and the poor, and for their faith in applied sciences as an instrument for achieving this compromise to promote national progress. They are also one important sector of the expanding middle class, a highly influential social stratum that historians have returned to examining recently after a hiatus of several decades. This renewed scholarly attention to the professional middle class offers a welcome opportunity for historians of technology to examine the cohorts of technically trained professionals who occupied influential positions in government bureaucracies and the private sector throughout Latin America across the twentieth century, and to insert the work and influence of these actors into existing political and social historiographies.

## NOTES AND REFERENCES

1. I wish to thank the two referees, Mikael Wolfe and an anonymous reviewer, for helpful commentary on an earlier version of this essay; all remaining errors are my own.

2. Mike Savage, 'Affluence and Social Change in the Making of Technocratic Middle-Class Identities: Britain, 1939–55', *Contemporary British History* 22, 4 (2008): 457–476.
3. Antoine Picon, 'French Engineers and Social Thought, 18–20th Centuries: An Archeology of Technocratic Ideals', *History and Technology* 23, 3 (2007): 197–208.
4. John D. Martz and David J. Myers, 'Technological Elites and Political Parties: The Venezuelan Professional Community', *Latin American Research Review* 29, 1 (1994): 7–27.
5. Angela de Castro Gomes et al., *Engenheiros e Economistas: Novos Elites Burocráticas* (Rio de Janeiro: Ed. FGV, 1994).
6. Pamela Murray, *Dreams of Development: Colombia's National School of Mines and Its Engineers, 1887–1970* (Tuscaloosa: University of Alabama Press, 1997): 54.
7. Andrés Valderrama, Juan Camargo, et al., 'Engineering Education and the Identities of Engineers in Colombia, 1887–1972', *Technology and Culture* 50, 4 (2009): 811–838. See also Frank Safford, *The Ideal of the Practical: Colombia Struggles to Form a Technical Elite* (Austin: University of Texas Press, 1976).

8. Eve Buckley, *Technocrats and the Politics of Drought and Development in Twentieth-Century Brazil* (Chapel Hill: University of North Carolina Press, 2017).
9. Juan C. Lucena, 'De Criollos a Mexicanos: Engineers' Identity and the Construction of Mexico', *History and Technology* 23, 3 (2007): 275–288. For more comprehensive coverage of Mexican engineers in the 1920s, see José Raúl Domínguez Martínez, *La ingeniería civil en México, 1900–1940: Análisis histórico de los factores de su desarrollo* (Mexico: IISUE, 2013).
10. Mikael Wolfe, *Watering the Revolution: An Environmental and Technological History of Agrarian Reform in Mexico* (Durham: Duke University Press, 2017): 72–73.
11. Eden Medina, *Cybernetic Revolutionaries: Technology and Politics in Allende's Chile* (Boston: MIT Press, 2011).
12. Patricio Silva, *In the Name of Reason: Technocrats and Politics in Chile* (University Park, PA: Pennsylvania State University Press, 2008): 17.
13. Michael A. Ervin, 'The 1930 Agrarian Census in Mexico: Agronomists, Middle Politics, and the Negotiation of Data Collection', *Hispanic American Historical Review* 87 (2007): 537–570.
14. D. Parker, 'The Making & Endless Remaking of the Middle Class', in David Parker and Louise Walker (eds.), *Latin America's Middle Class* (Lanham, MD: Lexington Books, 2013): 13.
15. B. Weinstein, 'Commentary', in A. Ricardo López and Barbara Weinstein (eds.), *The Making of the Middle Class: Toward a Transnational History* (Durham: Duke University Press, 2012): 109.
16. A. Ricardo López, 'Conscripts of Democracy: The Formation of a Professional Middle Class in Bogotá during the 1950s and Early 1960s', in López and Weinstein (eds.), *The Making of the Middle Class*: 188.
17. B. Owensby, *Intimate Ironies: Modernity and the Making of Middle Class Lives in Brazil* (Palo Alto: Stanford University Press, 1999): 206.
18. Owensby, *Intimate Ironies*: 4.
19. See Buckley, *Technocrats* for more detail on the agency's establishment and history.
20. J. A. Trinidade, *Os postos agricolas da inspetoria de sêcas* (Rio de Janeiro: Ministério de Viação e Obras Públicas, 1940): 7.
21. J. G. Duque, 'O fomento da produção agricola', *Boletim da Inspetoria Federal de Obras Contra as Sêcas* 11, 2 (1939): photo caption, n.p.
22. J. A. Trinidade, 'Os serviços agricolas da inspectoria de seccas', *Boletim da Inspetoria Federal de Obras Contra as Secas* 7, 1 (1937): 43.
23. Duque, 'O fomento da produção': 155
24. J. G. Duque, *Solo e agua no polígono das sêcas*, 2nd edn. (Fortaleza: MVOP-DNOCS, 1951): 14.
25. Duque, *Solo e agua*: 15.
26. Duque, *Solo e agua*:12.
27. Duque, *Solo e agua*: 111.

28. Duque, *Solo e agua*: 111.
29. J. G. Duque, *Solo e agua no polígono das sêcas*, 3rd edn. (Fortaleza: MVOP-DNOCS, 1953): 7.
30. Duque, *Solo e agua*, 3rd edn.: 201 and 291.
31. Duque, *Solo e agua*, 3rd edn.: 188 and 201.
32. Tirano Pires da Nobrega, 'Ensaio social-econômico de um setor do Vale do Rio São Francisco', *Boletim da Inspetoria Federal de Obras Contra as Sêcas* 16, 1 (1941): 14.
33. Nobrega, 'Ensaio': 10.
34. Nobrega, 'Ensaio': 12.
35. Nobrega, 'Ensaio': 14.
36. *Irrigante amigo! Seja bem vindo ao Projeto São Gonçalo*, undated pamphlet, Instituto Agronômico José Augusto Trinidade Collection (hereafter IAJAT), DNOCS library, Fortaleza.
37. Serviço Agro-Industrial, *Informações para irrigantes* (Fortaleza: Ministério de Viação e Obras Públicas/DNOCS, 1957), IAJAT, DNOCS library.
38. Carlos Bastos Tigre, *Catecismo do agricultor irrigante* (Fortaleza: DNOCS, 1954): 10.
39. Tigre, *Catecismo*: 12.
40. Tigre, *Catecismo*: 45.
41. Tigre, *Catecismo*: 51.
42. J. de Castro, *A Geografia da Fome* (Rio de Janeiro: Editora O Cruzeiro, 1946).
43. Tigre, *Catecismo*: 54.
44. Ervin, 'The 1930 Agrarian Census'.
45. Letters from J. G. Duque to drought inspector, 20 June 1944, Forquilha Reservoir file 7, Fundo: Açudes Públicos, Arquivo DNOCS (Fortaleza, Ceará).
46. Letters from J. G. Duque to drought inspector, November 1946, Forquilha Reservoir file 7, Fundo: Açudes Públicos, Arquivo DNOCS.
47. Seven memos from J. G. Duque to other DNOCS personnel, 10 April 1953 to 3 September 1954, Forquilha Reservoir file 7, Fundo: Açudes Públicos, Arquivo DNOCS.
48. Letter from Associação do Commércio, Sobral, to DNOCS director, 24 April 1957, Forquilha Reservoir file 7, Fundo: Açudes Públicos, Arquivo DNOCS.
49. DNOCS inspector Berredo to MVOP, 'Perspectivas mais animadores', 1948, Choró Reservoir file 1, Fundo: Açudes Públicos, Arquivo DNOCS.
50. Memo from José Nanges Campo discussing Delfino de Alencar case, March 1948, Choró Reservoir file 1, Fundo: Açudes Públicos, Arquivo DNOCS.
51. Pimentel Gomes, 'Solução agronômica do problema das sêcas', *Boletim do Departamento Nacional de Obras Contra as Secas* 19, 3 (1959): 113–124.
52. J. G. Duque, 'Agricultura do Nordeste e o desenvolvimento econômico', *Boletim do Departamento Nacional de Obras Contra as Secas* 19, 4 (1959): 52.
53. Duque, 'Agricultura do Nordeste': 64.

# Challenging Martial Masculinity: The Intrusion of Digital Computers into the Argentinian Armed Forces in the 1960s

DEBORA GERSTENBERGER

## INTRODUCTION

War is the father of all things; the armed forces have always stood at the forefront of technological innovation. These are common assumptions, and they also entail the belief that computer technology emerged from the military sphere and that the historical process labelled 'digitalization' that eventually spread across the world has its roots there. The idea of war and the military being the spearhead of technological innovation, including the digital revolution, certainly corresponds with the common belief that progress is a male phenomenon. Our Western culture locates (technological) innovation into male spheres and responsibilities. According to Claudia Honegger, Charles Darwin prominently ascribed progress to masculinity and stagnation to femininity, which heavily influenced the bourgeois view that developed around 1800 of men being rational and active and women being emotional and passive.[1] From the nineteenth century on, to be masculine meant to be competent in technological questions, while to be feminine meant to have nothing to do with technology.[2]

To be sure, the thesis of the military being the driving force behind all innovations has been challenged. Brazilian historian Érico Esteves Duarte goes as far as to state that the armed forces are and always have been 'hostile towards technological innovations'. According to him, the logic of technology is driven by efficiency and effectiveness, whereas the logic of war is not, because it relies heavily on well-known and established patterns. Interestingly, however, he claims that this does not hold true for the case of computer technology. In this field, he maintains, the military did indeed play a decisive role.[3] The master narrative of the computer – a technology that became masculinized in the course of the 1960s and 1970s[4] – emerging from the military sphere and conquering the world within a few years remains in place.

This article offers a different perspective on the history of military computing. It questions the story of the armed forces being the 'natural' leader in the process of digitalization, at least in Latin America. I will show that this narrative omits many aspects of the troublesome and slow process of the implementation of digital hardware and software. I will argue that rather than being at the forefront of the digitalization project, the military sphere was remarkably reluctant to accept new digital technology. In fact, the computer was seen as an alien, an intruder. The reason for this, I will argue, lay not least in the notions of 'martial masculinity' that prevailed within the armed forces.

The sources for this article consist mainly of Argentinian military journals. These texts shed light on the debates and controversies surrounding the introduction of computer technology into Latin American armed forces during the early 1960s. My guiding questions in this article will be: How did debates about computer technology within the military sphere evolve in the early 1960s? What arguments were in place, what positions can be singled out? How did the computer actually enter into the military sphere?

The relation between computers and the military in Latin America has not yet received much attention from historians. Whereas studies on the armed forces abound due to the fact that they played a crucial role in politics for many decades, the history of (computer) technology remains largely divided from 'general' military and political history. Interestingly, this holds true not only for Latin America but for other world regions as well. One finds some accounts on ARPANET, and discussions about the role of the US military in the Internet's development are still ongoing.[5] However, the history of military digital data processing prior to or beyond ARPANET seems to be astonishingly understudied.

Most accounts on the social and cultural history of computing that significantly consider gender roles and their transformations, in turn, focus on business or administration, not on the military sphere. Thus, my aim is to foster the entanglement of the history of computing with military history. I believe that this intersection should be analysed from the perspectives of social history, cultural history and not least gender history.

## THE FIRST DIGITAL COMPUTERS IN LATIN AMERICA AND ARGENTINA

In Latin America, digital computers (mainly imported machines from the US or Europe) were first installed in universities, industrial companies and public management facilities,[6] and not in military institutions. The very first digital computer in Latin America is said to have been an IBM 650 machine installed on 8 June 1958, at the *Universidad Autónoma de México* (UNAM). In Brazil, the first digital computer arrived in April 1960, at the Catholic University (*Pontifícia Universidade Católica* [PUC]) of Rio de Janeiro. President Juscelino Kubitschek himself, the self-ascribed modernizer of Brazil (who became famous for his formulation, 'fifty years in five'), was present when the Burroughs Corporation

B-205 machine was installed. Cardinal Giovanni Battista Montini of Rome, later Pope Paul VI, consecrated the newly established data processing centre at the PUC. A consortium consisting of the Brazilian Ministry of War, the National Research Council, the National Nuclear Energy Commission, the National Steel Company and the university itself had paid 400,000 US dollars for the equipment.[7]

In Argentina, an IBM computer was displayed at the exhibition to commemorate the 150th anniversary of the independence revolution on 25 May 1960. Another IBM machine and two Univac machines were installed at Argentinian public transport companies. A British digital computer, a Ferranti Mercury II, had been purchased for scientific use by the department of mathematics (*Departamento de Matemática de la Facultad de Ciencias Exactas y Naturales*) of Universidad de Buenos Aires (UBA) with financial support from the Argentinian *Consejo Nacional de Investigaciones Científicas y Técnicas* (CONICET) in 1958. However, the machine was only delivered in 1960 and could not be installed before Ferranti Mercury personnel arrived in 1961.[8] Nicolás Babini estimates that until 1965 some forty digital computers were imported to Argentina.[9] In 1962, the data processing institute (*Instituto de Cálculo*) was established at UBA. In 1960, the Argentinian Calculation Society (*Sociedad Argentina de Cálculo* [SAC]) was founded in order to channel all activities concerning the new electronic computation and to promote electronic computing in business companies and universities.[10]

As this short overview suggests, civil institutions in Latin America, such as universities and big companies, were eager to purchase digital computers at an early stage. The armed forces at first did not share this enthusiasm. To be sure, electronic calculating had long been part of military weapons like guided missiles or radar. However, digital computers of the above-mentioned type, huge machines that were not integral parts of other weapons or tools but constituted a system as an end in itself, were rather used for business and scientific purposes. How, then, was the computer received and evaluated by the members of the armed forces?

## THE ARGENTINE MILITARY AND ITS RELUCTANCE TO ACCEPT COMPUTERS IN THE EARLY 1960s

During the Cold War, Argentina was located at the 'periphery' of world politics; it was part of the 'Third World'. However, Argentina did not easily accept its role as an underdeveloped state. Argentinian elites had projects of modernization and industrialization.[11] Ever since the coup d'état against President Hipólito Irigoyen in 1930, the military had played a prominent role in Argentinian politics as a self-described countervailing force.[12] The armed forces at first supported Juan Domingo Perón, who acceded to his first presidential term in 1946, but then staged a coup against him in 1955 during the so-called *Revolución Libertadora*: the military bombed the central square of *Plaza del Mayo*, conducted violent attacks against civilians and staged public executions never seen before. In the turbulent times of the 1950s and 1960s, civilian presidents (who alternated with military men as the head of the state) remained dependent on the support of generals.[13] As 'anti-Peronists', the armed forces were opponents of Arturo Frondizi, who became president in 1958 but had to give way to José María Guido in 1962, after another military coup.

When the first digital computers were introduced to Argentina, the armed forces were confronted with severe domestic political problems. They believed themselves to be fulfilling a 'political duty' to the nation; their mission consisted in the repression of communism and all 'excessive leftist' movements. The commander of the Fuerzas Armadas, General Toranzo Montero, viewed the military as being responsible for preserving the republican order, and therefore for fighting against extremism and the 'totalitarian' Peronism. Because of the perceived failure of the civil authorities, he faced the task of restoring 'national unity' and 'public order'.[14] In sum: the Argentinian armed forces were, in the early 1960s, powerful, anti-communist, prepared to use violence and field-tested in coup d'états.

Given this historical setting, how are we to analyse the attitudes of the Argentinian military towards new digital technologies? Military journals are not frequently or systematically used for accounts on Latin American history in general, much less for accounts on the history of computing. Yet, according to J. Samuel Fitch, they are attractive sources: 'No other source offers such potential for analysing changes in military thinking in historical perspective.'[15] Most articles are written by military authors for their fellow officers. There might be some distribution of military journals outside of the military, but it is rare to find anyone outside the military who reads them.[16] The question whether the opinions expressed in some articles of military journals are in fact representative of 'military thinking' is a legitimate one. According to Fitch, however, they do tell something about military doctrine. Particularly, threats to national security are described and analysed, and strategies for dealing with those threats are recommended.[17] Furthermore, military journals are good sources for data on military myths and self-images.[18] Therefore, they are particularly relevant for understanding military self-positioning in the face of computer technology.

The basis of the present analysis consists of dozens of articles from the monthly journals *Revista de los Servicios del Ejercito* and *Revista del Círculo Militar* published between 1961 and 1965. Both journals were primarily written by and directed at (ex-)military personnel, although they considered themselves open to a greater readership and some subscribers were also civilians. The typical number of pages of each issue was about 90 (*Revista de los Servicios del Ejercito*) and 130 (*Revista del Círculo Militar*), the articles being between eight to twelve pages and four to twelve pages long, respectively.

In the 1960s, the topics of innovative military technology, nuclear war, military defence and supply crisis prevailed in the *Revista de los Servicios del Ejercito*. Only about one dozen articles between 1961 and 1965 dealt with computer technology. The *Revista del Círculo Militar* was divided into six subject areas, namely Military topics, History, Philosophy and other Sciences, Perspectives from the World and America, Miscellaneous, and Museology (*Temas militares, Historia, Filosofía y Otras Ciencias, Visión Mundial y Americana, Miscelanea, Museologia*). Especially in the 'Philosophy and other Sciences' and 'Perspectives from the World and America' sections there frequently were articles on computer technology, although decidedly military innovations remained of greatest interest. All in all, in the 1960s topics of war, weapons and military power prevailed.

Despite (or maybe because of) the fact that computer technology was not the top priority, one can clearly see that the 'technologization' of the military provoked debates within the armed forces, bringing to light different opinions and perspectives. Views sceptical of 'modern technologies' were expressed in the *Revista de los Servicios del Ejercito* in December 1962, when an article titled 'The soldier and the technological war' (*El soldado y la guerra tecnologica*) was published. The text by Ferdinand Otto Miksche, a French military theorist with German roots, was translated from French into Spanish.[19] In it, Miksche criticized the exaggerated belief in the benefit of technology in general. According to him, the 'material values' were over-estimated by those whose thinking was limited to the technical. Their reply to 'bombs with great destructive force' was bombs with an even greater destructive force, and their answer to 'missiles with a great range' was missiles with ranges of a hundred kilometres more. These people, Miksche claims, forgot that the human being was the measure by which 'all things' were to be measured. They ignored that the majority of all conflicts emerged independently from technological tools, and that therefore they could not be solved only with those tools.[20]

> 'Machines and other technological means are now part of all activities, but being excessively dependent on them could easily lead to fatal mistakes in the military field. The art of war is by nature fluid. It is no more probable to win battles solely with technological tools than it is to paint a valuable picture with a machine whose electronic brain choses the colours. Those who have an exaggerated faith in technological tools and methods are easily misled by the illusion that in war the devices determine the result and that war can be reduced to calculations of quantities and velocities.'[21]

According to Miksche, technological routine had taken over from 'creative intelligence' and art. Soldiers were no longer fighters but 'specialists in the use of certain instruments'. Armies were transforming into 'organisations akin to industrial companies' that functioned under the guidance of engineers. The troops were managed rather than commanded by a conceited military bureaucracy.[22]

Despite his harsh criticism of the belief in technology, Miksche did not promote the abolition of all technical equipment. 'We do not suggest to abolish the radar, guided missiles or air planes. The point is not this, but rather the fact that there is an ever increasing tendency towards technology which brings about a type of soldier which is inferior to the real fighter (*soldado inferior al verdadero combatiente*).'[23]

Argentinian chemist and military man Ricardo Mastropaolo, too, avoided the term 'computer' in his article about 'Efficiency and responsibility in modern war', published in 1963 in the journal *Revista del Círculo Militar*. However, in his text it becomes evident that he was criticizing digital processing. A good part of his article describes 'war as a business project' that is oriented towards efficiency, a tendency that Mastropaolo evaluates negatively. The 'modern armed forces', he maintained, 'have increasingly become technologised'. Even if this did not mean the 'death of the essentially human art' of the military, or the 'death of its ingenious incarnations', which could not be replaced by 'electronic brains', today's 'theatre of operations', with its 'horrible loneliness', had little to do with the 'nostalgic memories of the brilliant campaigns of the ancient leaders'. He further described 'old' war as 'colourful' due to the uniforms, 'passionate' due to the marching music, and 'monumental' due to the clash of swords and armour.[24]

Here, we have two opposing pictures: on the one hand, a contemporary commander who decides the next moves in 'terrible loneliness' (in a building, sitting at a computer?). On the other hand, the martial armed forces that 'in former times' clashed mightily with their 'real' weapons (swords) to the sound of marching music. It is not clear whether he is referring to Ancient Rome, the Middle Ages or the wars of independence. However, Mastropaolo expresses his conviction that the art of war would remain the preserve of humans in the future, and that the 'geniuses' could not be substituted by 'electronic brains'.

Despite his negative estimation of computer technology, he also wrote that the advent of the 'new technology' was a process that could – or should – not be stopped. The word 'nostalgic' indicates his conviction that the 'old art of war' had already become history, and that the 'new art of war' was the reality of the present and the future.

It becomes quite clear that the opponents of the 'technologization' of the armed forces pleaded against granting computers ('electronic brains') a dominant position in the military sphere. By no means should the 'electronic brains' substitute the 'ingenious' commanders, and one should not place too much faith in them. Otherwise the armed forces would become akin to an 'industrial company', and there would be no more 'real fighters' but only 'inferior specialists' trained to use specific tools. Other military weapons that had already been in use for some time (like the radar and the guided missile) were legitimate members of the collective. But in the face of the new technology, the 'electronic brains', Miksche and Mastropaolo remained decidedly sceptical.

It has been convincingly argued that computer work, particularly computer programming, *began* as women's work. It had to be *made* masculine, a process that, according to Nathan Ensmenger, accelerated only in the late 1960s.[25] However, from the late 1950s on, computing (at least the more sophisticated and better-paid aspects of it) was increasingly seen as a (masculine) management tool.[26]

What is remarkable here is that the type of masculinity associated with computer technology was interpreted differently in different spheres. Whereas the new and more managerial form of masculinity that came along with sophisticated computing was embraced in the business world and in administration (the ever-increasing importance of data processing promised ambitious men the chance to elevate their position within the corporate hierarchy),[27] this was not the case in the military sphere. Here, the successful manager and the 'masculinity of the organization man' that had evolved along with big business[28] had no appeal. Rather, a masculinity which can be termed 'martial masculinity' (a feature of the 'real fighter') remained the ideal conception of masculinity in the eyes of the opponents of computer technology.

## PROPAGATORS OF COMPUTER TECHNOLOGY WITHIN THE ARMED FORCES

The idea of electronic data processing being part of 'technological progress' that could not be stopped was also prevalent among the propagators of digital data processing. The very first article that was published about computers in the military journals under scrutiny was a text by Major José Javier de la Cuesta Ávila. After a brief description of technological progress in the twentieth century, he ranks the technologies that, in his view, had markedly changed the world. 'The twentieth century, in its progress and its struggle for improvement, clearly and unambiguously indicates three pillars of its future development'. In first place he names 'nuclear energy', in second 'biology' (especially medicine), and in third 'electronic data processing' (*cálculo electrónico*).[29]

Humankind does not appear in his short sketch of the history of the twentieth century. Rather, it is the century that acts: it *struggles* for improvement, it *enhances* progress. It, the century, thereby builds on the pillars of 'nuclear energy', 'biology', and the newest one, 'electronic data processing'. These three fields are imagined as solid, static and unbreakable 'pillars'. In other words, for de la Cuesta Ávila the solidity and sustainability of these fields were their most important characteristics – rather than dynamics, conflicts and controversies.[30]

The article paints a positive picture of technological progress. Technology and science are good, because they carry the century and facilitate progress and enhancement. De la Cuesta Ávila's article concludes with a sentence in italic face that again points to the connection between progress and the new machines: '*Electronic calculating machines are part of the avant-garde of progress and are, in the present, the key to the future.*'[31] Major José Javier de Cuesta Ávila, who strongly propagated electronic data processing within the military, is today celebrated in Argentina as a protagonist of computerization.[32]

In other articles on computer technology the topos of the 'second industrial revolution' is frequently mentioned,[33] and the development of this technology is seen as a 'logical' and 'automatic' consequence of other, previous innovations. In an article titled 'General remarks on the use of digital electronic computers', major and engineer Carlos Cesar Lopez wrote: 'The automatization is a natural extension or else a continuation of a sequence of technological changes that characterized the first industrial revolution, despite of the fact that concerning the appreciation of the person it is rather an inversion than an extension.'[34] In the following pages, he sketches the development of the computer as a process which has its roots in the 'beginning of civilization', and which was advanced by ingenious engineers, such Blaise Pascal ('inventor' of the arithmetic machine or 'Pascaline', 1642) and Leibnitz (sic!) (cogwheel, 1671).[35] Oscar A. Poggi, too, in his article published in 1966, traces the 'birth of the contemporary epoch' back to the 'technological revolution', which he connects with previous events, namely the Industrial Revolution (1760) and the French Revolution (1789).[36] He cites Italian industrialist Aurelio Peccei's view that a considerable part of humankind dedicates itself to the realization of 'permanent innovations' in the field of technology and science, and this process was currently in a stage of 'continuous and increasing acceleration'. Since all discoveries are based on the accumulation of previous knowledge, and all innovations are stimuli for progress in other fields, he reasons that progress has transformed itself into an 'automotive process'.[37]

Interestingly, in their sketches of a meta-narrative, the authors describe computer technology as part of the general history of humankind. They do not specifically discuss the military sphere; that is, they do not connect computer technology with the interests of the armed forces, although they were writing for soldiers and commanders. This is not self-evident. De la Cuesta Ávila, for instance, could have mentioned the atomic bomb instead of nuclear energy. However, in the face of the contemporary technological developments, the authors (who acted as spokespersons for the collective named 'armed forces'), did not see themselves as members of a specific professional field or as members of a specific (national) community in Latin America. Rather, the computer in their eyes had such importance that the relevant collective affected by computer technology was the 'whole of humanity in the twentieth century'.

The narrative according to which computer technology emerged 'automatically' and then 'fell upon' humankind is not the only concept developed in the military journals. In fact, the 'meta physics' of the computer accounts only for a few lines or paragraphs in each of the articles, which are five to sixteen pages long. The descriptions of the concrete functioning of computers, the work *with* computers and the alliance between men and computers account for the main body of the articles.

## ENVISIONING SUCCESSFUL ALLIANCES BETWEEN COMPUTERS AND MEN

In all the articles arguing in favour of computer technology, the authors dwell on the concrete work with computers, which they foresee as becoming a daily routine in the near future. In the above-cited article, after having sketched the meta-narrative of the history of the twentieth century and the computer de la Cuesta Ávila turns to much more practical matters. The larger part of his text describes what computers actually *are* and what they *do*. According to him, a computer was a 'scientific tool, reliable storage and element of confidence and great flexibility which saves time and enables unimaginable operations'.[38] The modern calculating machines, he explains, evidently to an audience unfamiliar with computers, were able to read, assemble data, print, perforate, copy, compare, store, select, add, subtract, programme, multiply and divide. These actions were possible if the following elements were in place: 'The human who orders what he wants', the programme card that translates and saves the order in its own language (the perforation language), the machine that works on the basis of electronic impulses, and the magnetic kernel that 'memorizes' (*los núcleos magnéticos, que 'memorizan'*).[39]

Here, human and non-human elements are conceived as parts of one unit: whereas the first element in his numeration is human, the following three elements are non-human, and all are integral parts of the calculating process. It is no longer 'the twentieth century' or 'the technology' that are the actors here, but a specific technological artefact, working *together* with humans. All the elements are described in detail in the rest of the text, in which an IBM machine serves as an example: the programming card, for instance, is described as an entity with a 'variable' number of columns, IBM's having eighty. In order to describe the 'vivid, infallible and immediate' memory (*memoria viviente, infalible e inmediata*) of systematizing machines, de la Cuesta Ávila in another article also draws on IBM hardware, namely the systematizing machine IBM 705 EDPM. This machine, he explains, could master 8,400 additions or subtractions (of numbers with five digits), 1,250 multiplications (of five times five digits), 550 divisions (of six dividends and four divisors) and '29,400 logical decisions' per second.[40]

According to the manufacturers, de la Cuesta Ávila explains, the machine was the 'most powerful tool' for solving 'all problems' of the management of a company. It encompassed the whole spectrum of data registration with the 'highest flexibility' and the 'fastest velocity'. It was able to store 'all information', to use at a convenient time for problem solving purposes. The only precondition was entering the 'right command'. Here, through the explicit reference, it again becomes apparent that the major had obtained his information from IBM (the manufacturer of the mighty systematization machine) and now passed it on to his readers. From 1958 on, de la Cuesta Ávila, then director of the military hospital in Buenos Aires, had participated in IBM trainings.[41] In these trainings he had accumulated knowledge about digital computing. In some cases, he explicitly mentioned his close connection and affinity to IBM. At the end of one of his articles, there is the following sentence: 'I wish to thank IBM WORLD TRADE CORPORATION for the cooperation and the instructions.'[42] Here, it becomes evident that IBM, a big US player, was already well connected to persons in the Argentinian armed forces – and with persons located in the right places.

Remarkably, but not surprisingly, the major usually speaks of *civilian use* of computers, praising the increased efficiency for industrial companies. Conveniently and probably not coincidently, there is an IBM advertisement in the same issue of the journal, placed prominently on the first page (verso of the cover), that says exactly the same. The wording is 'A company for companies' ('empresa para empresas', see Figure 1).

The fact that IBM primarily promoted the civilian, entrepreneurial use of computers also becomes apparent in another advertisement published in the following issue of the military journal 'Servicios del Ejercito' in July 1963. Here, the 'System 1401' is promoted in large letters: 'With the System 1401, IBM offers the MODERN company a new and powerful tool for data systematization. Capacity – Speed – Accuracy' (see Figure 2). Beneath the advertisement is the address of the originator, namely the IBM branch in Buenos Aires.

The IBM advertising department did not bother to draft advertisements specifically addressing the military sphere. Apparently, it was not necessary to praise the computer as a tool compatible with and useful to the armed forces. It was sufficient to highlight the usefulness of the System 1401 to *companies*. One can conclude that the computer was, at that time, seen both by IBM and the armed forces as a tool that mainly served the business sphere. It was thus not IBM that had to adapt to the other collective (the armed forces), but the other way around. IBM's self-confidence can be briefly explained with some figures: in 1960, IBM had about 100,000 employees worldwide,[43] some 1,000 working in Argentina.[44] IBM's sales volume was 1.81 billion US dollars, its profits were 205 million US dollars. The Argentinian armed forces, in contrast, consisted of some ten thousands of soldiers, with a budget of 321 million US dollars in 1960. In 1962, the budget increased to 368 million US dollars.[45]

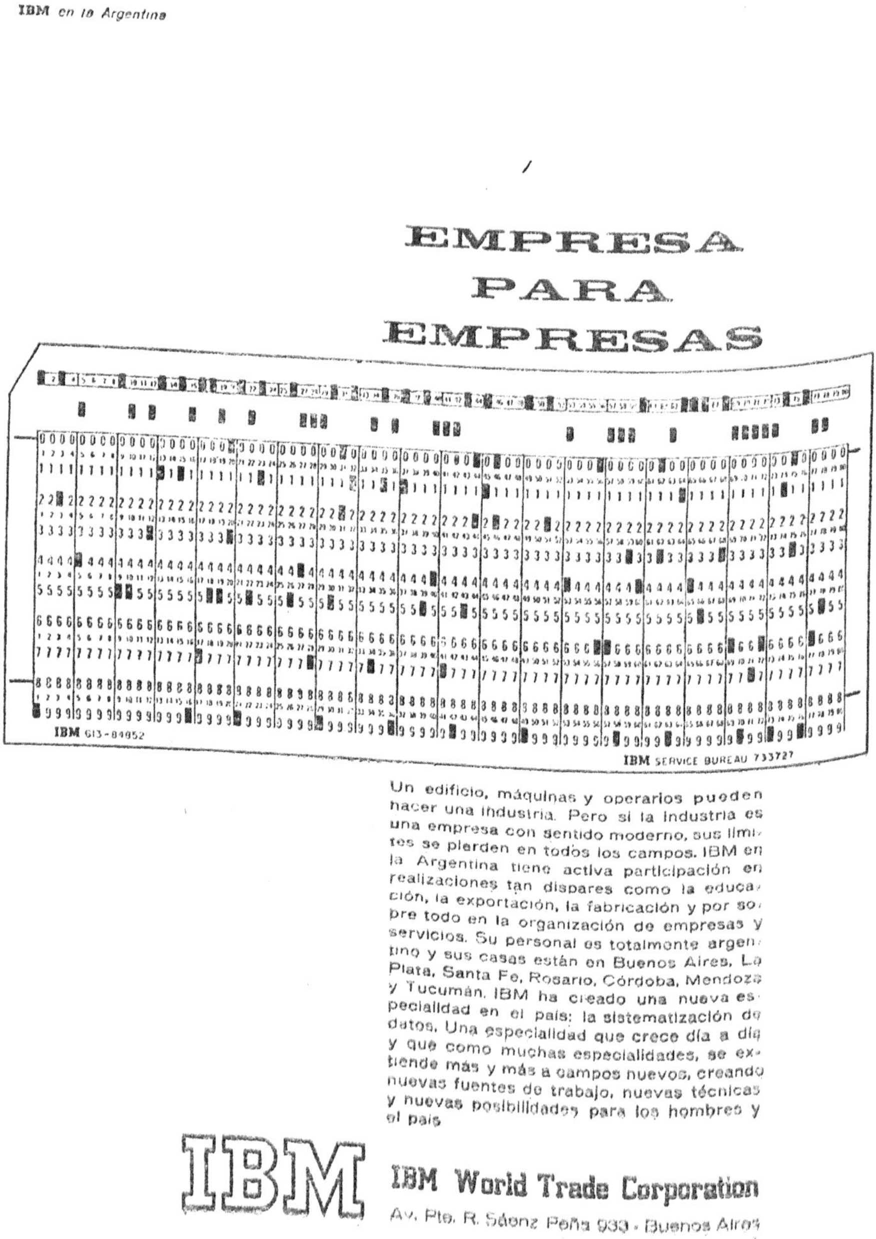
IBM en la Argentina
EMPRESA PARA EMPRESAS
IBM SERVICE BUREAU
Un edificio, máquinas y operarios pueden hacer una industria. Pero si la industria es una empresa con sentido moderno, sus límites se pierden en todos los campos. IBM en la Argentina tiene activa participación en realizaciones tan dispares como la educación, la exportación, la fabricación y por sobre todo en la organización de empresas y servicios. Su personal es totalmente argentino y sus casas están en Buenos Aires, La Plata, Santa Fe, Rosario, Córdoba, Mendoza y Tucumán. IBM ha creado una nueva especialidad en el país: la sistematización de datos. Una especialidad que crece día a día y que como muchas especialidades, se extiende más y más a campos nuevos, creando nuevas fuentes de trabajo, nuevas técnicas y nuevas posibilidades para los hombres y el país
IBM World Trade Corporation
Av. Pte. R. Sáenz Peña 933 - Buenos Aires

FIGURE 1

IBM
VINCIT
ofrece a la empresa MODERNA, con su sistema
1401
Un nuevo y poderoso equipo de Sistematización de Datos.
CAPACIDAD
VELOCIDAD
EXACTITUD ...
ELECTRONICAS
IBM World Trade Corporation
Av. Pte. R. Sáenz Peña 933, Bs. As

FIGURE 2

The authors of the articles in the military journals saw the computers as tools to increase efficiency, a tool that could also be used in the military field. The business-like and modern image of the machines was, in the eyes of the propagators of computer technology in the military, not a disadvantage but rather a selling point.

In April 1962, de la Cuesta Ávila published an article about the 'Military Programmer'.[46] Remarkably, he begins his text with an extensive description of general processes of computerization. 'The data processing machines or systems of data processing (SCD) slowly but surely invade all spheres of human activity.' According to him, science benefited due to the speed of the processes, which saved the scientists 'years of life', industry gained 'solutions for production problems', and traders resolved their problems with balances of account by receiving statistics for current and future developments. 'Archives of all sorts', too, benefited from having direct access to data. In short: computers brought humans tools they had 'never imagined before'.[47]

Two things deserve attention here. First, de la Cuesta Ávila believed that computers were currently *invading* all human spheres. Therefore, the human spheres were not solely human any more – they were inhabited by computers, which took over important tasks and became constitutive elements of the respective collectives. Second, the mentioned collectives were all of a civilian nature. Science, industry, trade, archives – these all had already been profiting from computer technology. Here de la Cuesta Ávila implicitly makes a plea for the use of computer technology in the military field. In his eyes, the armed forces were late-comers who should adapt to the new technology as soon as possible.

In 1962, Javier de la Cuesta Ávila moved from the post of director of the military hospital to that of director for reservists in the Human Resources Department (*Jefe de la División Reservas del Departamento Potencial Humano*). Together with military engineer and general lieutenant Aníbal H. Aguiar, who was chief of the technical division of the Human Resources Department, he developed the basis for the RIPOM register (*Registro Integral de Personal con Obligación Militar*), which served for the registration and administration of military personnel.[48]

Aguiar, a close ally of de la Cuesta Ávila, went even further in his interpretations of the interconnections between men and machines. In an article published in October 1963, Aguiar describes, under the subtitle 'Effects of the technology of the electronic data processing on humans, organisations and norms of procedures', the characteristics of the 'systems of electronic data processing' as systems in which a 'real amalgamation of humans, machines and processes' took place (*verdadera amalgama de hombres, máquinas y procedimientos*). In this system, the relation between man and machine was not fixed but dynamic. While the human directed the system, and the activities of the system normally served human purposes, there still was a stage in which the human had to subordinate his will to the exigencies (*exigencias*) of the machine if he wished to get what he asked for. Therefore, 'paradoxically', some situations required that at certain moments the machine not be in the service of men but, on the contrary, men be in the service of the machine (*en ciertos momentos, no sean las máquinas las que están al servicio del hombre, sino que es a la inversa*).[49] In the end, the whole unit had to act 'like a single, perfectly synchronized body', and from this moment on the machine held control (*todo el conjunto debe actuar como un solo cuerpo perfectamente sincronizado y desde ese momento es la máquina la que manda*). This also implied that in the system, a 'complex made out of humans and machines' (*complejo formado por hombres y máquinas*), all individuals must understand the general concept and be 'deeply convinced' by it. The actions of each individual were as significant as those of all others. Nothing was to be left to personal interpretations; all humans in the organism, the centre or the installation, had to subordinate themselves to clearly determined norms and processes in order to ensure a correct procedure.[50]

The main and oft-repeated argument for the use of computers was, for Aguiar, its 'infallibility': '[T]he computer executes what it was ordered, and if there is an error in the system, it is in 99.9% of cases not caused by the machine but by the deficiency of the programmer'.[51] The human being was, according to Aguiar, not perfect, whereas the computer in fact was. 'We always have to consider that the human is not perfect. . . . Even the most perfect and most concentrated man is prone to failure.' The machine, by contrast, is 'perfect and does not commit errors'.[52] On the other hand, Aguiar describes the computer as an element that – again contrary to the human – could not think. '[The computer] is not able to execute something which has not been ordered, it has no capacity for analysis, logic and reasoning, it is not the electronic brain that it is commonly called, it does not think, it cannot think.'[53]

The incapacity to think independently and to give orders is frequently used to address the fear of the computer replacing humans. 'Computers and automatization do not replace the human, plain and simple (*las computadoras y la automación no reemplazan al hombre, lisa y llanamente*)', wrote Carlos Cesar Lopez. They were rather applied 'far away' from the places of military operations. In addition, they freed humans from calculating and saved them time (*ganancia de tiempo*). But 'always, at least for the time being', they followed the instructions that were elaborated 'directly or indirectly by a human mind'.[54] According to the articles, the final authority – a central issue for the military – remained in human hands.

However, in the eyes of the contemporary observers electronic data processing would have an impact on the work of humans. In the face of computer technology, explains Lopez, it was necessary to redistribute tasks and 'readapt' workers in order to forestall severe social implications as computers took over routine and administrative tasks. In addition, humans had to change their 'current perspective of time' if they wanted to profit from the advantages of computers. They would have to measure time in thousandths and millionths of seconds in order to appreciate the quantity of data and information that was now manageable.

## THE COMPUTER AS AN ALLY OF THE ARGENTINIAN ARMED FORCES?

Despite the fact that all analysed articles were published in military journals, they speak remarkably little about the specific military use of computers. Instead, they frequently emphasize civilian use in companies or in science, and always mention these uses *before* any military use. However, in the oldest article examined (1961), de la Cuesta Ávila mentions military operations accomplished by computers.

According to him the electronic machine accomplished, within the military sphere (*dentro del campo militar*), a 'wide spectrum of activities'. 'Modern armed forces', he emphasizes, could not afford leaving data collection, calculation of resources and planning activities to the 'memory and the preparation of their officers' (*a la memoria o preparación de sus cuadros*), but had to rely on calculating machines.[55]

In the last page of the same article, the author provides an example of the specific use of the computer for a 'military problem' (*problema castrense*). The perforator machine was able to issue a file (*tarjeta*) on all citizens, containing all personal data. If a man left the army, this file would be complemented by information about his specific trainings, qualifications and possible use in different units.[56] After this, the *máquina intercaladora* could organize the documentation of the personnel according to register number, qualification, types of units, etc. and process it for the archive. The *máquina clasificadora*, then, would be able to organize the personal data according to groups or units and so to search for persons that combined specific characteristics. Finally, the *máquina sistematizadora de datos*, as the 'permanent memory of the whole system', would be able to assemble a strike force whenever needed. The machines were ready to fulfil this and similar tasks on short notice and with 'reliable safety' and a 'continuous and infallible memory' (*memoria constante e infalible*).

What de la Cuesta Ávila sketches here, as a field of application for computer technology, is the administration of human resources. His main argument was that personal data and information could be collected, processed and systematized through digital data processing. At the outbreak of a war, for instance, the computer could create units out of a mass of former conscripts, forming them according to the required characteristics, within a very short time. With its specific capacities and attributes (large and infallible memory, speed, accuracy), the computer was an important tool to create the collective 'strike force' in the first place.

In November 1961, de la Cuesta Ávila published another article about the 'Military application of the systems of data processing by electronic machines'.[57] In it, he chooses another strategy. In the very first sentence, he states that the *sistema de computación de datos* was 'another tool for the work of the commanders', not a substitute for them (*una herramienta más de trabajo para los Comandos, no un substituto de los mismos*). This was apparently a reaction to fears within the armed forces that the machines would take over. De la Cuesta Ávila replied that 'the system' was not able to take decisions and added: 'The command decision will remain the exclusive activity of the leader, as a man.' (*La resolución del comando es y seguirá siendo una actividad exclusiva del conductor, como hombre.*)[58] What the military should expect from electronic data processing was rather 'accessibility, control, distribution, security and secrecy, automatic destruction, simultaneity in processing'. Interestingly, security and secrecy were, in the eyes of the major, guaranteed because of the reduced number of persons involved in the documentation, and because of the ease of storing and encrypting data. Furthermore, the automatic destruction of data within a fraction of a second would allow the command to use and to save data 'until the last second' (before a hostile takeover).

Altogether, de la Cuesta Ávila evaluates electronic data processing as multifaceted and useful for all areas within the military sphere. 'The military use of electronic data processing is practically infinite, only limited by imagination and needs.'[59]

Frequently, however, he does not give many answers to the question *how* computer technology was to be used by militaries. Even in a long article entitled 'The military data processing centre' (1963),[60] only the fourteenth and last paragraph is dedicated to specific military uses. The rest is given up to the description of a common, that is, a civilian, data processing centre.[61] In some of his articles published in military journals, de la Cuesta Ávila dedicated himself explicitly to the 'technology and schematization of forms'.[62] In these texts, there is not a single mention of a specific military use; the described processes solely concerned the administration of personal files in human resources management.

Throughout his articles, de la Cuesta Ávila consistently relates computer technology to administration and bureaucracy. According to him, bureaucratic matters (notifications, schedules, receipts, etc.) accounted for 'half of all tasks' within an operation. All jobs of this sort – 'control, administration, supply, infrastructure, cost calculations, automatic notifications on stocks, units etc., distribution of military personnel and the commands, civil activities, knowledge about the personnel, tasks, statistics, assessment and production of reports etc.' – were to be left to electronic data processing. This kind of data processing, he sums up, would enable a reduction in administrative personnel, freeing them up for operation in other, more important areas.[63]

Quite telling with regard to the omission of a precise definition of the benefit of computer technology for the military sphere is Aguiar's article titled 'Electronic data processing – an indispensable ally for national defence' (see Figure 3).[64] Remarkably, the text contains many arguments in favour of computer technology (especially the speed, infallibility, and perfection of the machine), but does not say anything about specific military use. Despite the title, the use of computer technology for 'national defence' purposes is not even mentioned in the text. What is mentioned, however, is the change caused by computers within the armed forces. According to Aguiar, the 'amalgamation' of humans, machines and procedures requires comprehensive knowledge about the installation among the 'entire personnel' and within 'all parts of the system'. In other words, not just the director general but every member, down to the 'very last employee' or 'subaltern', had to be familiarized with the whole organism, even if they had no direct contact with the system. Only then would the 'mystic of the systematization' function.[65] Aguiar here envisions nothing less than a new distribution of knowledge among the military institutions. Implicitly, a more horizontal form of organization is given preference because all elements were believed to play an essential role in 'the system'.

From Aguiar's perspective, what had to take place was an encompassing transformation of the armed forces and the military field. He was aware of the fact that, especially within the military sphere, there was much scepticism and a lack of comprehension, because electronic data processing was a 'type of discipline' that differed 'notably' from the 'nature of military tasks' (*tipo de disciplina que difiere notablemente de la índole de las tareas castrenses*). All in all, the author is confident about the future. With his article, he wanted to provoke the reader's mind. In his opinion, this was to be the beginning of a 'fierce desire' and a further step on the path which at present was only a beaten track but which would in time transform into a 'motorway'.[66]

LA SISTEMATIZACION ELECTRONICA DE DATOS – UN ALIADO INDISPENSABLE PARA LA DEFENSA NACIONAL

Por el Tcnel. (RE) Ing. Mil. D. Aníbal H. Aguiar

PREFACIO:

*Desde hace tiempo, se ha hecho común escuchar en distintos medios algunas expresiones que se refieren a los procesos de computación o sistematización de datos, realizados por máquinas, y, salvo algunas personas ligadas estrechamente a este tipo de trabajo, pocas son las que saben a ciencia cierta en qué consisten y qué significan.*

*Esta situación se ha hecho más evidente, en especial en las Fuerzas Armadas, por tratarse de un tipo de disciplina que difiere notablemente de la índole de las tareas castrenses.*

*El autor de este trabajo pretende hacer llegar al Cuadro de Oficiales y de Suboficiales de las Fuerzas Armadas, algunas ideas que permitan aclarar el verdadero significado y el alcance efectivo que involucra la aplicación de los procesos Electrónicos de Sistematización de Datos, en el convencimiento de que se hace necesario no retardar por más tiempo la*

56

FIGURE 3

## CONCLUSION

The article aimed at questioning the common notion of the military being the driving force behind processes of digitalization. The debates and controversies about computer technology within the Argentinian armed forces in the early 1960s suggest that the computerization of the military was by no means a process that ran silently and smoothly. Much less did the Argentinian *Fuerzas Armadas* see themselves as at the forefront of digitalization. Instead, members of the armed forces discussed the value and advantages, as well as the dangers and disadvantages of the new digital machines, which were at first used in civilian contexts (science, business) and which were seen as 'un-military' tools. What emerged from the importation of the first digital computers to Argentina was a struggle about and ultimately a new positioning of the military sphere within other fields, namely business, trade, science and public institutions. Propagators and opponents alike judged the computers to be part of a 'revolution' that could not be stopped any more. Both groups believed that electronic data processing would permanently and pervasively 'invade' all human spheres.

The advent of computers led to discussions and negotiations about what was 'soldierly' and manly – and what was not. In other words, the appearance of computer technology kicked off the transformation of the spheres in which gender was 'performed'. The opponents of computer technology expressed 'nostalgic' sentiments in the face of the technologization of the armed forces, remembering the good old times when soldiers were 'real fighters' instead of 'inferior' specialists trained to use a specific tool. Gendered visions and divisions of different spheres became particularly apparent in the talk about commanders making their decisions in 'terrible loneliness' (at the desk, sitting at the computer, inside a building) and not – as in former times – directly on the battlefield, with marching music to accompany the 'colourful' activities of the fighting soldiers. For the conservative stakeholders within the armed forces, the desk and 'the inside' were apparently heavily linked to 'unmanly' work. The more managerial form of masculinity that had arisen in conjunction with sophisticated and well-paid computing jobs in business and administration from the late 1950s on was not in great fashion within the military sphere. In short: the computer did not comply with the armed forces' picture of themselves as a particularly martial institution.

The propagators of computers, too, used the metaphors of a 'technological revolution' when referring to the advent of electronic data processing. In fact, they describe the 'invasion' of computers into all human spheres as a *fait accompli* and as a tremendous event, as inevitable as a natural disaster. In contrast to the sceptics, they praised the machine as being part of the 'modernization' process. Technological innovation, to their mind, took place 'automatically', with auto-motion. Humans could only *react*.

One of the main arguments of the 'computerizers' within the military was that all other spheres, namely those of 'business', 'commerce', 'industry' and 'science', had already adapted to computer technology – with tremendous success. The armed forces, a late-comer in this respect, had to participate in the process of technologization. 'Modern societies' and 'modern armed forces', they maintained, had no choice but to join this process. Additionally, they argued that, contrary to humans, the machines were 'perfect' and 'infallible'. They would provide time-saving, efficiency and excellent administration capabilities, all useful attributes for the military.

Important protagonists of electronic data processing included persons like José Javier de la Cuesta Ávila and Aníbal H. Aguiar, who, from 1962 on, worked together in the Human Resources department of the Argentinian armed forces. They were, for sure, no battle-hardened combatants. Rather, they were military administrators who were used to acting behind the scenes. From 1958 on, de la Cuesta Ávila had attended training sessions offered by IBM and obtained practical skills with IBM machines. In his article, he emphasized the daily work with computers – which he was perfectly acquainted with. In de la Cuesta Ávila and Aguiar's conception, there should no longer be a strict division between the human sphere and the computer sphere, or between human and non-human labour. Rather, they propagated the establishment of one single organism, consisting of men and machines. The processes and the success of 'the system' were only attainable if men and machines worked together. Aguiar explicitly talked about a 'real amalgamation of men, machines, and procedures'.

Apparently, gender stereotypes did not have to change radically for the integration of computers into the military sphere to take place. At least the 'masculinity of the organization man' (Thomas Haigh) that had been successfully established in the business and administration spheres was not praised (or even mentioned) by the proponents of computer technology within the military sphere. Within their own institution, they avoided arguing against 'martial masculinity'. Remarkably, they did not even bother to praise the computer as a particularly 'manly', 'masculine' or military tool. They simply described it as a useful tool for management and administrative purposes, and emphasized its efficiency and time-saving capabilities. The transformation of the armed forces into an institution 'akin to industrial companies' (a process feared so much by the opponents) was not neglected by de la Cuesta Ávila or Aguiar or seen as a disadvantage. Rather, they favoured and propagated this tendency. Interesting in this regard are also the advertisements by IBM. In the military journals under scrutiny, IBM published ads under the heading 'A company for companies', promoting the System 1401 as an 'almighty tool in the hands of the administration'. Regarding the new computer technology, IBM (and the business sphere as a whole) was not inclined to adapt itself to the military sphere – on the contrary, the military sphere had to adapt to the business sphere. Both the proponents of computer technology within the Argentinian military and members of IBM's public relations office apparently were very confident that this would eventually happen – and probably in the near future. How exactly the history of computerization of the military (in Latin America) happened in the 1960s and thereafter, and how 'martial masculinity' was (or was not) thereby contested still requires more differentiated accounts, based on empirical evidence.

## NOTES AND REFERENCES

1. Claudia Honegger, *Die Ordnung der Geschlechter. Die Wissenschaften vom Menschen und das Weib, 1750–1850* (Frankfurt/New York: Campus, 1991).

2. Cynthia Cockburn, *Die Herrschaftsmaschine. Geschlechterverhältnisse und technisches Know-How* (Berlin/Hamburg: Argument, 1988).
3. Érico Esteves Duarte, *Tecnologia militar e desenvolvimento econômico: uma análise histórica* (Rio de Janeiro: Ipea, 2012): 16f.
4. Nathan Ensmenger, '"Beards, Sandals, and Other Signs of Rugged Individualism": Masculine Culture within the Computing Professions', *Osiris* 30, 1 (2015): 38–65.
5. For an overview, see Roy Rosenzweig, 'Wizards, Bureaucrats, Warriors, and Hackers: Writing the History of the Internet', *The American Historical Review* 103, 5 (1998): 1530–1552; Janet Abbate, 'Government, Business, and the Making of the Internet', *The Business History Review* 75, 1 (2001): 147–176.
6. Ivan da Costa Marques, 'History of Computing in Latin America', *IEEE Annals of the History of Computing* 37, 4 (2015): 10–12; Jaquelinne Dominguez Nava, Juan C. Acosta-Guadarrama, Rosa M. Valdovinos Rosas, Víctor H. Solis Ramos, Nely Plata César and Leticia Quintanar Rebollar, 'A Brief History of Computing in Mexico', *IEEE Annals of the History of Computing* 37, 4 (2015): 76–86, 77f.
7. Steven R. Beck, *Computer Bargaining in México and Brazil 1970–1990: Dynamic Interplay of Industry and Politics* (London, PhD dissertation at the London School of Economics, 2012): 76f.
8. Nicolás Babini, 'Modernización e informática: Argentina, 1955–1966', *Quipu* 9, 1 (1992): 89–109, 93; Nicolás Babini, *La Argentina y la computadora. Crónica de una frustración* (Buenos Aires: Ed. Dunken, 2003): 29ff.; Nicolás Babini, 'La llegada de la computadora a la Argentina', *LLULL* 20 (1997): 465–490.
9. Babini, *La Argentina y la computadora*: 28; Pablo Miguel Jacovkis, *De Clementina al siglo XXI: breve historia de la computación en la Facultad de Ciencias Exactas y Naturales de la Universidad de Buenos Aires* (Buenos Aires: Eudeba, 2013).
10. Ernesto García Camarero, 'Algunos recuerdos sobre los orígenes del cálculo automátio en Argentina, y sus antecendentes en Espana e Italia', *Revista Brasileira de História da Matemática* 7, 13 (2007): 109–130, 117.
11. Guillermo O'Donnell, 'Modernization and Military Coups', in María Gabriela Nouzeilles and Gradela Montaldo (eds.), *The Argentina Reader: History, Culture, Politics* (Durham, NC; London: Duke University Press, 2002): 399–420, 399.
12. Robert A. Potash, 'The Changing Role of the Military in Argentina', *Journal of Inter-American Studies* 3, 4 (1961): 571–578.
13. David Huelin, 'Conflicting Forces in Argentina', *The World Today* 18, 4 (1962): 142–152, 143. Of sixteen Argentine presidents between 1955 and 1983, seven were civilians and nine were generals.
14. O'Donnell, 'Modernization and Military Coups': 403.
15. J. Samuel Fitch, 'Military Attitudes toward Democracy in Latin America: How Do We Know If Anything Has Changed?', in David Pion-Berlin (ed.), *Civil–Military Relations in Latin America. New Analytical Perspectives,* (Chapel Hill, NC: University of North Carolina Press, 2001): 59–87, see especially 71–76.
16. Fitch, 'Military Attitudes': 72.

17. Fitch, 'Military Attitudes': 73.
18. Fitch, 'Military Attitudes': 75.
19. Ferdinand Otto Miksche, 'El soldado y la Guerra tecnológica', *Revista de los Servicios del Ejercito* 25, 800 (1962): 1057–1063.
20. Miksche, 'El soldado y la Guerra tecnológica': 1058.
21. Miksche, 'El soldado'. All translations are the author's.
22. Miksche, 'El soldado'.
23. Miksche, 'El soldado': 1059.
24. Ricardo A. Mastropaolo, 'Eficiencia y Responsabilidad en la Guerra Moderna', *Revista del Círculo Militar* 669 (1963): 112–117, 115.
25. Nathan Ensmenger, 'Making Programming Masculine', in Thomas J. Misa (ed.), *Gender Codes: Why Women Are Leaving Computing* (New York, NY: John Wiley & Sons 2010): 115–141.
26. Marie Hicks, 'Meritocracy and Feminization in Conflict: Computerization in the British Government', in Thomas J. Misa (ed.), *Gender Codes: Why Women Are Leaving Computing* (New York, NY: John Wiley & Sons 2010): 95–114.
27. Thomas Haigh, 'Masculinity and the Machine Man: Gender in the History of Data Processing', in Thomas J. Misa (ed.), *Gender Codes: Why Women Are Leaving Computing* (New York, NY: John Wiley & Sons 2010): 51–71, 56.
28. Haigh, 'Masculinity and the Machine Man': 51, 58.
29. José Javier de la Cuesta Ávila, 'Computadoras Electronicas', in *Revista del Círculo Militar* 201–203, 661 (1961): 124–129, 124.
30. For proponents of ANT, technological innovation and scientific facts are always outcomes of complex processes and conflicts; see Bruno Latour, *Science in Action. How to Follow Scientists and Engineers Through Society* (Cambridge, MA: Harvard University Press, 2003).
31. 'Las máquinas de cálculo electrónicas están en la avanzada del progreso y son en el presente la llave del futuro', Cuesta Ávila, 'Computadoras Electronicas': 129.
32. The national computer science museum in Buenos Aires (Museo de la informática) offers a comprehensive biography of de la Cuesta Ávila on its website: http://www.museoinformatico.com.ar/~museoinfor/articulos/personalidades-informaticas/27-tcnl-cuesta-Ávila (last accessed 26 September 2016).
33. Aníbal H. Aguiar, 'La sistematizacion electronica de datos – un aliado indispensable para la defensa nacional', *Revista del Círculo Militar* 228–230, 670 (Oct. 1963): 56–69, 57.
34. Carlos Cesar Lopez, 'Generalidades Sobre Empleo de las Computadoras Electrónicas Digitales', *Revista de los Servicios del Ejercito* 26, 311 (Nov. 1963): 917–932, 923.
35. Lopez, 'Generalidades': 924.
36. Oscar A. Poggi, 'Trascendencia de uno de los hechos más característicos de nuestro tiempo', *Revista del Círculo Militar* 678 (1966): 80–87, 80f.

37. Poggi, 'Trascendencia': 81f.
38. Cuesta Ávila, 'Computadoras Electronicas': 124.
39. Cuesta Ávila, 'Computadoras Electronicas': 125.
40. Cuesta Ávila, 'Computadoras Electronicas': 127.
41. http://www.museoinformatico.com.ar/~museoinfor/articulos/personalidades-informaticas/27-tcnl-cuesta-Ávila (last accessed on 12 September 2016). The history of IBM in Latin America is not yet well known. First insights into IBM in Chile are provided by: Eden Medina, 'Big Blue in the Bottomless Pit: The Early Years of IBM Chile', in: *IEEE Annals of the History of Computing* 30, 4 (2008): 26–41.
42. Cuesta Ávila, 'Computadoras Electronicas': 124.
43. IBM-Highlights 1885–1969, 22. This document is available at the IBM online archive: https://www-03.ibm.com/ibm/history/documents/pdf/1885-1969.pdf (last accessed 26 September 2016).
44. Babini, *La Argentina y la computadora*: 50.
45. David L. Feldman, 'Argentina, 1945–1971: Military Assistance, Military Spending, and the Political Activity of the Armed Forces', *Journal of Interamerican Studies and World Affairs* 24, 3 (1982): 321–336, 328.
46. José Javier de la Cuesta Ávila, 'Sistema computador de datos. Progamador (sic!) militar S.C.D.', *Revista de los Servicios del Ejercito* 25, 292 (1962): 281–286.
47. Cuesta Ávila, 'Sistema computador de datos': 281.
48. http://www.museoinformatico.com.ar/~museoinfor/articulos/personalidades-informaticas/27-tcnl-cuesta-Ávila (last accessed on 12 September 2016).
49. Aguiar, 'La sistematizacion electronica de datos': 65.
50. Aguiar, 'La sistematizacion electronica de datos': 65.
51. Aguiar, 'La sistematizacion electronica de datos': 67.
52. Aguiar, 'La sistematizacion electronica de datos': 70.
53. '[L]a Computadora . . . no tiene capacidad de análisis, de lógica, ni de razonamiento, no es un Cerebro Electrónico como vulgarmente suele llamárselo, no piensa, no sabe pensar', Aguiar, 'La sistematizacion electronica de datos': 67.
54. Lopez, 'Generalidades': 926.
55. Cuesta Ávila, 'Computadoras Electronicas': 124.
56. Cuesta Ávila, 'Computadoras Electronicas': 128.
57. José Javier de la Cuesta Ávila, 'Aplicación Militar del Sistema de Computación de datos mediante maquinas electrónicas', *Revista de los Servicios del Ejército* 24, 287 (1961): 933–937.
58. Cuesta Ávila, 'Aplicación Militar': 933.
59. Cuesta Ávila, 'Aplicación Militar': 934.
60. José Javier de la Cuesta Ávila, 'Sistema de computación de datos. Centro SCD Militar', *Revista de los Servicios del Ejército* 26, 304 (April 1963): 265–267.
61. Cuesta Ávila, 'Sistema de computación de datos': 267.

62. José Javier de la Cuesta Ávila, 'Sistema de Computación de Datos. Técnica de Diagramación de Formularios', *Revista de los Servicios del Ejército* 26, 306 (June 1963): 449–458.
63. Cuesta Ávila, 'Aplicación Militar': 934f.
64. Aguiar, 'La sistematizacion electronica de datos'.
65. Aguiar, 'La sistematizacion electronica de datos': 68.
66. Aguiar, 'La sistematizacion electronica de datos': 57.

# Transnational Astronomy: Science, Technology and Local Agenda in Cold War Chile

BÁRBARA SILVA[1]

Nowadays, Chile has a well-known reputation as a world astronomical centre. Although this prestige seems to be a recent acquisition, this paper argues that the development of Chilean astronomy is the result of a transnational effort during the 1960s, specifically connected to the Cold War. In 1962, American, European and Soviet scientists chose Chile as a strategic place to build their own large-scaled astronomical observatories. The country's political stability, the relative small size of the institutions involved in the negotiations and governmental reliability, enhanced the opportunities Chile's physical landscape provided to astronomical enterprises.

Local and international scientists were at the crossroads of political connections, while science legitimized those ideological exchanges. For local politicians, to take part in projects related to avant-garde technology was a key aspect to demonstrating their active commitment to progress and modernization. Despite their lack of expertise in the field, national authorities saw a one-time opportunity in these projects, and established alliances in order to develop astronomy, astrophysics and astronomical engineering. Within a broader framework, these local actors managed to create public policies regarding astronomy, and to get involved with both European, American and Soviet scientists.

By studying the development of Chilean astronomy, this article questions the intersection between power and large-scale scientific and technological projects, such as astronomical observatories, at a time of international political polarization in Latin America. Furthermore, through the study of the convergence between science and politics, this article suggests novel ways to explore other circulations of expert knowledge and proposes a new understanding of Latin American historical processes throughout technology.

# TRANSNATIONAL ASTRONOMY: SCIENCE, TECHNOLOGY AND LOCAL AGENDA IN COLD WAR CHILE

A visit to the Atacama Desert can be an intense experience, in different ways. Geographically, this location in northern Chile is the Earth's driest place. This – almost – absolute absence of water and humidity relates to a sensation a visitor can get when walking around the area: the feeling of emptiness. There are few inhabited areas, with communities struggling to bend nature's condition because of the lack of water. For this, scientific imagination and inventions have tried to solve the scarcity of water and to contribute towards a more habitable environment in different ways. One of the most amazing technologies developed there is *Atrapaniebla*, which collects the water drops from the morning fog.[2] Other technologies developed in the Atacama Desert are solar energy panels, which have led to the establishment of one of the world's largest photovoltaic plants that has to deal with the overload of sunlight.[3] Every effort to get water or energy, and how these basic needs enhance and promote new and imaginative technologies can be easily understood when in the desert itself.

Nevertheless, in the desert we also find other enterprises and technologies that have challenged the soil's features. Mining has a long history in the area and, of course, the extraction of nitrates, copper and other minerals presumed new technologies. But besides the mining industry, and water and energy initiatives, another advanced technology can be seen in the middle of the desert, and indeed using in its favour the desert's 'emptiness'. At first sight, it has nothing to do either with subsistence or for economic growth. Suddenly, enormous antennas and monumental astronomical domes interrupt the different colours of the sand, rocks and mountains. The red, brown and yellow shades of the soil, and the deep blue of the skies contrast with the white and silver infrastructure of high-tech telescopes.

The convergence of scarcity of life and cutting-edge technology draw a curious landscape, which resembles science fiction images. These antennas, mirrors and domes are part of the most advanced astronomical observatories in the world, aiming to unveil the mysteries of the universe. First built by transnational holdings between 1963 and 1969 in the southern area of the desert, these strange settlements are now part of the silent vibrations of Atacama, and are growing at an accelerated pace. By 2020, approximately 70 per cent of the world's astronomical observation will be collected from Chile, along more than 1,000 kilometres of the desert.[4]

This vertiginous scientific development can be explained by the specific climatic and geographical conditions, usually described as the cleanest sky in the world: an astronomer's paradise. Nowadays these features are indeed essential for successful astronomical research, but there is also a historical trajectory we cannot leave out. International holdings arrived in Chile in the 1960s, with projects of building enormous astronomical observatories. In the middle of the global Cold War, Americans, Europeans and Soviets landed in the country and, almost at the same time, settled in the same place: the semi-arid area of the Coquimbo Region. From this perspective, the futurist landscape that the observatories provide today, open up questions about the role of political conditions for scientific development in a Third World country, as well as how the country itself managed to secure and negotiate with different and opposing international powers, simultaneously.

Through the case of astronomy in Chile during the 1960s it is possible to open new perspectives of the global Cold War, considering that 'the most important aspects of the Cold War were neither military nor strategic, nor European-centred, but connected to political and social development in the Third World'.[5] To this we can add that politics, ideologies and even culture have been, to different extents, studied regarding this historical period,[6] but we still don't know 'to what degree Cold War conditions changed the content or the character of the scientific knowledge that was produced (or not produced)'.[7] Even if scholars have addressed this topic in the so-called developed world,[8] we still need a deeper understanding on this topic for regions such as Latin America. The Cold War enhanced scientific development, and also determined new relationships with the Third World, where the 'superpowers' competed and measured their power. For Latin American countries, the Cold War logic implied they had to negotiate in an apparently binary-structured world. Paradoxically, within this power struggle they found small openings where they could grasp their own limited power quotas and, at the same time, promote the scientific development they were lacking.

This paper aims to present the complex network of the scientific and political relationship between Chile and the Cold War major powers regarding astronomical development in the southern country. Furthermore, this paper addresses how we can understand what technology and science meant for Latin America at a time of extreme political polarization, and how social actors acted and reacted to these new challenges. Many history studies, such as gender history, cultural history, nation-building studies and even diplomatic history – only to mention a few – have tried to reverse the top to bottom analysis of historical processes. But science is still assumed as part of the developed world, and little agency is given to Third World countries,[9] even less in a period of intense polarization such as the Cold War. This paper aims to contribute in shifting that perspective on science and technology in Latin American history.

## THE 1960s IN THE SOUTHERN CONE: WHY TECHNOLOGY AND SCIENCE MATTERED

By the 1960s, Latin American societies were experiencing diverse changes. Different social actors were more conscious of their political role in society and the states were assuming a broader responsibility in the economical and socio-political transformations. Through the 1930s and 1940s in Chile, new social actors integrated the political scenario, and enhanced the gradual widening of citizenry.[10] The spreading of mass culture, and access to new technologies such as the radio or the cinema, were modernizing and globalizing the society's imaginary. The strengthening of a new middle-class intellectuality was transforming the ways in which Chileans thought and imagined their nation. The CORFO model, which started in 1939, was shaping an industrialization process whose results, although real, were far from the expectations politicians hoped for. Other factors contributed in elaborating these changes, such as modernization theory, that assumed a 'common and essential pattern of "development", defined by progress in technology, military and bureaucratic institutions, and the political and social structure'.[11] Chileans were experimenting more than three decades of constitutional stability, which had become an important identity reference for the country since the nineteenth century onwards.

In the Chilean presidential election of 1958, the right-wing candidate, Jorge Alessandri, barely won, closely followed by the socialist leader, Salvador Allende. The Christian Democrats also had a strong electorate behind them, and became the political centre. This election was the ultimate expression of Chilean politics organized in what has been called the 'three thirds'; a pattern with scarce possibilities of alliances and which led to a process of deep polarization. At the same time, the Cold War characterized the international context, in terms of blocs' alignment, and a dichotomic difference between them. The Cuban Revolution and the American response of the Alliance for Progress had a critical impact in the Latin American region, contributing to the social actors' political commitment.[12]

Nevertheless, that election of 1958 was a turning point as the socialist candidate, Allende had managed to secure a 'third' of the votes, and a year after, with the triumphant Cuban Revolution already established, this was seen as a risky situation for the US. As Nixon would say years later, the US could not have a 'red sandwich' in Latin America, framed by Cuba in the north and Chile in the south.[13] Hereafter, the US could not jeopardize its natural influence zone, and thus considered its strong presence in the continent as necessary.

In Chile, each political 'third' had to demonstrate that they had the appropriate project to finally transform the country into the modern nation it was supposed to be. In this sense, Chile and also Latin America had a long history of modernization desires, in which scientific and technological development related to the ideal of progress. Having industrial technologies and scientific practices legitimized and measured that ideal of progress.[14] This was an idea widely present at the time, acknowledged and followed by the different Latin American countries. As stipulated in the Special Latin American Coordination Commission: 'It is every country's responsibility and duty to contribute to humanity's progress through the development and outreach of science and technology, to the extent of their own capacity.'[15]

On the other hand, the Cold War allowed a vertiginous technological development whose exploratory phases were usually promoted and financed by military imperatives.[16] Besides the scientific progress itself, technology was a very important dimension through which the world measured the superpowers' hegemony. Therefore, science and technology were relevant not only in military terms, but also for the superpowers' political and cultural influence, interwoven with their global projects and concepts of future and supremacy.[17] This showed how the power struggle was also a conflict between modernization models, extending and moving throughout the entire world, from the capitalist model to the Soviet one.[18]

In a Cold War context, the calibration of forces through science took place in different projects. Among them, the space race produced a particular sensation of novelty and uncertainty. The space race showed how a global confrontation moved towards a multilateral and extra-atmosphere arena,[19] bringing together science, politics, military power and ideologies. We can find different examples of the space race, such as the 1957 Sputnik,[20] the creation of NASA,[21] the Apollo missions of the 1960s and 1970s, and even their expression in sci-fi literature and movies.[22]

These scientific and technological initiatives related to the space race echoed on astronomical development, whose efforts projected even further, and aimed to reverse the borders of the unfathomable. Even if both were working beyond the Earth's atmosphere, the space race focused in the 'near-by' universe, while astronomy aimed to unveil questions about farther away star systems. However, scientists and engineers developed cutting-edge technology that was functional to both astronomy and the space race, such as how to deal with the speed of light or which materials suited astro-engineering purposes, among others.

Astronomical knowledge could be a turning point for space race performance, and was also a 'new' domain of knowledge yet to be conquered. From this perspective, the analysis of astronomy during the global Cold War allows us to connect its technological and scientific importance with its meaning in Latin American culture, identities and politics. It thus relates to the ancient longing of dominating space, and understanding what is going on 'up there', where human eyes could no longer see. The powerless feeling of looking up to the skies, probably for the first time, was about to change, and it would become another dimension of power.

## FOREIGNER SCIENTISTS AT THE END OF THE WORLD

From the beginning of the twentieth century, the enormous potential the Southern skies offered for astronomical research was almost unquestioned.[23] From the Southern hemisphere it was possible to observe and study the Magellanic Clouds, and other celestial phenomena, with enormous potential for astronomical knowledge.[24] Simultaneously, as observation technologies progressed, astronomers looked for places with less humidity and luminescence pollution. This led scientists to explore depopulated semi-arid places in Australia, South Africa and South America.[25] These site-testings show that there were other locations in sight; it was neither mandatory nor obvious that the place for investing in astronomical observation ought to be the Atacama Desert; other options were on the table.

In the case of Chile, there was a previous relationship with American observatories, which had begun in the mid-nineteenth century and the early twentieth century.[26] These precedents were mostly temporary expeditions in Chilean territory; the first one studied the solar parallax, and the other measured radial velocities of brighter stars. In reality, the development of Chilean astronomy was extremely limited, but this was to drastically change in the 1960s.[27] In the late 1950s, scientists working for the Association of Universities for Research in Astronomy (AURA), an American organization, were exploring areas of the Southern hemisphere looking for the best site for building an astronomical observatory.

The explorations were particularly intense in Australia and South Africa. After previous experience in the Southern hemisphere, an English-speaking community was highly regarded, for this would facilitate administrative work. Although cultural reasons such as language were important, there were also political reasons to decide where to get settled. These were high-tech investments, and scientists could not afford to risk losing the sophisticated equipment, due to unstable political conditions. The relationship Australia and South Africa had with the British government also made agreements even more difficult; Americans would have to deal with another power, and not only with a Third World country.[28] This partially explains the need to look for other places, such as South America.

AURA established some networks with scientists at *Cerro Calan*, the National Astronomical Observatory of Chile (hereafter OAN, by its name in Spanish), by then directed by Federico Rutllant. In an official speech in 1953, highlighting some of the achievements of the OAN, Rutllant said 'how much we wished that, in the future, similar events would resonate again the name of Chile in the international scientific environment'.[29] His words show how, years before the arrival of AURA, the director of OAN already had in mind the potential international relationships, and the positive impact it could offer to the small Chilean scientific community. Shortly after that, Rutllant went to the US to negotiate with American scientists, particularly with Gerard Kuiper, and to materialize an exploration for astronomical development in Chilean territory.[30]

Through agreements with the Chilean government and the partial funding of the National Science Foundation (NSF), AURA developed a plan to install a large-scale observatory in the country. Jürgen Stock, a German astronomer working for the University of Chicago and afterwards for AURA was exploring the semi-arid area of Chile from April to December 1959, and then from 1960 to 1962.[31] In that expedition, Stock reported about *Cerro El Morado*, *Cerro Tololo* and *Cerro Robles*, among other places. In 1962, AURA decided to build the Cerro Tololo Interamerican Observatory, in the Coquimbo region.[32] As Stock would recall decades later, 'as a result [of this trip] the world's largest collection of astronomical instruments is now in Chile'.[33]

Simultaneously, in 1954, several Western countries in Europe signed the *Carte de Janvier*, something that can be described as the birth certificate of the European Southern Observatory (ESO).[34] This organization formalized its existence on 5 October 1962.[35] Since its establishment, ESO started exploring South African territories, mostly in the Karoo region.[36] ESO was informed about the activities of AURA regarding its explorations in South America, but the reports of the 1959 Committee stated: 'This project will have little influence on the development of ESO.'[37] Soon after that however, in 1961, ESO sent an expedition for site testing in Chile. A year after, J. H. Oort, a prestigious Dutch astronomer stated, 'having at sight that Americans have recently obtained favorable indications for the placement in Chile, we will have to seriously review the possibility of a radical change regarding the location of our observatory'.[38]

ESO was an autonomous organization, but had strong connections with American scientists, and even with Stock himself. This becomes very clear when we observe the support ESO got from the Ford Foundation in 1959, of one million US dollars.[39] Besides funding, when ESO decided to explore sites in Chile, they set a convergent plan which, as we shall now see, was never completed. In June 1963, authorities and astronomers from ESO and AURA attended an important meeting at the summit of *El Morado*. From that meeting onwards, it was absolutely clear that both Americans and Europeans would settle in Chile for astronomical observation. However, the relationship, although continuously polite, shifted from collaboration to political and scientific tension.[40] On the one hand, the efforts of combining the work of AURA and ESO 'were dropped because of certain diplomatic problems and the complexities of dealing with an organization which was itself a combination of the scientists of five European countries'.[41] On the other hand, and regarding those 'diplomatic problems', we cannot forget that both were dealing with a Chilean government that was about to change: from the right-winged liberals to the Christian Democrats led by Eduardo Frei. In 1964, almost at the same time of the presidential election, ESO decided to buy lands in *La Silla*, and settle in Chile.[42]

Beginning with the first explorations in 1959, and then the construction of the AURA and ESO observatories in 1963 and 1964 respectively, astronomy in Chile became a drastically global enterprise, a feature which only accentuated. To these AURA and ESO initiatives we have to add the arrival of the Carnegie Southern Observatory (CARSO) – another American institution – that also settled their observatory *Las Campanas* in the same area near Coquimbo. Not long after that, ESO projected a newer and bigger observatory in Paranal, a location approximately 800 kilometres north of Coquimbo, in the middle of the Atacama Desert.

The global dimension of astronomy in Chile would become even more complex. In 1962, astronomers from the Soviet Púlkovo Observatory were collaborating with scientists working in *Cerro Calan*, at the OAN, whose director at the time was Claudio Anguita. The task of the Soviet scientists, led by Mitrofan Zverev, was to measure the position of several stars. These astronomers were, in the words of Frank Edmonson – the AURA Director – 'among the world leaders in fundamental positional astronomy'.[43] An agreement to set a southern station of the Púlkovo Observatory in Chile, and specifically to work from *Cerro Calan*, was announced in 1960 in a conference in La Plata Observatory.[44] The Soviet work in Chile would expand from that initial project to the building of an observatory and an astrograph with a Maksutov telescope in *Cerro Robles*[45] which happened in 1967–1968. *Cerro Robles* had indeed been one of the first sites Jürgen Stock had analysed for the AURA mission in 1959.

The relationship between the Soviet, American and European scientists in Chile was a complicated one. On the one hand, the astronomers knew each other, and had some kind of professional relationship, despite their nationalities. They were all in Chile for scientific tasks. But on the other hand, they all knew the specific national and global context and political condition they were living in. As Edmonson would recall, 'several members of the AURA Board expressed concern about the Soviet presence in Cerro Calan'.[46] The construction of *Cerro Robles* started at the same time as the opening of the Cerro Tololo Interamerican Observatory (CTIO). Americans were actually paying attention to the Soviet astronomers in Chile, for this was indeed a sensitive matter related to the space race: 'To date the Soviet activities in Chile seem limited to astrometry and are apparently tied to the USSR's space programs'.[47]

## CHILEANS AND FOREIGNERS DEALING WITH SCIENCE

By the mid-1960s, European, American and Soviet astronomers could be found working in Chile. At the time, and considering the shift in the balance of power due to the Cuban Revolution of 1959, and the American response of the Alliance for

Progress, the Cold War had become quite 'warm' in Latin America. The understanding of national and international politics was especially sensitive in Chile: the identity features the country had promoted in the past, as an exceptional, trustworthy, modern nation,[48] and the good relationship it had with the United States,[49] was coming to a halt with the growing polarization of Chilean politics throughout the 1960s. In this context, the presence of multiple international scientists in the country provided a possibility to engage with the scientific dimension of modernity that the country had almost always sought. From this perspective, it could have been risky to sign only with the Americans, since presumably the Soviets were the first to collaborate with Chile in the twentieth century – or at least that was what the Chilean press published.[50] Most likely, and in conjunction with the privileged conditions that the Chilean geography offered for astronomical research, it was the Soviet presence that would have triggered the Americans' arrival and decision to remain in the country. Therefore, the first step of the Chilean government was to sign agreements with the different international parties.

In this matter, the University of Chile, although it did not have an Astronomy Department, was a key institution, as it was legally in charge of the OAN. However, academic international relationships with the scholars at the University were not at their best; even if Rutllant made one of the first attempts in bringing international science to Chile, when these projects became real, by 1963, 'questions had been raised about his handling of the funds sent to him from abroad by AURA and other institutions'.[51] Because of this, Rutllant was no longer associated with the OAN. Despite this incident, the agreements were still standing, and the next director, Claudio Anguita, continued developing them. In addition to the agreements with the University itself, the Chilean government was also involved in these negotiations with AURA, as well as with the other research projects.

In a Cold War context, there was a significant difference between the agreements signed with Americans and Europeans, on one side, and Soviets, on the other. For 'Western Powers', the Chilean government tried to secure the foreigners settlement in Chile. In the case of AURA, the University was in the front line of the negotiations, but a Chilean–American holding was also established for land ownership.[52] In the case of Europeans, the negotiation diverged. To secure the investment meant providing a safety environment for the assets Europeans were to bring to Chile, as well as to protect the scientists themselves. ESO and the Chilean government signed an official contract on 8 November 1963. In that first agreement, ESO members and any other scientist, teacher or engineer working in the ESO facilities would hold full diplomatic status: 'The government shall grant ESO the same immunities, prerogatives, privileges and facilities as the Government applies to the United Nations Economic Commission for Latin America and the Caribbean as grants in the Agreement signed in Santiago on 16 February 1953.'[53]

The model for diplomatic status was set by the United Nations, but specifically by the institution created for Latin America. Moreover, this established that the agreement itself was set between ESO and the government of Chile, leading to an extraterritorial status for the Observatory. If at any point Europeans and Americans thought of working together in the same place, to opt for a contract at governmental level was one of the gaps opened between ESO and AURA. The latter were working on agreements with the University of Chile, and accessing the governmental level from there, while Europeans worked the other way round.[54] The different models were described as an inefficient one by the Americans: 'The existence of three separate multimillion dollar research centre and support systems seems a tragic waste of funds and intellectual potential.'[55]

However, and revealing a further complexity in the establishment of international astronomical centres in Chile, both ESO and the North Americans (AURA and CARSO) would establish agreements with the University of Chile as well as with the government through the Ministry of Foreign Affairs. This included several benefits for them, as an expression of the government's interest in securing that the astronomers would stay and invest in Chile. Chileans authorities explicitly declared this interest. Regarding the first exploration of ESO in the country, the Ministry of Foreign Affairs wrote to Otto Heckmann – the ESO Director – of 'the bright interest of the Chilean Government of being chosen as headquarters of the Observatory, for which ESO can rely on the widest cooperation of the national authorities'.[56]

Shortly after that first statement, the Minister himself wrote to Heckmann, confirming the country's interest, but also installing the idea that if Chile was chosen, 'The Chilean government would consider adhering itself to the respective International Convention, as long as it was invited to.'[57] Following this first attempt of the government to be included in the scientific project, the Minister stated the government would be available to give ESO members all the benefits given to international bodies working in the country. In concrete, this meant a full diplomatic status, tax exemption and waiving customs fees.[58]

These negotiations, which had started during the right-wing government of Alessandri, were finally ratified during the Christian Democrat government of Eduardo Frei. Both presidencies had a close relationship with the US government, mostly under the Alliance for Progress.[59] This new framework for US–Chile relations would mean a more effective and strong bond with the political centre, identified with the Christian Democrats.

The collaboration process with the Soviets from Púlkovo was significantly different: they did not get the same benefits when arriving in Chile. They signed some agreements with the OAN and the University of Chile, but they did not get diplomatic status. However, to forbid their settling in Chile would have been even more risky, since the delicate political scenario could have led the left-winged forces to demand equal treatment with the Europeans and Americans. Also, it is probable that the Soviet arrival in the country motivated Americans to settle in Chile; to leave all the potential the Southern skies offered to their opponent was not an option. But the anti-communist spirit of the government made it impossible to host them in the same way they had received ESO and AURA. Although at the Archives of the Ministry of Foreign Affairs there is no sign of the Soviet astronomers ever being in Chile during the 1960s, the Chilean press did register their presence in the country. A few years later, when their observatory was almost ready for observation, the press reported: 'Five Soviet technicians are installing in Cerro Robles, near Rungue, about 70 kilometres from Santiago, a Maksutov telescope, the only one in the world.'[60] The news report also made explicit that this was a collaboration agreement between the University of Chile and the Soviet Academy of Sciences.

The press also printed news related to the openings of *La Silla* and *Tololo*, and showed all the potential these scientific infrastructures could have. Regarding the American installations, the press reprinted a US news report, with details of how the initiative of *Tololo* had brought together funding from the Ford Foundation, the National Academy of Science, and expected the Federal Government to be included.[61] Politicians were not left out of the scientific arena. For example, Europeans invited Chilean authorities for the *La Silla* opening, and got the President himself to attend: 'President of the Republic, Eduardo Frei Montalva, will officially launch the astronomical observatory La Silla in a ceremony at an altitude of 2500 meters.'[62]

The complex relationship between astronomy and politics, both national and international, also had another angle. This was an opportunity to drive through a turning point reform for Chilean science. It was a one-time opportunity to form astronomers in Chile, and to get them involved in these international experiences. It was not by chance that the University of Chile created the Astronomy degree and the Astronomy Department in 1965.[63] With a broader scope, but with the same perspective of public policy and enhancing scientific research in Chile, the government created CONICYT, the Science and Technology National Commission, in 1967.[64]

Overall, the relationship between Americans and the Chilean scientific community was much closer than the one with the Europeans. Authorities at AURA stated this point: 'Chileans will have the opportunity to use the telescope and even if in Chile there are not many astronomers, the equipment will be loaned for a useful association in this sense.'[65] Therefore, this was the chance for Chileans to learn how to work with astronomy's technology. The Carnegie Institution, which arrived in 1967 and settled in *Las Campanas*, also offered scholarships for Chileans, in order to 'contribute, in Chile, to the progress of this science, to improve scientific education, and to the professional development of young astronomers'.[66]

The Chilean–American cooperation included working with the novel Chilean community of astronomers, an agreement that was not reached with the Europeans. As Anguita – the OAN director – explained:

> Paradoxically, Chilean researchers will have access only to 25 per cent of the research in *Cerro Tololo*, and to a few chances to access *La Silla* Observatory; 'With the Associations of Universities for Research on Astronomy of the United States (AURA) we have a contract that stipulates for Chilean researchers a quarter of the experiments to be done. But with European institutions it was not considered a possibility for Chilean researchers to work there.'[67]

But ESO would elaborate a response regarding this difference, and consequences for Chile would also involve other matters. Commenting on the leverages for Chile, Harold Hyslop, working for ESO referred that 'the observatory will be some sort of "Public Relations Officer". The outcome of every work will be accompanied with Chile's name. Therefore, Chile will be mentioned in scientific and scholar media all the time.'[68] Also he addressed the role of travellers, and tourists, as well as administrative personnel and workers. This was the turning point for the development of Chilean astronomy, although still – and up to today – with multiple shortcomings.[69]

Regardless of the deficiencies, and in only a few years, Chile went from almost having no expertise in Astronomy, to becoming a central player in the field. The country soon integrated astronomical development in some of its national discourses, as the press stated: 'Chile is in its way to becoming one of the astronomical centres of most importance in the world, due to its areas of exceptional quality and stability of its atmosphere.'[70] However, it was not only about clear skies, but also about the conditions and the political base Chileans could offer to their foreigner counterparts. It was as if the nation's stable institutions were reflected in its atmosphere and skies. And the Southern country would take a chance to make the best of that crossroads: it took the chance to simultaneously sign agreements with Europeans, Americans and Soviets, as if in that intersection there was a small opening for Chileans to have a place in astronomy. The converging factors of geography, politics, ideologies and science thus led to a key moment for both Chile and astronomical research.

## ASTRONOMY AND POWER: FINAL REMARKS

Chilean development in astronomy was born in this Cold War scenario. And yet this development can be questioned if it is really Chilean: it is still in foreign management, and currently Chilean researchers only get about 10 per cent of the observation time in the newest telescopes built in the following decades.[71]

Despite the limited participation, nowadays astronomy is a relevant knowledge area in the country for different reasons: it constantly promotes the creation of new technologies, and increases the country's production in astro-engineering; it also enhances the scientific and scholar production in the area; and, additionally, it has become part of the current identity narrative the country has built of Chile as the world's astronomical centre.

This story began during the global Cold War, and from the astronomical perspective, it looked indeed as a 'cold' conflict: everybody seemed to be collaborating, and even sharing a non-territorial space. But behind that cold relationship, there was a warm war happening. There was tension not only between Soviets and Western scientists, but also between Americans and Europeans. In the centre of that tension, an almost non-existing Chilean scientific community, and a government in a highly polarized environment, managed to negotiate with each one of these foreign powers, and found a gap that would allow Chileans to get involved in astronomy's development.

That convergence of factors, and how a Third World country arranged contracts with powerful nations shows that scientific development was not only promoted by scientific conditions or negotiations, but also by other factors. Civic epistemology points to those frequently not spoken or not written rules that are relevant when forming a knowledge governance.[72] In this matter, we have to consider the role of the state, not only as manager of the country's decision, but also as a mediating power regarding science and technology issues. Even with the criteria of truth or authenticity science is supposed to have, what a society believes is interwoven with concepts of authority, power struggles and mentalities.[73]

The Cold War period shows us how these complex relations in scientific development are to be considered not only regarding a national dimension, but also an international one. In the context of multiple tensions where developed nations were trying to progress in astronomy, as a sensible dimension for Cold War politics, a small country actually had an opportunity and took it. Chileans managed to negotiate with all powers in conflict simultaneously regarding a specific question: whoever would dominate the knowledge of the skies would most likely win the space race, the area where the superpowers were at the time measuring their strength.

As a result of the global – and ideological – conflict, Chile played a role in astronomical science. The opening was born in the desert: in the emptiness of Atacama, a 'micro cold war' unfolded; that silent power struggle opened the possibility for Chile to get into avant-garde science. Telescopes are the future in technology, and are looking at a distant past, thousands of light years ago. By analysing the telescope's own past, we are contributing to unveiling unknown parts of the history of science and technology in Latin America.

## NOTES AND REFERENCES

1. This paper is a product of the FONDECYT postdoc nº 3170099 'Astros y Galaxias desde el sur. Chile y su inserción científica en la Guerra Fria Global, 1962–1973'.
2. Pablo Osses, Robert Schemenauer, Pilar Cereceda, Horacio Larraín and Cristóbal Correa, 'Los atrapanieblas del Santuario Padre Hurtado y sus proyecciones en el combate a la desertificación', *Revista Norte Grande* 27 (2000): 61–67.
3. Aleszu Bajak, 'Bright Future', *Nature* 552, 14 December 2017: 53–55.
4. Michele Catanzaro, 'Big Players / Chile: Upward Trajectory'. *Nature* 510, 12 June 2014: 204–205.
5. Odd Arne Westad, *The Global Cold War. Third World Interventions and the Making of Our Times* (Cambridge: Cambridge University Press, 2005): 396.
6. The literature in this respect is wide, but we can provide some examples, such as Benedetta Calandra and Marina Franco (eds.), *La guerra fría cultural en América Latina. Desafíos y límites para una nueva mirada de las relaciones interamericanas* (Buenos Aires: Biblos, 2012); Yale Ferguson and Rey Koslowski, 'Culture, International Relations Theory, and Cold War History', in Odd Arne Westad (ed.), *Reviewing the Cold War. Approaches, Interpretations, Theory* (London: Frank Cass Publishers, 2000): 149–179; Patrick Iber, *Neither Peace nor Freedom. The Cultural Cold War in Latin America* (Cambridge: Harvard University Press, 2015); Gilbert Joseph, Catherine LeGrand and Ricardo Salvatore (eds.), *Close Encounters of Empire: Writing the Cultural History of U.S.–Latin American Relations* (Durham: Duke University Press, 1998); Patrick Major and Rana Mitter, 'East is East and West is West? Towards a Comparative Socio-cultural History of the Cold War', *Cold War History* 4, 1 (2003): 1–22; Gilbert Joseph and Daniela Spenser (eds.), *In From the Cold. Latin America's New Encounter with the Cold War* (Durham: Duke University Press, 2008).
7. Naomi Oreskes and John Krige (eds.), *Science and Technology in the Global Cold War* (Cambridge: MIT Press, 2014): 2.
8. Mark Solovey, 'Science and the State during the Cold War: Blurred Boundaries and Contested Legacy', *Social Studies of Science* 31, 2 (2001): 165–170.
9. Lissa Roberts, 'Situating Science in Global History. Local Exchanges and Networks of Circulation', *Itinerario* 33, 1 (2009): 9–30.

10. Bárbara Silva and Rodrigo Henríquez, 'El pueblo del Frente. Representaciones sobre la ciudadanía en Chile: 1930–1950', *European Review of Latin American and Caribbean Studies* 103, January–July (2017): 91–108.
11. Nils Gilman, *Mandarins of the Future. Modernization Theory in the Cold War America* (Baltimore: The Johns Hopkins University Press, 2007): 3.
12. Jeffrey Taffet, 'Implementing the Alliance for Progress', in Jeffrey Taffet (ed.), *Foreign Aid as Foreign Policy: The Alliance for Progress in Latin America* (New York: Routledge, 2007): 29–46.
13. Richard M. Nixon, *The Memoirs of Richard Nixon* (London: Arrow, 1979): 490.
14. Gabrielle Hecht (ed.), *Entangled Geographies. Empire and Technopolitics in the Global Cold War* (Cambridge: MIT Press, 2011): 5.
15. 'Reunión Extraordinaria. Comision Especial de Coordinación Latinoamericana. Nivel de expertos', Doc 67/rev 1, Anexo VIII: Principios en materia de ciencia y tecnología. *Archivo del Ministerio de Relaciones Exteriores de Chile*, Fondo Organismos Internacionales, vol. 494, May 1969. Translated by the author.
16. David Reynolds, 'Science, Technology, and the Cold War', in Melvyn P. Leffler and Odd Arne Westad (eds.), *The Cambridge History of the Cold War*, vol. 3: Endings (Cambridge: Cambridge University Press, 2010): 399.
17. Gilman, *Mandarins of the Future*: 7.
18. Tobias Rupprecht, *Soviet Internationalism after Stalin. Interaction and Exchange between the USSR and Latin America during the Cold War* (Cambridge: Cambridge University Press, 2015): 22.
19. John Krige, 'Embedding the National in the Global: US–French Relationships in Space Science and Rocketry in the 1960s', in Naomi Oreskes and John Krige (eds.), *Science and Technology in the Global Cold War* (Cambridge: MIT Press, 2014): 230.
20. George Reisch, 'When Structure met Sputnik: On the Cold War Origins of the Structure of Scientific Revolutions', in Naomi Oreskes and John Krige (eds.), *Science and Technology in the Global Cold War* (Cambridge: MIT Press, 2014): 371.
21. Erik M. Conway, 'Bringing NASA Back to Earth: A Search for Relevance during the Cold War', in Naomi Oreskes and John Krige (eds.), *Science and Technology in the Global Cold War* (Cambridge: MIT Press, 2014): 251.
22. David Seed, *American Science Fiction and the Cold War. Literature and Film* (Chicago: Fitzroy Dearborn Publishers, 1999).
23. 'Convention Portant la Création d'une Organisation Européenne pour des Recherches Astronomiques dans l'Hémisphère Austral, signée à Paris, 5 Octobre 1962'. *Archivo del Ministerio de Relaciones Exteriores de Chile*, Fondo Organismos Internacionales, Vol. 400.
24. Victor M. Blanco, 'Aportes científicos del Observatorio de Cerro Tololo', in Arturo Aldunate Phillips, *Chile mira hacia las estrellas* (Santiago: Ediciones Gabriela Mistral, 1975): 251.
25. Hilmar W. Duerbeck, 'National and International Activities in Chile 1849–2002', *Astronomical Society of the Pacific*, Conference Series 292 (2003): 9.

26. James M. Gilliss, *The US Naval Expedition to the Southern Hemisphere* (Washington: A.O.P. Nicholson Printer, 1855); Herbert D. Curtis 'Report on Astronomical Conditions in the Region about Copiapo', UA ser. 4, Box 8, Folder 12: D. O. Mills expedition: Report on site survey near Copiapo. 17 April 1909. *Lick Observatory Records*, Mary Lea Shane Archives at University of California Santa Cruz.
27. Hernán Quintana and Augusto Salinas, 'Cuatro siglos de astronomía en Chile', *Revista Universitaria* 83 (2004): 55.
28. 'Letter J. W. Joyce, Acting Deputy to Paul A. Siple, Scientific Attaché, American Embassy Canberra', 28 February 1966, *National Archives College Park*, United States, Record Group 59, Entry 3008D, Box 20, Folder SCI 21 Visits, Missions. Astronomy.
29. Federico A. Rutllant, 'Discurso del profesor Federico Rutllant, director del Observatorio', *Anales de la Facultad de Ciencias Físicas y Matemáticas*, Universidad de Chile 10, 10 (1953): 16. Translated by the author.
30. Frank K. Edmonson, *AURA and its US National Observatories* (Cambridge: Cambridge University Press, 1997): 137.
31. Jürgen Stock, 'Chile Observatory Project. Seeing Expedition'. *Stock Reports*, w/d, Cerro Tololo Interamerican Observatory Library, Coquimbo, Chile.
32. *El Mercurio*, Santiago, 24 November 1962: 25.
33. Jürgen Stock quoted in Dirk Lorenzen, 'Jürgen Stock and his Impact on Modern Astronomy in South America', *Revista Mexicana de Astronomía y Astrofísica* 25 (2006): 71–72.
34. Adriaan Blaauw, *ESO's Early History. The European Southern Observatory from Concept to Reality* (München: ESO, 1991): 2.
35. 'Convention Portant la Création d'une Organisation Européenne pour des Recherches Astronomiques dans l'Hémisphère Austral, signée à Paris, 5 Octobre 1962'. *Archivo del Ministerio de Relaciones Exteriores de Chile*, Fondo Organismos Internacionales, Vol. 400.
36. Govert Schilling and Lans Lindberg Christensen, *Europe to the Stars. ESO's First 50 Years of Exploring the Southern Sky* (Weinheim: ESO, 2012): 25.
37. Quoted in Blaauw, *ESO's Early History*: 44.
38. J. H. Oort, *ESO Historical Archives*; Archives A. Blaauw, 'Documents Pertaining to the Secretariat of the ESO Committee', quoted in Blaauw, *ESO's Early History*: 48.
39. Frank Edmonson, 'The Ford Foundation and the European Southern Observatory', in Blaauw, *ESO's Early History*: 255.
40. Edmonson, *AURA and its US National Observatories*: 203–214.
41. 'Letter from Herman Pollack to Ralph A. Dungan, American Ambassador in Santiago', 29 March 1966, *National Archives College Park*, United States, Record Group 59, Entry 3008D, Box 20, Folder SCI 21 Visits, Missions. Astronomy.
42. Duerbeck, 'National and International Activities in Chile': 16.
43. Edmonson, *AURA and its US National Observatories*: 199.
44. *Transactions of the International Astronomical Union*, vol. XI–A (Berkeley, 1961): 24.

45. Duerbeck, 'National and International Activities in Chile': 18.
46. Edmonson, *AURA and its US National Observatories*: 202.
47. 'From Amembassy Santiago to Department of State, Science: Soviet Astronomy Activities in Chile', 19 June (1968), *National Archives & Records Administration*, United States. Record Group 59, Central Foreign Policy Files 1967–1969, Box 2934, Folder SCI CHILE.
48. Bárbara Silva, *Identidad y Nación entre dos siglos. Patria Vieja, Centenario y Bicentenario (*Santiago: Lom, 2008).
49. Stefan Rinke, *Encuentros con el Yanqui. Norteamericanización y cambio sociocultural en Chile. 1898–1990* (Santiago: Dibam, 2013).
50. 'First it was the University of Chile that through a contract with specialized observatories of the Soviet Union managed to build, in the mountains' buttress, modern and powerful telescopes', 'Chile, Centro de Interés Astronómico', *El Mercurio*, Santiago, 21 March 1969: 3. Translated by the author.
51. Phillip Keenan, Sonia Pinto and Héctor Álvarez, *The Chilean National Astronomical Observatory 1852–1965* (Santiago: Facultad de Ciencias Físicas y Matemáticas, Universidad de Chile, 1985): 64.
52. 'Apoyo a la investigación astronómica', *El Mercurio*, Santiago, 6 May 1967: 1.
53. 'Article IV: Agreement between the Government of Chile and the European Organisation for Astronomical Research in the Southern Hemisphere for the Establishment of an Astronomical Observatory in Chile', 6 November 1963, in ESO, *Basic Texts, Authoritative and Original English Texts and English Translations* (2012): 38.
54. Schilling and Lindberg Christensen, *Europe to the Stars*: 25.
55. 'Letter from Ralph A. Dungan to Herman Pollack, Director of International Scientific and Technological Affairs, Department of State', 2 March 1966, *National Archives College Park*, United States, Record Group 59, Entry 3008D, Box 20, Folder SCI 21 Visits, Missions. Astronomy.
56. 'Carta de Fernando Orrego, Subsecretario de Relaciones Exteriores a Otto Heckmann', 4 June 1963, *Archivo del Ministerio de Relaciones Exteriores de Chile*, Fondo Organismos Internacionales, Varios, vol. 308, f. 187. Translated by the author.
57. 'Carta de Carlos Martínez, Ministro de Relaciones Exteriores a Otto Heckmann', 18 June 1963, *Archivo del Ministerio de Relaciones Exteriores de Chile*, Fondo Organismos Internacionales, Varios, vol. 308, f. 205. Translated by the author.
58. 'Carta de Luis Cubillos, Ministerio de Relaciones Exteriores a Otto Heckmann', 27 November 1963, *Archivo del Ministerio de Relaciones Exteriores de Chile*, Fondo Organismos Internacionales, Varios, vol. 308, f. 366.
59. The Alliance for Progress was Kennedy's strategy to contain the spread of communism in Latin America, after the Cuban Revolution. It was officially announced in the 1961 Punta del Este Conference, as a collaboration plan between the US government and Latin American states. It sought for strategic reforms for the region, such as Agrarian Reform, Educational Reform, housing, trading modernization,

among others. It assumed that if living conditions were improved, it was less likely that communism would have a nurturing social environment to grow.

60. 'Moderno Observatorio', *Las Últimas Noticias*, Santiago, 3 June 1967: 32. Translated by the author.
61. 'Apoyo a la investigación astronómica', *El Mercurio*, Santiago, 6 May 1967: 1.
62. 'Frei Inaugurará Nuevo Observatorio "La Silla"', *El Mercurio,* Santiago, 22 March 1969: 38. Translated by the author.
63. 'Creación del Departamento de Astronomía y de Licenciatura en Astronomía', *Anales de la Universidad de Chile* serie 4, octubre–diciembre 136, 123 (1965): 264.
64. 'Decreto 13123, Crea Comisión Nacional de Investigación Cientifica y Tecnologica', 10 May 1967: https://www.leychile.cl/N?i=1038389&f=1967-05-10&p=
65. 'Apoyo a la investigación astronómica', *El Mercurio*, Santiago, 6 May 1967: 1.
66. 'Panorama. Beca para astrónomos', *Las Últimas Noticias*, Santiago, 20 March 1969: 6. Translated by the author.
67. 'Centro Astronómico en Cerro Tololo', *El Mercurio*, Santiago, 19 June 1967: 23. Translated by the author.
68. 'Relaciones públicas desde un observatorio sentado en "La Silla"', *Las Últimas Noticias*, Santiago, 20 March 1969: 15. Translated by the author.
69. Bajak, 'Bright Future': 53–55.
70. 'URSS construye Centro de Astronomía en Chile', *El Mercurio*, Santiago, 3 June 1967: 1. Translated by the author.
71. Javiera Barandiaran, 'Reaching for the Stars? Astronomy and Growth in Chile', *Minerva* 53 (2015): 141–164.
72. Lorrae Van Kerkhoff and Victoria Pilbeam, 'Understanding Socio-cultural Dimensions of Environmental Decision-making: A Knowledge Governance Approach', *Environmental Science and Policy* 73 (2017): 29–37.
73. Sheila Jasanoff, *Science and Public Reason* (New York and London: Routledge, 2012): 15–18.

# The Curious Case of Cuba's Biotech Revolution

HELEN YAFFE

In late September 2018, the first US–Cuban biotech joint venture was established to trial CIMAVax-EGF, an innovative Cuban lung cancer immunotherapy treatment, with the intention of delivering it to patients in the United States. Innovative Immunotherapy Alliance SA was set up by Buffalo-based Roswell Park Comprehensive Cancer Center and Havana's Centre for Molecular Immunology (CIM); two institutions which benefited from the tentative rapprochement between the US and Cuba in mid-December 2014. The sheer fact that such an entity should exist is scientifically and politically ground-breaking for several reasons: First, it testifies to the extraordinary development of biotechnology in Cuba, a case largely overlooked in medical science and business history literature on the field.[1] Havana's Scientific Pole, a complex in Western Havana which hosts over fifty major integrated research, education, health and commercial biotech institutions, is not included in references to twenty global biotech regions.[2] Lara Marks' masterful 2015 history of the development of monoclonal antibodies (MABs) omits Cuban advances that have placed the island at the forefront of MAB based immunology therapies.[3]

Second, despite the fact that for nearly sixty years the United States blockade has obstructed Cuba's foreign trade, external financing, technology transfers and scientific exchange, including in the medical field, it is the Cubans who have contributed the innovative science to this joint venture; they have cracked a difficult nut – using immunology therapy to combat cancer. Third, while global biopharma is associated with speculative, mostly private, capital, the Cuban industry is entirely state owned and financed. The emergence of the Scientific Pole for example, was the result of state planning, not market forces attracting private interests to a given location. Cárdenas points out that 'the State is focused not only on fast returns, but also on long term socioeconomic targets'.[4]

Finally, Cuban 'exceptionalism' does not lie in the mere fact of government support and/or public funding for the biotech industry; several analysts have pointed to this feature in the United States and elsewhere. The exceptionalism lies in the political economy context; Cuba is a small island nation that can boast 'developed-country health outcomes despite its developing-country economy'.[5] Arguably this has been possible because it has a centrally planned, state-controlled economy and a development strategy which has prioritized healthcare, education and research into science and technology from the early 1960s. The universalization of (free) education, beginning with the Literacy Campaign of 1961 and the University Reform of 1962, generated a 'critical mass' of scientists, many of whom initially were sent abroad to study. The Revolution also committed to provide free public health and pursued a policy of 'medical internationalism' from the outset. But the global pharmaceutical industry is dominated by the United States, the country imposing the comprehensive blockade of the island. Thus, the development of a domestic capacity to produce medical drugs and supplies was an imperative forced on Cuba, particularly after the loss of Cuba's socialist trade partners around 1990. Beyond the political economy framework, also important is the historical context of Cuban medical science, with its focus on parasitology and immunotherapy and the fact that the biotech sector was initiated with researchers who came from biophysics and nuclear physics.[6] While Cuban scientists claim that their biotech revolution 'demonstrates the feasibility of developing a new industry in a developing country under foreign economic pressures', the specific historical and political context within which the sector developed means that the Cuban case is difficult to emulate.[7]

Biotechnology is the application of biological knowledge and techniques pertaining to molecular, cellular and genetic processes to develop products and services.[8] Biological research into genomes and cells does not constitute biotechnology unless it involves industrial production. In that sense, biotechnology is a manufacturing process, but the transformation of the raw material into a final product is done inside a living cell.[9] The world's first biotechnology enterprise was established in the United States in 1976. Just five years later, in 1981, the Biological Front was set up to develop the industry in Cuba. This was the first time in its economic history that Cuba had incorporated itself into an emerging industrial sector. While most developing countries had little access to the new technologies (recombinant DNA, human gene therapy, biosafety), Cuban biotechnology expanded and took on an increasingly strategic role in both the public health sector and the national economic development plan.[10] It did so despite the US blockade obstructing access to technologies, equipment, materials and even knowledge exchange.

Among Cuba's biotech innovations are the world's first Meningitis B vaccine, treatment for diabetic foot ulcers, vaccines for Hepatitis B and dengue, and the world's first human vaccine with a synthetic antigen (a vaccine against *Haemophilus influenza* type b, or Hib).[11] Today, Cuba is among the world leaders in oncology drugs. In 2012, CIM patented the first therapeutic lung cancer vaccine. In examining the curious case of Cuba's biotech revolution, this article starts with an overview of the industry in the advanced capitalist countries, particularly the United States, to better appreciate what makes the Cuban case so distinctive.

## THE RISE OF THE BIOTECHNOLOGY SECTOR

California hosted the first and the largest biotech companies: Genetech was founded in San Francisco in 1976 by venture capitalist Robert A. Swanson and biochemist Herbert Boyer; and AMGen was set up in Los Angeles in 1980. Boston and San Francisco were leading biotech hubs early on and continue to be dominant. The location was no coincidence. These cities each have three of the nation's twenty top-ranked medical research institutions.[12] Biotech firms set up in close proximity to medical science research institutions (universities, hospitals, research centres) hosting the best scientists and facilities, 'and a globalising market where the best and the brightest from all over the world come to study'.[13] Those institutions receive government funding that runs into the billions of dollars annually. Studies have also shown the importance of access to local venture capital as a determinant of biotech firm location.[14]

In the early twentieth century the US pharmaceutical industry developed in clusters between New York and Philadelphia, and these regions went on to develop substantial concentrations of biotech activity in the late twentieth century.[15] San Diego, Seattle and Raleigh-Durham subsequently emerged as hubs. By 2002, these three cities had received 'an average of $500 million annually in National Institutes of Health (NIH) funding (in 2001 dollars) for more than a decade and $750 million new venture capital investment during the past six years, and each area also has one or more of the nation's 20 top ranked medical research universities'.[16] By the outset of the twenty-first century, the US biotech sector was concentrated in nine of the country's fifty-one largest metropolitan areas.

Biotechnology took off in Europe in the 1990s, and subsequently in Japan, Singapore and China. By 2009, there were more biotech firms in Europe than in the US.[17] The character of the industries differed in that the US had twice as many publicly listed companies than Europe and more access to venture financing than the European industry. In 2008, three decades after the industry began, there were 1,754 biotech firms in the US, 371 of them publicly listed, and sector employment was hovering below 200,000.[18] But 'progress' was often evasive. Most small biotech firms were losing money. Half of the US biotech firms formed since the 1970s had folded or were merged into other companies by 2000.[19] Internationally, the biotech sector did not achieve profits from product sales until 2009.

Nonetheless, billions of dollars poured into the industry. Lazonick and Tulum refer to this as the 'Pisano Puzzle', after Harvard Professor Gary Pisano's 2006 book questioned why money from venture capital and big pharma flowed into an industry in which profits are so hard to come by.[20] The answer is the role of financial mechanisms such as Initial Public Offerings (IPOs), Special Purpose Entities (SPEs), Special Purpose Corporations (SPCs) and patents licenses, in permitting profits to be made from a high-tech sector with low productivity. Start-up firms typically depend on venture capital investment to underwrite their initial costs. Cortright and Mayer explain that once some promising products are developed, venture capitalists and other early-stage investors seek to recoup their investment (or a portion of it) by having the firm issue stock to the public in an IPO.[21] However, biotechnology products can take up to twenty years to commercialize, and many will never reach that point. By 2002, only about 100 biotech-related drugs had reached the market in thirty years, with the top ten accounting for nearly all of the sales.[22] In 2011, Lazonick and Tulum pointed out that virtually all biopharma companies that do IPOs are product-less: 'Rather than waiting 10–20 years to see whether a commercial drug will in fact be produced, the existence of a speculative stock market provides [venture capital and big pharma] with a mode of exit from their investments by mean of an IPO'.[23] Without such mechanisms, capital may never have been made available.

The 1971 establishment of the NASDAQ stock exchange, for financing risk-intrinsic high-tech businesses, and the rising power of financial capital since the early 1980s, shaped the emerging biotechnology sector. 'The stock exchange has become an instrument for subordinating corporations to the management norms and profitability standards requested by the shareholders . . . the evolution of the biotechnology industry clearly illustrates these findings', concludes Christian Zeller.[24] However, this relationship hides the vital role of US government policy in almost every stage of the biotech industry.

## PUBLIC MONEY FOR PRIVATE PROFIT?

Cortright and Mayer list the forms this takes: the government heavily subsidizes the training of medical researchers; US patent policy is set by Congress and administered by the US Patent and Trademark Office; the Food and Drug Agency (FDA) has to approve most biotech products and regulates conditions for manufacturing and advertising to consumers; decisions on healthcare and what drugs or therapies to include on national healthcare programmes like Medicare and Medicaid are decisive.[25] Lazonick and Tulum conclude: 'The biopharmaceutical industry has become big business because of big government [and] remains highly dependent on big government to sustain its commercial success.'[26]

The US government funds basic science through the NIH. Between 1978 and 2004, NIH spending on life sciences totalled $365 billion in 2004 dollars.[27] US institutions provide legislative support for the private biotech sector; the 1980 Bayh-Dolye Act gave universities and hospitals clear property rights to new knowledge that resulted from federally funded research. Also in 1980, the Supreme Court decision in Diamond vs. Chakrabarty ruled that genetically engineered life forms are patentable. The Orphan Drug Act of 1983 provided generous tax credits for pharmaceutical companies that develop drugs for rare diseases. By 2008, orphan drugs accounted for 74 per cent of total revenues and 75 per cent of product revenues of the six leading companies.[28]

Lazonick and Tulum argue that these financial mechanisms and the wide scope of IP protection in biotechnology has become an obstacle to follow-on innovation, by limiting the new entrants who would compete to produce better and cheaper drugs, while financial speculation permits stock-market investors to reap huge rewards by trading biopharma stock, even in the absence of a commercial product. A similar view is strongly held by the Cuban medical scientists. Certainly, the absence of such financial mechanisms has contributed towards the distinctive character of their biotechnology industry. However, biopharma developments on the island also build upon a long history of infectious disease vaccinations. And that story begins in the nineteenth century.

## THE HISTORICAL TRAJECTORY OF CUBAN MEDICAL SCIENCE

Three private science institutes were set up in nineteenth-century Cuba,[29] but the best-known medical scientist of the century is Carlos Finlay who was born in colonial Cuba in 1833 to a Scottish father and a French mother. In 1881, Finlay presented a ground-breaking theory that the transmission vector, or carrier, of yellow fever was the mosquito. The following year, Finlay identified the *Aedas aegypti* mosquito as the culprit, and recommended controls to halt the spread of the disease. His finding was described as the greatest advance in medical science since the discovery of the smallpox vaccine in 1796.[30] Following Cuba's formal independence, between 1902 and 1909, Finlay served as Cuba's chief health officer.[31]

There were other outstanding individuals in the decades that followed, but the period between Cuban 'independence' in 1902 and the Revolution of 1959 were austere years for medical science. The country had only three universities; in Havana, Oriente (Santiago) and Villa Clara (set up in 1952), and they conducted little medical research. In 1937, Dr Pedro Kouri privately founded the Institute of Tropical Medicine, which did conduct investigations and earned a good international reputation among parasitologists and other specialists of Tropical Medicine.

Private medical clinics thrived, largely by offering US clients' services at lower cost than at home, or services not available in the US.[32] There was a tradition of eminence in surgery. The big money spinner was cosmetic surgery which generated $5 million a year between 1948 and 1958. The main causes of childhood death on the island were parasitic infestation, gross malnutrition and enteric infections, leading to diarrhoea and dehydration. But the paediatrics department at the University of Havana Medical School barely addressed these ailments. Instead it specialized in hyperactivity and leukaemia.

There was also a medical focus on the diagnosis and treatment of cancer. Created in 1925, the League against Cancer in Cuba secured private funding to set up the hospital 'Calixto Garcia' where prestigious Cuban medics held private clinics. In 1929, the Institute of Cancer was set up, the first to treat malignant growths. Subsequently two more oncology centres were founded, mainly to treat patients in the advanced stages of cancer. Most of the doctors did not receive payment; this was philanthropic work. They did teach, however, and exchanged scientific information and experiences with oncology centres in developed countries.

The 1950 report by the Truslow Commission of the International Bank of Reconstruction and Development declared that 'the Mission could not find any suitable applied research laboratory, public or private, in Cuba'.[33] Three years later, the 1953 census recorded that 60 per cent of the population had between three years and no education.[34] Just over 1 per cent of Cubans had university education, and of those only 1.7 per cent were science students who mostly graduated without practical experience.

## REVOLUTIONARY CHANGE FROM 1959

Speaking at the Cuban Academy of Sciences in mid-January 1960, one year after the Rebel Army took power, Fidel Castro declared: 'The future of Cuba will be a future of men of science.' This must have seemed like a pipe dream, given the backward state of Cuban scientific research and generally low level of education. The Revolution, declared Castro, was sowing opportunities for intelligence. It needed thinking men who would put their intelligence to 'good', on the side of 'justice', in the interests of the nation.

Initiated the following January, the literacy campaign of 1961 reduced illiteracy from 24 per cent to 4 per cent in Cubans over-ten years old within one year. It was followed by the University Reform Law in January 1962, which removed their traditional autonomy and, by eliminating fees and facilitating access at all levels, opened the universities to the children of workers, peasants and non-white Cubans. Courses were introduced to train the specialists required for the Revolution's economic development plans. In 1962 the Cuban government created the National Commission for the Academy of Sciences of Cuba. New schools, colleges and universities were built, new teachers trained. Thousands of Cuban students studied in the socialist bloc countries, while others received scholarships from the West. Free, universal public healthcare provision was introduced.[35] From 1962 a national immunization programme provided all Cubans with eight vaccinations free of charge. Infectious diseases were rapidly reduced, then eliminated, including polio (eliminated 1962); malaria (1968); diphtheria (1971); measles (1993); pertussis (1994); and rubella (1995).[36]

## GUEVARA: REVOLUTIONARY, MEDIC, INDUSTRIALIST

In 1959, Cuba was dependent on US pharma for medicines. The market was dominated by two firms which made exorbitant profits. The industry was expropriated by revolutionary decree, putting production and distribution of medicines into government hands. It fell under the jurisdiction of Ernesto 'Che' Guevara as Minister of Industries. In his pre-revolutionary life, Guevara had graduated as a medic and researched allergies, asthma, leprosy and nutritive theory. As Minister, he set up nine research and development institutes, including, in 1963, the Institute for the Development of the Chemical Industry to foster the industrial application of human and animal antibiotics.[37] While progress was limited, the institute established a research methodology which later became a distinctive feature of Cuban biotechnology. 'The idea was excellent', claims Tirso Sáenz, then Guevara's Vice Minister of Science and Technology, 'to make an institution with what they call a complete cycle of innovation. The institute develops products at a scale where it can build pilot plants which, if successful, are turned into production plants.'[38]

Guevara commandeered an abandoned farm, to use for socio-productive and botanical experiments.[39] The personnel were students from the Rebel Army School, and they were joined by Chinese medical scientists, a Cuban post-doctoral researcher and three agronomy engineers. From the farm two dozen varieties of medicinal plants were supplied to scientists conducting laboratory experiments in the Hospital of Oncology in Havana. The fourth floor of the Hospital had forty scientists working under Guevara's directives on laboratory experiments with plants, animals and raw materials.[40]

Guevara left Cuba in 1965 and the farm was transferred to the newly established National Centre for Scientific Research (CENIC), set up to initiate biological studies and establish a new scientific infrastructure. The directors in the post-1980 biotechnology institutes all began as students in CENIC.[41] Throughout the 1960s and 1970s, thousands of Cubans trained as scientists and engineers. Cuba achieved 1.8 researchers per 1,000 inhabitants, well above the Latin America mean (0.4) and close to that of Europe (2.0).[42] Among the many institutions created was the National Council of Science and Technology, in 1975, the same year that a new national scientific policy was approved at the First Congress of the Cuban Communist Party.

By the 1980s Cuba had the health profile of a highly developed country, having eliminated most infectious and poverty-related diseases, so that ailments like cancer, diabetes and heart disease became priorities, on a par with the developed capitalist world. These conditions are expensive to treat. Additionally, a new law passed by US President Reagan in 1982 prohibited foreign nations from exporting goods and equipment to Cuba if any part or process in its manufacture had been mediated by US companies or individuals. Cuban subscriptions to US science and technology journals could not be honoured.[43]

## THE CURIOUS ORIGINS OF CUBAN BIOTECHNOLOGY

Cuba's biotechnology sector emerged independently from both the Soviet Union and the corporate capitalist model in the US and Europe.[44] Driven by public health demand, it has been characterized by the fast track from research and innovation to trials and application. This is illustrated by the story of how interferon was used to arrest a deadly outbreak of the dengue virus in 1981.[45]

Interferons are 'signalling' proteins produced and released by cells in response to infections (viruses, bacteria, parasites and tumour cells) which alert nearby cells to heighten their anti-viral defences. They were first identified in 1957 by Jean Lindenmann and Aleck Isaacs at the National Institute of Medical Research in London during their research into 'viral interference' – the process by which a cell that is infected by one virus can produce an immune response which protects it from another virus. Following this breakthrough, in the 1960s Ion Gresser, a US-researcher in Paris, showed that interferons stimulate lymphocytes that attack tumours in mice.

In the 1970s, US oncologist Randolph Clark Lee, Director of a Cancer Hospital in Houston, Texas, took up this research. Catching the tail end of US President Carter's improved relations with Cuba, Clark joined a delegation to visit the island's health facilities. During the trip Clark met with Fidel Castro and convinced him that interferon was *the* wonder drug. Clark offered to host a Cuban researcher at his hospital. Castro persuaded him to take two. Shortly afterwards, a Cuban doctor and a haematologist spent time in Clark's laboratory. He gave them the latest research about interferon and put them in contact with Kari Cantell, the researcher who, in the 1970s, had isolated interferon from human cells. Cantell's commitment to global health led him to share his breakthrough without patenting his interferon procedure. In March 1981, six Cubans spent twelve days in Finland learning to produce large quantities of interferon. They were from the first generation of medical scientists entirely trained under the Revolution of 1959.

In April 1981, the day after returning from Finland, the Cubans moved into 'House 149', a former mansion converted into an interferon laboratory, which became the Centre for Biological Studies. Fidel Castro visited them frequently, securing them the resources they required. Within just forty-five days the Cubans had produced their first Cuban batch of interferon. Safety and sterility tests were performed on mice before three of the scientists inoculated themselves. They experienced a slight rise in temperature, nothing worse. The laboratory in Finland confirmed the quality of their interferon.

Just in time, it turned out. Weeks later Cuba was struck by an epidemic of dengue, another disease transmitted by mosquitos. Notably, it was the first time this particularly virulent strand, which can trigger life-threatening dengue haemorrhagic fever, had appeared in the Americas. The epidemic affected 340,000 Cubans with 11,000 new cases diagnosed every day at its peak.[46] One hundred and eighty people died, including 101 children. The Cubans suspected the CIA of releasing the virus. Castro announced: 'We share the people's convictions and strongly suspect that the plagues that have been punishing our country, especially the hemorrhagic [*sic*] dengue, could have been introduced into Cuba, into our country, by the CIA.'[47] The US State Department flatly denied it, although a recent Cuban investigation claims to provide evidence that the epidemic was introduced from the US.[48]

At the height of the epidemic, Cuba's Ministry of Public Health (MINSAP) authorized the scientists in House 149 to use interferon to halt it. This was done at great speed. They found that in advanced cases of dengue, interferon was not useful, but in recent infections in children, it cut short cases of haemorrhagic dengue shock. Mortality declined. In their historical account, Cuban scientists Caballero Torres and Lopez Matilla claim: 'It was the most extensive prevention and therapy event with interferon carried out in the world. Cuba began to hold regular symposia, which quickly drew international attention.'[49] The first international event in 1983 was prestigious; Cantell gave the keynote speech and Clark attended with Albert Bruce Sabin, the Polish American scientist who developed the oral polio vaccine that has helped to nearly eradicate the disease globally.

Convinced about the contribution and strategic importance of innovative medical science, the Cuban government set up the 'Biological Front' to develop the sector. In January 1982, the Centre for Biological Studies moved from House 149 into a newly built and better-equipped laboratory, with eighty researchers. Cuban scientists went abroad to study, many in Western countries. Their research took on more innovative paths, as they experimented with cloning interferon. By the time Cantell returned to Cuba in 1986, the Cubans had developed a second generation interferon cloned in yeast.[50]

In 1982, the United Nations Industrial Development Organisation (UNIDO) launched a competition for an internationally funded project to foster biotechnology in the 'Third World'. UNIDO's International Centre for Genetic Engineering and Biotechnology was to facilitate North–South knowledge transfer and cooperation in science. Reagan's 1982 measures tightening the US blockade gave Cuba an additional incentive to apply. In 1984, the fund was awarded to a joint application by India and Italy. However, the Cubans, and most emphatically Fidel Castro, decided to proceed without support. Construction immediately began on Cuba's showpiece Centre for Genetic Engineering and Biotechnology (CIGB), to work in biology, chemical engineering and physics. It opened just two years later in 1986. By then Cuba was submerged in another health crisis, a serious outbreak of Meningitis B, which further spurred Cuba's biotechnology sector.

## CUBA'S MENINGITIS MIRACLE

In 1976, Cuba was struck by Meningitis B and C outbreaks.[51] Since 1916 only a few isolated cases had been seen on the island. At that point, internationally, vaccines existed for Meningitis A and C, but not for B. Cuban health authorities secured a vaccine from a French pharmaceutical company to immunize the population against type C Meningitis. However, in the following years, cases of type B Meningitis began to rise. With infections and fatalities on the increase, in 1983 MINSAP established a team of specialists from different medical science centres, led by a woman biochemist, Concepción Campa, to work intensively on finding a vaccine. By 1984 Meningitis B had become the main health problem in Cuba. After six years of working around the clock, in 1988, Campa's team produced the world's first successful Meningitis B vaccine. Again, the scientists tested the vaccine on themselves, and their own children, before beginning clinical trials. Between 1987 and 1989, a randomized, double-blind controlled trial of the vaccine efficacy took place, with over 100,000 students aged ten to fourteen years. The results showed the vaccine to be 83 per cent effective. Another member of Campa's team, Dr Gustavo Sierra recalled their joy: 'this was the moment when we could say it works, and it works in the worst conditions, under pressure of an epidemic and among people of the most vulnerable age.'[52]

MINSAP decided that more lives would be saved by starting nationwide vaccinations immediately with an 83 per cent effective vaccine than delaying until a more effective one was produced (or not). During 1989 and 1990, three million Cubans, those most at risk (children and young people), were vaccinated. In the roll out, the efficacy ranged from 83 per cent to 94 per cent in difference provinces. No severe reactions occurred and another severe disease outbreak had been halted. Subsequently, 250,000 young people were vaccinated with the VA-MENGOC-BC vaccine, a combined Meningitis B and C vaccination. It recorded 95 per cent efficacy overall, with 97 per cent in the high-risk three months to six years age group. Cuba's Meningitis B vaccine was awarded a UN Gold Medal for global innovation. This was Cuba's meningitis miracle.[53]

'I tell colleagues that one can work 30 years, 14 hours a day just to enjoy that graph for 10 minutes,' says Agustín Lage, Director of the Centro for Molecular Immunology (CIM), referring to an illustration of the rise and sudden fall of Meningitis B cases in Cuba. 'Biotechnology started for that. But then the possibilities of developing an export industry opened up, and today, Cuban biotechnology exports to 50 countries.'[54] This possibility came about after an outbreak of Meningitis B in Brazil a few years later. 'The Brazilians bought the Cuban vaccine. It was a huge purchase and that money was invested in expanding the biotechnology industry here', says Lage.

By 1986, Cuba had 39,000 scientific workers; one for every 282 people. Twenty-three thousand were involved in research. Thousands had been trained abroad, mainly in the socialist countries, but also in Western Europe. The revolutionary government's investments in education and public health had created the 'critical mass' necessary for further progress in medical science. The Biological Front invested $1 billion to develop a biotechnology industry between 1981 and 1989, including establishing the Western Havana Scientific Pole, known as Science City, between 1986 and 1991.[55]

## SCIENCE CITY

Science City is a cluster of biotechnology institutions that coordinate and integrate their work. Centre directors met monthly to discuss projects and exchange information in meetings attended by top government officials, including Fidel Castro. Thousands of housing units were constructed locally to enable the institution's employees, working daily shifts of 14 hours, to walk to work. At the centre of Science City is the CIGB, which received the greatest investment in a Cuban science institution. Other institutions followed:

- 1987, the Centre for Immunoassay: to manufacture computerized and automated equipment for biochemical tests and screenings to detect pathologies.
- 1989, the National Centre for Meningococcal Vaccines: for research and production of the VA-MENGOC-BC vaccine and other human vaccines. It was renamed the Finlay Institute (to honour Carlos Finlay) in 1991.
- 1990, the Cuban Centre for Neuroscience: for the diagnosis and treatment of brain diseases.
- 1992, the Centre of Biopreparados: to produce Cuban biologicals.
- 1994, Centre of Molecular Immunology.

Along with the pace of these state investments, the astonishing fact is that they took place in the midst of Cuba's most acute economic crisis: the Special Period which began in autumn 1990, as the case of Cuba's Centre for Molecular Immunology shows.

## ONCOLOGY MEETS BIOTECHNOLOGY

In the 1980s, biotechnology and oncology began to converge globally. In Cuba's National Institute of Oncology and Radiobiology (which emerged from the Hospital of Oncology that Guevara had once commandeered), Lage was among a group of young scientists working on an experimental project into the role of immunology to fight cancer. In the early 1980s, INOR developed and trialled the first Cuban monoclonal antibodies (MABs) – clones of single antibody cells – with multiple medical uses. By the late 1980s, MABs were used for detecting malignant tumours and preventing organ rejection in Cuban transplant patients.[56] After a visit to INOR in 1989, Fidel Castro recommended expanding the institution, integrating it into the Scientific Pole, and providing it with industrial production capacity and authorization to export. Construction began in 1991. In December 1991, when the USSR collapsed, only the prefabricated columns of the new Centre of Molecular Immunology (CIM) had been laid. Nonetheless, Fidel would not allow the project to be halted. Lage describes this as: 'a very audacious decision, when the country had no financial resources to say "this centre has to be completed". It was a Fidel's decision, a kind of offensive defence.'[57] The fact that the sector was entirely state owned and controlled made that decision possible, and necessary, if CIM was to continue.

## THE SPECIAL PERIOD – BIOTECHNOLOGY EXEMPT!

The collapse of the USSR and Eastern European socialism had a traumatic economic impact on Cuba. From having 85 per cent of its trade conducted under planned agreements with socialist countries (unimpeded by the US blockade), Cuba was suddenly dependent on an international capitalist market dominated by the US, the country pursuing a merciless blockade of the island. One-third of world pharmaceutical production took place in the US. From where could Cuba get medical equipment and medicines?

In 1993, as Cuban socialism tottered on the edge of an abyss after GDP plummeted 35 per cent in two years, elsewhere there were drastic cuts, things fell apart, production and transport were halted, belts were tightened and scarcity appeared in almost every sector and space. In this context, the inherently high-risk nascent biotechnology sector was selected as one of three strategic economic sectors for investment, along with tourism and food production.

Between 1990 and 1996, another US $1 billion (1.5 per cent of Cuba's GNP) was invested into the Scientific Pole. It functioned as an incubator of medical science enterprises which were protected directly by the President's office during the Special Period. When the institutions began to export, that money was reinvested into them; a 'closed economic cycle' was created. Thus it was in the most difficult economic period that Cuban biotechnology flourished. The motivation was domestic public health benefits, explains Lage. 'And because of the US blockade. We could not afford expensive drugs, sometimes we could not get them even if we had all the money in the world, because they would not sell them.'[58] Elsewhere in Latin America, neoliberal structural adjustment programmes saw public health provision privatized and rolled back in the 1990s, but the Cuban government held steadfast to its social-welfare orientated centrally planned economy.[59]

Gradually the dust covering the country's pharmacy shelves was replaced by Cuban manufactured 'biosimilars' – copies of traditional and biotechnological medicines, as Cuba's medical science sector was channelled to meet this need. Copying biotech products involves high-level science; to synthesize genes and introduce them into cells to clone them. Cuba's interferon, erythropoietin, the vaccine against Hepatitis B have all been 'copy products'. Homeopathic alternatives also claimed shelf space. The need for cheap generic drugs was international, so Cuba increased its pharmaceutical exports. By the mid-1990s, they were earning $100 million a year.

## BIOTECHNOLOGY WITH CUBAN CHARACTERISTICS

The president of a multinational pharmaceutical company once told Lage that he was bound by his shareholders' interests. The director of CIM asked how many shareholders his company had. The answer was 300,000. Lage replied, 'well, I have 11 million. Our shareholders are all 11 million Cubans!'[60] Founded solely through state investment, with financing guaranteed through the state budget, the Cuban biopharma sector is state owned, with no private interests or speculative investments. Profit is not sought domestically, because the sector is completely integrated into the state-funded public health system. National health needs are prioritized. Medicines that Cuba cannot afford, or cannot get access to because of the US blockade have to be produced domestically. Today, 517 of the 800 or so medicines consumed in Cuba are produced domestically, close to 70 per cent.

Cooperation prevails over competition as research and innovations are shared between institutions. Teams of scientists are established to take a project through from basic science, to product-oriented research, to manufacturing and marketing; activities that are carried out by different businesses in most countries. Dr Kelvin Lee is the Chair of Immunology in Roswell Park Comprehensive Cancer Centre in Buffalo, New York, and is currently leading clinical trials on CIM's CIMAVax-EGF, a therapeutic cancer vaccine for patients with advanced lung cancer.[61] He highlights these 'striking' and 'unique' characteristics of the Cuban biotech sector: 'They start with identifying a need, then figure out the science to develop that in the lab, manufacture their agent, test it in the Cuban medical system and then commercialize it and sell it overseas. Their system is particularly nimble in that ability.' The disadvantage the Cubans face, he adds, is that they can't pursue thousands of good ideas and write off those which don't work as sunk costs. 'They don't have the resources to do that.'[62] Their access to capital is extremely limited. So what, then, have been fruits of this distinctive Cuban system?

## THE CUBAN CURE

In 2015, the World Health Organization announced that Cuba was first in the world to eliminate mother-to-child HIV transmission. Cuba has prevented an AIDS epidemic with domestically produced antiretroviral medicine that halts patient transmission, as if it were a vaccine. Cuba's mortality curve for AIDS continues to fall. The universal use of the CIGB's Hepatitis B vaccine on newborns means Cuba should be among the first countries free from Hepatitis B. This is one of the eight vaccines (out of eleven vaccines for thirteen diseases) administered to Cuban children which are produced in Science City. Within ten years, 100 million doses of Cuba's Hep B vaccine were used around the world.

By 2017, CIGB employed 1,600 people and sold twenty-one products internationally. CIGB's portfolio of innovations with major public health implications includes Heberprot-P for diabetic foot ulcers, affecting some 422 million people worldwide, which reduces the need for amputations by 71 per cent.[63] In ten years, 71,000 patients in Cuba have been treated with Heberprot-P plus

130,000 people in twenty-six other countries. The CIGB also has products for the agricultural and food production sectors.

Cuba was second in the Americas to achieve a complete congenital hypothyroidism screening programme, after Canada and before the United States. Cuba's Immunoassay Centre developed its own Ultramicroanalytic System (SUMA) equipment for prenatal diagnosis for congenital anomalies. Among other things, nearly four million babies have been tested for congenital hypothyroidism, which effects the production of thyroxine, a hormone needed for normal growth and development.[64] Treatment is easy and cheap. 'Since this system was introduced some children who would have had problems with mental development are in the universities,' said Lage.[65] In 2017, the Immunoassay Centre had 418 workers producing 57 million tests per year for nineteen different conditions, including Hepatitis B and C, dengue fever, cystic fibrosis, Chagas disease, leprosy and HIV.

Cuba's Centre for Neuroscience is developing cognitive and biomarker tests for early screening of Alzheimer's disease. They have developed a hearing aid for children that costs just (US) $2, a fraction of the cost in the US and Europe, made to individual specification using a 3D printer.[66]

Cuban professionals have received ten gold medals from the World Intellectual Property Organization (WIPO) over twenty-six years. The first was in 1989 for the Meningitis B vaccine; another was for the *Haemophilus influenza* type b (Hib) vaccine,[67] the result of a collaboration with the University of Ottawa; also for Heberprot-P; and Itolizumab for treating psoriasis.[68] By summer 2017, the Cuban biotech sector boasted 182 inventions with 543 patents granted in Cuba, 1,816 patents abroad and 2,336 patent applications.[69] Its products were marketed in forty-nine countries and it had partnerships with nine countries in the Global South. Cuba's pharmaceutical industry has the capacity for large-scale production of Cuban and generic drugs for export cheaply to developing countries.

CIM's focus is on biotechnology of mammalian cells; monoclonal antibodies and cancer vaccines. Cancer is the biggest cause of death for under sixty-five year olds in Cuba, and second only to heart disease for over sixty-fives. By summer 2017 CIM had 1,100 employees, four manufacturing facilities, twenty-five products in the pipeline, six registered products, exported to thirty countries, had five joint-venture companies abroad, forty-five patented inventions, and 750 patents abroad. Over 90,000 Cubans had been treated by CIM products. Among CIM's most exciting innovations is CIMAvax: EGF, the lung cancer immunotherapy.

The term 'cancer' refers to a group of diseases involving abnormal cell growth with the potential to invade or spread to other parts of the body. Epidermal growth factor, or EGF, is a cellular protein that stimulates cell growth by binding to cells via epidermal growth factor receptors (EGFRs) on the cell surface. Back in 1984, Lage and the scientists at INOR were first in the world to describe the role of EGFRs in breast cancer: EGFRs were over-expressed in 60 per cent of human breast tumours. They discovered that EGF was rapidly distributed, reached tumour cells and recognized specific cell membrane receptors. Lage reported: 'These results suggest that high doses of EGF could eventually be used for inhibition of the cell proliferation in some tumours.'[70] Up until that point, EGF had been seen as part of the cancer problem; it nourishes tumours yet is natural to the body so it does not trigger the immune system. The Cubans were proposing to use human EGF as part of the solution: as an active agent that could be used to interfere with the normal, cancer-producing binding of EGF to its receptor – the EFGR.[71]

It is because the immune system struggles to recognize cancer as foreign to the body that immunology therapy to combat cancer had proven so difficult. CIM wanted to use EGF to 'train' the body to respond to EGF and so produce cancer-specific antibodies. No other cancer researchers had managed it. This therapy would require one or two doses of a vaccine which would be cheap to develop and could be delivered through primary healthcare. This suits for the Cuban medical system. However, because it does not involve a lengthy course of expensive therapy in high-tech institutions, it is arguably antithetical to the profit-motivated interests of the global biopharma corporations. CIMAVax built on the Cuban therapeutic expertise of the vaccinologists at the Finlay Institute (Meningitis B) and the CIGB's work on recombinant protein from Neisseria meningitides bacteria P64 K. By using P64 K as a carrier protein with which to introduce EGF into a patient's body, the researchers at CIM broke the body's tolerance to its own EGF.[72] The results are a vaccine that helps the body to help itself. Dr Kelvin Lee points out that the Cubans 'designed their lung cancer vaccine to actually be useful in things like colon cancer, head and neck cancer, breast cancer, pancreatic cancer. It has broad applicability.'[73] The Roswell Park/CIM joint venture, Innovative Immunotherapy Alliance, will be investigating some of these additional potentials.

These biotech achievements are accompanied by demographic ones. Most employees in Cuba's medical science centres are the children and grandchildren of workers and peasants, beneficiaries of Cuba's free education system. Today, 66 per cent of Cuba's science and technology personnel are women.[74] The inordinate additional burdens imposed on the sector by the unrelenting and extensive US blockade have undoubtedly impoverished Cuba, denying the island access to resources, markets and knowledge transfers. But it has also fostered resilience and creativity in Cuban scientists.[75]

## BIOCUBAFARMA

In 2012, BioCubaFarma was created as a kind of 'holding company' for the pharmaceutical and biotechnology sectors. It integrated thirty-eight companies, sixty manufacturing facilities and 22,000 workers, almost one-third of them scientists and engineers. By 2017, BioCubaFarma was exporting to nearly fifty countries with over 2,000 patents granted abroad. This reorganization of Cuban medical science is integral to the broader restructuring of the Cuban economy under the 'guidelines for updating the economic and social model'. The economic reforms, which were initially introduced in 2011, confirm a key role for the biopharma sector in the national development plan. But only 1 per cent of the nearly 3,000 Cuban state enterprises export innovative scientific products. How can this sector expand without the kind of speculative private investment and profit-motivated competition which characterizes the capitalist biotech industry?

Agustín Lage conceptualizes 'high-tech socialist state enterprises' in Cuba, requiring a distinct regulatory framework from the 'budgeted sector' (health, education and other social provisions funded entirely by state budget) and the state enterprise sector (state owned and expected to contribute towards the national coffers).[76] There is commitment to maintain biotechnology institutions under state control, with state investment for scientific research, with the sector increasingly contributing to economic growth through the export of high value-added products in a context of a state monopoly over foreign trade. This requires strengthening of the integration between science and production, research institutions and universities and the promotion of the close-cycle of production.

## CUBAN BIOTECH EXPORTS

Several global biopharma corporations have demonstrated interest in Cuban biotech products, but the US blockade, and the threat of receiving a huge fine, has largely proven an effective disincentive. Cuban collaborations with the Global South have been more successful, in terms of exports and joint ventures. The knowledge economy in small countries has to be developed on the basis of exports, because domestic demand is insufficient. Between 2008 and 2013, Cuban biotech sales earned US $2.5 billion.[77] By 2017, this had risen to $500,000 annually. The low cost of Cuba's biopharma exports is an incentive for governments in the Global South to pursue deals. CIMAvax costs $1 per shot to manufacture, much cheaper than alternative lung cancer treatments.

The sector has also established joint ventures with foreign (state) companies. By 2017, the Cuban biotech sector had joint ventures operating in Brazil, China, Venezuela, Algeria, South Africa, Vietnam, India, Thailand and Iran. CIM alone has three joint ventures in China, manufacturing, variously, monoclonal antibodies and therapeutic cancer vaccines, recombinant proteins and biotech products for agriculture. Lage boasts that China – a country with 1 billion inhabitants, which produces aeroplanes and heavy industry – made its first monoclonal antibodies with Cuban technology.[78] The CIGB also has two joint ventures with China. A Cuba–China workgroup oversees the collaboration between the two countries. CIM also has two marketing joint ventures, in Spain, for therapeutic cancer vaccines, and in Singapore for monoclonal antibodies. Another mixed enterprise is being set up in Russia.

The first Cuban patents are expiring, and the race is on to produce new innovative products. Globally, the average annual costs of research and development for new drugs is increasing. Between the 1980s and 2006 it increased 7.4 per cent.[79] A 2014 report calculated the cost of developing a prescription drug that gains market approval at $2.6 billion, a 145 per cent increase over their previous estimate in 2003.[80] Cuba is now bidding for substantial foreign investment in its biopharma industry.[81] However, there is no intention to relinquish state ownership of the sector, or to privatize it in any other form. The Cuban government does not want foreigners to fund innovations, thus claiming a stake in them. It wants investment in creating its industrial production capacity and it needs partners to help insert Cuban products into the global market, subverting the US blockade and increasing export earnings.

In the process of unravelling the US–Cuban rapprochement, on 8 November 2017, US President Trump's administration published a list of Cuban entities and sub-entities with which US citizens and businesses are prohibited from engaging. This placed the brakes on new cooperation with the US, and because of the extraterritorial character of the blockade, with its European allies too. However, the insertion of a 'grandfather' clause meant that collaborations licensed under Obama, like the one between CIM and Roswell Park, can continue. There are ways and means around these regulations, even for US businesses, if they are determined. And given Cuban medical science breakthroughs on some key global health issues, such as cancer, HIV and AIDS, hepatitis and infectious diseases, it is likely they will find increasing incentives to do so.

## CONCLUSION

With increasing frequency, Cuba's biotech breakthroughs are making global headlines. Cuba's biotechnology success can be understood in the context of the post-1959 commitment to high-standard, universal free public health and education, which created a 'critical mass' of medical scientists. It also reflects the political will of the country's leadership, particularly Fidel Castro, to develop science and technology as a means to socio-economic development. It builds on a long history of infectious disease research and vaccinations. The US blockade also imposed the need to meet domestic demand for medical products with domestic production. The question of state control in a planned economy is also key. The flow between universities, research centres, medical science enterprises and the public health system is far more fluid than elsewhere, precisely because there is no proprietor-based conflicts of interest. The state controls production and distribution, equally of scientists and researchers as of medical products and innovations.

Research demonstrating the 'inseparability of the biotechnology industry and the state . . . in the US and elsewhere',[82] highlights that the curious case of Cuba's biotech revolution is about far more than government support for the sector. The US government supports the biopharma sector with public-research funding, legislation and patenting laws. It has also fostered the venture capital sector, on which biotech firms principally rely, through favourable monetary and fiscal policies and regulations. In the US, the danger is that the private interests of venture capitalists obstruct medical science innovations which would bring public health benefits globally. In Cuba, the challenges are accessing capital and navigating the US blockade which obstructs access to foreign biopharma partners, export markets and patents and resources from around the world.

## NOTES AND REFERENCES

1. Notable exceptions include work by historians of science Angelo Baracca and Rosella Franconi, for example, 'Cuba: The Strategic Choice of Advanced Scientific Development, 1959–2014', *Sociology and Anthropology* 5, 4 (2017): 290–302; political geographer Simon M. Reid-Henry, *The Cuban Cure: Reason and Resistance in Global Science* (Chicago: The University of Chicago Press, 2010); and in the area of innovation theory, Andrés Cárdenas, *The Cuban Biotechnology Industry: Innovation and Universal Health Care*, 2009, https://pdfs.semanticscholar.org/df8b/95006fb835075a7b50a51cf3f61273b00304.pdf. Cuban medical scientists contribute to international journals, but these provide technical rather than historical accounts.

2. For example, Christian Zeller, 'The Pharma-biotech Complex and Interconnected Regional Innovation Arenas', *Urban Studies* 47, 13 (2010): 2867–2894.
3. Lara V. Marks, *The Lock and Key of Medicine: Monoclonal Antibodies and the Transformation of Healthcare* (New Haven and London: Yale University Press, 2015).
4. Cárdenas, *Cuban Biotechnology*: 7.
5. Cárdenas, *Cuban Biotechnology*: 1.
6. I am grateful to the anonymous reviewer who highlighted this element as decisive.
7. Ernesto Lopez Mola, Ricardo Silva, Boris Acevedo, José A Buxadó, Angel Aguilera and Luis Herrera, 'Biotechnology in Cuba: 20 Years of Scientific, Social and Economic Progress', *Journal of Commercial Biotechnology* 13, 1 (2006): 1–11.
8. Joseph Cortright and Heike Mayer, *Signs of Life: The Growth of Biotechnology Centres in the US* (Massachusetts: The Brookings Institution Center on Urban and Metropolitan Policy, 2002): 6.
9. Agustín Lage, Director of Cuba's Centre for Molecular Immunology, interview with the author in Havana, Cuba, 7 July 2017.
10. The focus in this article is biotechnology for human healthcare, not animal healthcare or agriculture where Cuba also boasts some innovative developments.
11. Halla Thorsteinsdóttir, Tirso Saenz, Uyen Quach, Abdullah S. Daar and Peter A. Singer, 'Cuba – Innovation through Synergy', *Nature Biotechnology* 22 (2004): 21.
12. Cortright and Mayer, *Signs of Life*: 14.
13. Zeller, 'Pharma-biotech Complex': 2889.
14. Cortright and Mayer, *Signs of Life*: 3.
15. Zeller, 'Pharma-biotech Complex': 2883.
16. Cortright and Mayer, *Signs of Life*: 33.
17. William Lazonick and Öner Tulum, 'US Biopharmaceutical Finance and the Sustainability of the Biotech Business Model', *Research Policy* 40 (2011): 1182.
18. Lazonick and Tulum, 'US Biopharmaceutical Finance': 1170.
19. Cortright and Mayer, *Signs of Life*: 8.
20. Lazonick and Tulum, 'US Biopharmaceutical Finance': 1170; Gary Pisano, *Science Business: The Promise, the Reality, and the Future of Biotech* (Boston, MA: Harvard Business School Press, 2006).
21. Cortright and Mayer, *Signs of Life*: 19.
22. Cortright and Mayer, *Signs of Life*: 9.
23. Lazonick and Tulum, 'US Biopharmaceutical Finance': 1172.
24. Zeller, 'Pharma-biotech Complex': 2870.
25. Cortright and Mayer, *Signs of Life*: 9.
26. Lazonick and Tulum, 'US Biopharmaceutical Finance': 1180.
27. Lazonick and Tulum, 'US Biopharmaceutical Finance': 1176.
28. Lazonick and Tulum, 'US Biopharmaceutical Finance': 1178.

29. The Institute of Research in Chemistry (1848); the Observatory of Meteorology and Physics (1856); and the Royal Academy of Medical, Physics and Natural Sciences (1861), founded by the Spanish queen's decree. The word 'Royal' was dropped in 1902 following Cuban 'independence'.
30. US President Obama acknowledged Finlay's contribution when he announced rapprochement between the US and Cuba, 17 December 2014, https://obamawhitehouse.archives.gov/the-press-office/2014/12/17/statement-president-cuba-policy-changes
31. Many Cubans refer to this period as the 'Pseudo-Republic' as US domination of the island effectively turned it into a semi-colony or dependency.
32. This and following information draws on Theodore MacDonald, *Hippocrates in Havana: Cuba's Health Care System* (Mexico: Bolivar Books, 1995): 15–79.
33. International Bank for Reconstruction and Development (IBRD) in collaboration with the Government of Cuba, *Report of the Mission to Cuba* (Washington, DC: Office of the President, 1951): 223.
34. 31 per cent of Cubans over six years old had no schooling; another 29.4 per cent had three years' schooling or less. In rural Cuba, 41.7 per cent of over-ten year olds were illiterate.
35. C. William Keck and Gail A. Reed, 'The Curious Case of Cuba', *American Journal of Public Health* 102, 8 (2012).
36. Baracca and Franconi, 'Cuba: Strategic Choice': 9; MacDonald, *Hippocrates in Havana*: 28.
37. See Helen Yaffe, *Che Guevara: The Economics of Revolution* (London: Palgrave Macmillan, 2009): 163–198.
38. Tirso Sáenz, interview with the author in Havana, 20 February 2006.
39. The farm was called Ciro Redondo after a fallen Rebel Army captain from Guevara's column.
40. Yaffe, *Che Guevara*: 188–190.
41. José Luis Rodríguez, former Minister of the Economy, interview with author in Havana, 20 December 2016.
42. Lopez Mola et al., 'Biotechnology in Cuba': 2.
43. MacDonald, *Hippocrates in Havana*: 143.
44. Reid-Henry, *Cuban Cure*: 26. Biological sciences stagnated in the USSR from the 1920s to the mid-1960s, under the influence of Trofim Lysenko. In 1948, the study of genetics was outlawed. See Reid-Henry, *Cuban Cure*: 26.
45. Information from Baracca and Franconi, 'Cuba: Strategic Choice'; Idania Caballero Torres and Lien Lopez Matilla, *La historia del CIM contada por sus trabajadores*', unpublished paper, 2017. Lage, interview; and Reid-Henry, *Cuban Cure*.
46. *New York Times*, 'Epidemic in Cuba Sets Off Dispute with US', 6 September 1981. www.nytimes.com/1981/09/06/world/epidemic-in-cuba-sets-off-dispute-with-us.html (last accessed 29 November 2018).

47. Fidel Castro, 26 July 1981, cited by *New York Times*, 'Epidemic in Cuba'.
48. Marieta Cabrera, 'La ciencia desnuda un crimen contra Cuba', *Bohemia*, 29 January 2016, http://bohemia.cu/ciencia/2016/01/la-ciencia-desnuda-un-crimen-de-ee-uu-contra-cuba-en-1981/. The study states: 'Cuban researchers were able to amplify and sequence the full genome of the original strains obtained in different moments of the epidemic in 1981, using bioinformatic tools . . .'.
49. Caballero Torres and Lopez Matilla, *Historia del CIM*: 10.
50. Reid-Henry, *Cuban Cure*: 47.
51. Meningococcal disease is one of a group of bacteria responsible for the life-threatening infections meningococcal meningitis and meningococcal septicaemia. Untreated, these conditions can kill within 24 hours. Ten per cent of survivors suffer serious, long-term disabilities, including brain damage. Meningococcal disease is among the top ten global causes of death due to infection. There are thirteen different forms of the disease but serogroups A, B and C are by far the most common. According to the World Health Organization there are up to 25,000 meningococcal deaths every year in Africa.
52. Dr Gustavo Sierra cited in 'Meningitis B – Cuba', documentary posted by Journeyman Pictures, posted 25 January 2008, www.youtube.com/watch?v=rgQZhTg04IM
53. Despite this, Cuba's achievement has been ignored or censored in Britain. In September 2015 when the NHS introduced a new Meningitis B vaccine for babies it claimed the vaccine: 'makes England [sic] the first country in the world to offer a national, routine and publicly funded MenB vaccination programme' (NHS Choices https://assets.publishing.service.gov.uk/government/uploads/system/uploads/attachment_data/file/448820/PHE_9402_VU230_June_2015_12_web.pdf). The then-British Health Secretary Jeremy Hunt repeated the claim. Until 2014, only Cuba had developed a safe and effective Meningitis B vaccine and millions of people in Cuba and around the world had benefited.
54. Agustín Lage, interview, 2017.
55. This occurred during a period in Cuba known as Rectification from the mid-to-late 1980s, during which Fidel Castro pulled Cuba away from the Soviet economic management model, fostering innovative science and technologies instead of the heavy industries the Soviets recommended.
56. The first monoclonal antibody registered for cancer treatment in the US was 1997.
57. Lage, interview, 2017.
58. Lage, interview, 2017.
59. Helen Yaffe, 'Cuban Development: Inspiration to the ALBA-TCP', in Thomas Muhr (ed.), *Counter-Globalization and Socialism in the 21st Century: The Bolivarian Alliance for the Peoples of Our America* (London: Routledge): 101–118.
60. Lage, interview, 2017.
61. Roswell Park received a US government licence to pursue collaboration with Cuba and trials were authorized in the brief period of improved relations towards the end of Obama's presidency.

62. Dr Kelvin Lee, Chair of Immunology in Roswell Park Comprehensive Cancer Centre in Buffalo, New York, interview with the author via Skype, 3 October 2017.
63. Given this ratio, 52,000 amputations which take place annually in the US due to diabetic foot ulcers could be prevented.
64. Orfilio Peláez, 'The Jewel that Fidel Conceived', *Granma*, 7 September 2017, http://en.granma.cu/cuba/2017-09-07/the-jewel-that-fidel-conceived
65. Lage, interview, 2017
66. Sara Reardon, 'Can Cuban Science go Global?', *Nature*, 29 September 2016, www.scientificamerican.com/article/can-cuban-science-go-global/
67. *Haemophilus influenza* type b (Hib) is a bacteria responsible for severe pneumonia, meningitis and other invasive diseases almost exclusively in children aged less than five years.
68. Similarly to Heberprot-P, over 100,000 patients worldwide have benefited from using Itolizumab.
69. This pales into insignificance in comparison to the US, however, where some 5,500 patents are awarded annually according to Cortright and Mayer, *Signs of Life*: 9.
70. Lage, cited by Reid-Henry, Cuban Cure, 99.
71. Reid-Henry, *Cuban Cure*: 99–100.
72. Subsequently, they sought a more efficient carrier for EGF.
73. Dr Lee, interview.
74. In Britain women make up only 12.8 per cent in the science, technology, engineering and maths workforce. See www.theguardian.com/news/datablog/2015/jun/13/how-well-are-women-represented-in-uk-science
75. This, and low salaries in Cuba, have prompted many Cuban medical scientists to seek employment overseas. However, there is no evidence that the 'brain drain' from Cuba is greater than in other developing countries.
76. Agustín Lage Dávila, *La Economia del Conocimiento y el Socialismo: Preguntas y Respuestas* (La Habana: Editorial Academia, 2015).
77. Reardon, 'Cuban Science'.
78. Lage, interview, 2017.
79. Lopez Mola et al., 'Biotechnology in Cuba': 3.
80. Rick Mullin, 'Tufts Study Finds Big Rise In Cost Of Drug Development', Tufts Center for the Study of Drug Development, 20 November 2014, https://cen.acs.org/articles/92/web/2014/11/Tufts-Study-Finds-Big-Rise.html
81. Total biopharma investments sought in 2017/8 were at least $850 million, nearly one-tenth of the $9 billion sought in the annual portfolio.
82. Sharmistha Bagchi-Sen, Helen Lawton Smith and Linda Hall, 'The US Biotechnology Industry: Industry Dynamics and Policy', *Environment and Planning C: Government and Policy* 22 (2004): 199.

# A Song of Water and Fire: The Brief Coming of Age of the Venezuelan Oil Industry's R&D Programme at the turn of the Twentieth Century

SAUL GUERRERO

## INTRODUCTION

The geographical location of the world's largest oil reserves (South America, Middle East), has rarely corresponded to the sources of the technical know-how required to monetize efficiently said reserves. The tension this had created in many of the major oil-producing countries was reflected in 1977 in the words of the Secretary General of OPEC (Organization of Petroleum Exporting Countries), Ali Mohammed Jaidah:

> The development of an indigenous technological base, backed by domestic research institutions [is] . . . a central pivot in the process of industrialization . . . we have been buying technology at exorbitant prices, whilst turn-key projects have proved to be tied to the suppliers of technology for patents, spare parts, operations, research . . . the present terms of transfer of technology are a source of deep concern to us . . . because they are of a grudging nature.[1]

Even the owners of oil-related technology have changed over the last three decades. Up to the 1990s, eleven oil majors carried out 80 per cent of Research & Development (R&D), but by 2013 it was oil field services (OFS) companies, mainly based in the USA, that issued 80 per cent of all the patents related to upstream oil activities.[2] By outsourcing drilling technology, the oil majors opened the door to a very lucrative and increasingly complex technical expertise now dominated by the OFS companies.[3] The leader, Schlumberger (USA), has grown from revenues of US$7 billion in 1990 to US$30.4 billion in 2017, with a market capitalization of US$89.7 billion.[4]

With such a moving target for a National Oil Company (NOC), having to choose whether to invest in research like a major independent oil company or to purchase bespoke technology in the marketplace, the decision to proceed with an in-house R&D programme has not been a straightforward one. The cost of acquiring technology is always high, but so is the political opportunity cost of diverting oil revenues to long-term R&D, by an NOC that is by nature subject to strong demands by its sole owner, the State, to divert most of its revenues to prop up government spending. Yet two years prior to the Vienna seminar cited above, Venezuela had already announced its intention to provide its oil industry after statization with its in-house R&D centre, a decision that singled it out among OPEC members at the time.[5] On 6 August 1975, a key figure in the formulation and execution of Venezuela's oil policy rose in the Senate to deliver a speech to mark this historic milestone. Rómulo Betancourt, the powerful and controversial elder statesman of the Acción Democrática (AD) political party, and twice President of Venezuela, announced that 'the current government [AD]. . . has acquired . . . "Villa Pignatelli" . . . the Instituto Tecnológico Venezolano del Petróleo [INTEVEP] will be built there'.[6] Betancourt had just provided the most powerful political backing to a research institution born in 1974 under former President Rafael Caldera's tenure, in a show of continuity of policy between the two major political parties of the period.[7]

In January 1976 the immediate task for the nascent Petróleos de Venezuela, S.A. (PDVSA) was to guarantee production, while in the longer term, new reserves had to be found and quantified, and technology created to upgrade or refine its heavy (HC) and extra-heavy (XHC) crudes, which had no immediate outlet in existing refineries.[8] The oil majors that had previously operated in Venezuela continued to provide technical assistance contracts to their Venezuelan successors, joined later by OFS companies.[9] In the light of available technical support, albeit at a high cost, the decision to create INTEVEP placed PDVSA quite ahead of its NOC peers, but not without major risks.[10] If the main argument in favour of developing in-house technology was the perception that third-party providers would never guarantee innovations best adapted to local crudes, or the degree of dependency it created together with a heavy financial and strategic cost, there was no certainty that INTEVEP would ever be able to develop such a high level of R&D output, across the spectrum of oil and energy challenges facing Venezuela. In the following sections I will follow some of the salient aspects of the life cycle of INTEVEP from its creation in 1976 to its watershed years at the turn of the twentieth century, and the fortunes of its most iconic technical creations in the field of catalysis and emulsions. Driven by Venezuela's need to optimize the market potential of its world-leading reserves of heavy and extra-heavy crudes, the coming of age of INTEVEP played against a backdrop of sharply contrasting ideologies on Venezuela's oil policy, that ultimately defined the destiny of the institution.

## A MARRIAGE OF CONVENIENCE

INTEVEP was incorporated in 1979 as a mercantile society within the PDVSA corporate umbrella, with a Board of Directors that included the heads of PDVSA's operating companies. Its revenues would be generated by providing services to the PDVSA group, together with contributions from head office to support long-term research. Its budget allocation was relatively flat over the years, as indicated in Figure 1, with a total expenditure from 1979 to 2000 of US$ 2.3 billion (2000 US$), lower than the $2.8 billion (2000 US$) paid by PDVSA to third parties for technical support during the same period, though the latter is distorted by the major disbursements incurred during the first years after statization. The average ratio of INTEVEP budget to total revenues of PDVSA in this period was 0.4 per cent.[11] The R&D intensity of oil companies reported for a more recent period (2009 to 2015) show a range between 0.4 per cent and 0.5 per cent.[12] On average, oil and gas producers show a low R&D intensity in general, spending less than 1 per cent on research.[13]

INTEVEP began as a marriage of convenience between two quite different partners. On the operational side there was a core Venezuelan experience from which to extract a nucleus of oil engineering staff for INTEVEP.[14] The other partner came from the Venezuelan scientific establishment, that seized the opportunity to nurture future research in Venezuela under the aegis of the main industrial activity of the country. The magnet of better salaries, equipment and budgets created a career pull very difficult to ignore in a country where R&D by the private sector was insignificant, leaving mainly the State universities, with their more fragile budgets, as the only other option. It was therefore inevitable that the creation of INTEVEP

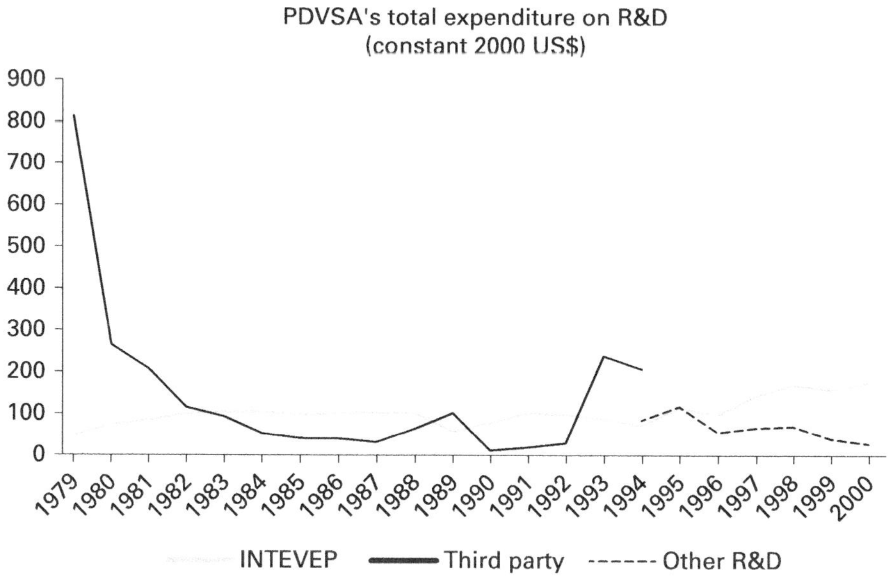


FIGURE 1: INTEVEP budget and third-party R&D expenses, 1979 to 2000, in constant 2000 US$. Source of raw data: note 11.

would create waves in the very small pond of the Venezuelan scientific research community. According to Requena, its operational budget was on average five times greater than the budget assigned to Venezuela's foremost scientific research facility, IVIC (Instituto Venezolano de Investigaciones Científicas, with origins in 1951), and twice that of CONICIT (Consejo Nacional de Investigaciones Científicas y Tecnológicas, founded 1969), the government body that coordinates Venezuela's scientific and technical policy, and provides funding.[15] From 1984 to 2000, 0.39 per cent of Venezuela's GNP had been spent on Science and Technology. INTEVEP absorbed 31 per cent of this amount, greater than the total assigned to all the Venezuelan universities (27 per cent), to CONICIT (18 per cent) or IVIC (12 per cent).[16] INTEVEP could also now lure the cream of available talent which had expanded in recent years, thanks to the scholarship programmes fostered initially by CONICIT, then after 1974 by FONINVES (Fondo para la Investigación y Formación de Personal en el Area de los Hidrocarburos) and the Fundación Gran Mariscal de Ayacucho (FGMA), flush with oil bonanza funds.[17] While CONICIT figured prominently in the early planning stage of INTEVEP, the latter ultimately became too large and independent of the former. The priority for the lines of research for INTEVEP were now to be set uniquely by PDVSA, outside the centralized planning of the scientific and technical sector of Venezuela. As a result, INTEVEP virtually disappears from the historiography on the government policies for the scientific and research sector of Venezuela.[18]

Two major divisions were initially created: the engineering pole of Exploration and Production, and the chemical pole of Refining and Petrochemicals. In addition, two career paths within INTEVEP had to be established, so that researchers could focus on R&D and still ascend within the corporate hierarchy, while managers with no experience in research, assigned to INTEVEP as part of PDVSA's human resources policy, could still use the posting as another stepping stone in their career. An anonymous ditty reflected with wry humour this corporate and cultural *melange*: 'we the unwilling/ led by the unqualified/ have been doing the unbelievable'.[19] For researchers coming from university backgrounds, the new world of rank expressed in colour-coded name tags, separate dining rooms, areas of restricted access, segregated levels of plane travel and hotel accommodation, guaranteed annual training options overseas, represented a culture shock, though with attractive benefits. By the end of the twentieth century INTEVEP had assembled an impressive pool of Venezuela's intellect: a total workforce of 1,580 employees, of which 334 worked in administrative roles, and 985 professionals in R&D and technology, 164 with PhDs, 241 with Master's degrees and 577 with engineering and other university first degrees. In addition, as of 1984 INTEVEP had funded 108 PhDs overseas.[20] INTEVEP had assembled a sizeable proportion of the approximately 1,400 researchers reported to have been active at the time in Venezuela.[21] As General Alfonso Ravard, the blunt-spoken first President of PDVSA stated in a prescient interview, it was not the buildings or the equipment, but the discipline and the intellect in the institute that would make the difference.[22] The steadfast implementation of INTEVEP's plan took place against the backdrop of seven democratic transfers of government between parties of different ideology, a major currency devaluation in 1983, two failed military coups in 1992, an impeached

President of Venezuela in 1993, and low oil prices from the mid-1980s that would remain at just over $10 per barrel by 1999.

When attempting to measure quantitively the degree of success of an institute of applied research such as INTEVEP, the immediate benchmark at hand is the number of patents applied for on a yearly basis (Figure 2).[23] The maturing of INTEVEP is evident in the period 1976 to 2002, when it filed a total of 266 patents, placing it as the most prolific R&D institute focused on oil in Latin America of the period, with regards to patents filed in the U.S. Patent and Trademark Office (USPTO). During the same period, Petrobras filed 147 patents with USPTO, while the Instituto Mexicano del Petróleo filed ten patents in USPTO, preferring to file its patents in Mexico (425 patents awarded by the Instituto Mexicano de Propiedad Industrial (IMPI) between 1973 and 1993). By the year 2000 INTEVEP had filed 1,046 patents in twenty-four countries.[24] Patents are certainly valuable indicators, though the importance of INTEVEP to PDVSA passed foremost as first responder to the operational problems of its affiliates, who supplied the core revenues to fund INTEVEP's activities. It is also true that useful knowledge can be accumulated and used within the R&D facility through original research or adaptation of existing technology, without having created a patent. However, since patents are like seeds in a parable, where many can be issued but very few sprout into commercial realities that can generate revenues to the parent company, it is precisely in the successful husbanding of these patents to commercial fruition that an R&D institute can demonstrate to its peers that it has come of age. The challenges faced by INTEVEP are highlighted by the contrasting commercial fortunes of the R&D innovations from two of its main areas of activity, catalysis and emulsions. Their history illustrates why the marketing success for INTEVEP's creations ultimately depended not only on the intellectual depth of the institute, but on the synchronicity required between long-term R&D projects and PDVSA's business plans, together with the willingness

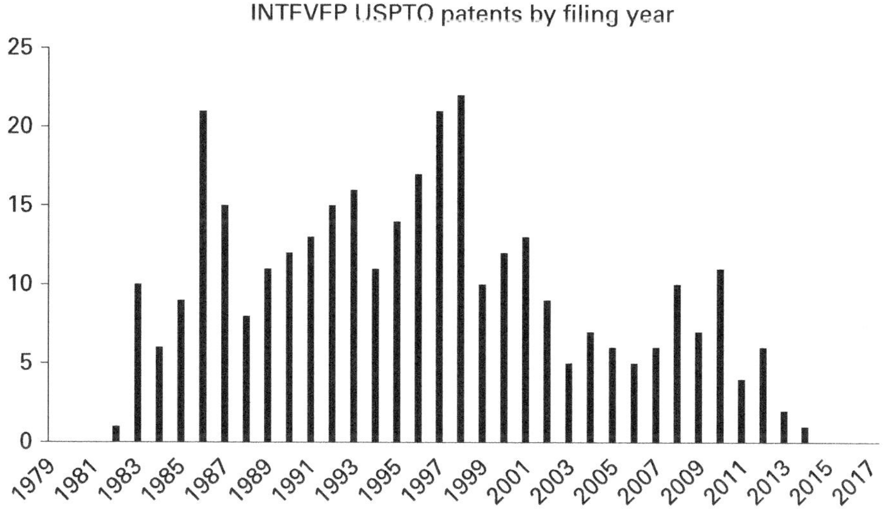


FIGURE 2: Patents filed in the USPTO by INTEVEP, 1979–2017. Source: note 24.

and ability of the corporation to back its own R&D products with the whole resources of the corporation.

## HDH PLUS™ AND AQUACONVERSION™

For Venezuela it is not enough to claim the world's largest reserves of hydrocarbons, since they consist mostly of very heavy and high sulphur oil that has to be transformed into commercial lighter products, that need to comply with environmental legislation, at a profit.[25] The established route to upgrade heavy crudes remains delayed coking, a well-tested process that uses temperature to break down the heavy hydrocarbons into a lighter (less viscous) synthetic crude oil, which can then be used as feedstock in a conventional refinery. Its drawback is that one of the main products of the process is coke, a carbon-rich solid, with very high metal and sulphur content, a waste product with very restricted market outlets for environmental reasons. For PDVSA it was important that INTEVEP focus on finding a catalytic route to process more efficiently its HC and XHC reserves, and avoid the formation of coke. In the words of one of the main protagonists:

> The development of catalytic refining processes for heavy crudes started with four independent programs, led in parallel by two senior managers. Two of the initial four would be chosen after a long period of trials at the laboratory and pilot-plant scale, subject to the recommendation from a panel of independent experts. This selection took place over a period of approximately eight years. The process known as HDH™ would be the ultimate choice, and a scale-up test, to 150 bpd (barrels per day), was planned in the installations of Veba Oel in Germany. The trial also allowed a comparison with Veba's own process (VCC). The project then passed on to a new team of research leaders, who introduced a synthetic nano-catalyst dispersed in an emulsion, a very advanced technical solution for the period. This improved process was named HDH Plus™. It was transferred to the IFP (Institut Français du Petrole) / Axens for its licensing and commercialization.
>
> María Magdalena Ramírez, Researcher, Catalysis Group, ex-INTEVEP[26]

HDH™ scaling tests were completed and improved into HDH Plus™ by the early 1990s. By 1997 INTEVEP had also developed AQUACONVERSION™, another catalytic process to lower the viscosity of the extra-heavy crudes to at least 14° API. This would solve the critical problem of transporting the heavy crudes without having to mix them previously with lighter fractions as a diluent, and the product of AQUACONVERSION™ could also feed the HDH Plus™ process.[27]

INTEVEP addressed the upgrading of the reserves of the Orinoco Oil Belt in a global manner, using minerals from nearby mines as catalysts in the case of the HDH™ process, or by integrating its in-house knowledge on emulsions in the case of AQUACONVERSION™. However, none of these INTEVEP innovations had been sufficiently scaled up before PDVSA implemented its business plan of constructing four delayed coking units at Jose (Eastern Venezuela), each one a Joint Venture (JV) with different overseas partners, to produce synthetic crudes that would be supplied to conventional refineries.[28] The total investment came to over

US$12 billion, of which approximately half was financed with loans. The limitations in sourcing sufficient natural gas has been cited as the reason why delayed coking was chosen as the technology for all the joint ventures built at Jose.[29] Nevertheless, it is doubtful that either the JV partners or the financing banks would have taken the risk of pioneering on an industrial scale INTEVEP's new catalytic processes, given their state of development in the 1990s. PDVSA had to proceed on its own if it wanted to showcase INTEVEP's technology for the upgrading and/or refining of heavy crudes. There would be no better market credentials than their use by the owner of the world's largest HC and XHC reserves, yet even at present neither process has been implemented on an industrial scale by PDVSA.

Does the absence of a commercial application for these amply cited INTEVEP icons imply a failure of its long-term R&D research in the area of catalysis? Not so, if a broader perspective is adopted:

> the technical knowledge base [in catalysis within INTEVEP] was rich, but its commercial success very poor . . . however the knowledge generated has allowed a more advantageous negotiation of technology by the refineries . . . and to resolve their technical problems . . . in the case of catalytic technology, its commercial value may not reside in net sales . . . [but] in providing support for the [core operations of the] business . . . in the matter of [INTEVEP] patents, catalytic technologies and of absorbents account for 18%, while . . . inventions are 30%.[30]

## ORIMULSION®, THE NEW HYDROCARBON FUEL IN THE MARKET

Global commercial success for INTEVEP was the result of PDVSA's wide search for the optimal use of its vast reserves in the Orinoco Oil Belt.

> In the mid 1980's it was LAGOVEN's turn to work on PDVSA's long-term plan. This was a two-year effort that had to evaluate a global energy horizon up to the early 21st century, in order to avoid a scenario where Venezuela's heavy crude reserves of the Orinoco Oil Belt would be as useless in the future as they had been in the past. We looked at all options, including the use of natural bitumen as a fuel for power generation, since even then we knew that electric cars would be vying with combustion engines in the transport market. At the time there were two immediate problems: how to extract the reserves, one option being the in-situ emulsification of the heavy crudes and natural bitumen, and then how to transport them via pipelines to the traditional outlets for such heavy feeds, the delayed cokers.
>
> Eugenia Vásquez, ex-Managing Director of Bitor Europe Ltd[31]

As a result, LAGOVEN provided the funding and guidance for the research effort undertaken by INTEVEP on multiple fronts to tackle these issues, with emulsions as the common thread to all the activity. Staff from both companies first carried out field tests on the *in situ* emulsification of heavy crudes in the reservoir.[32] In a separate effort, the Emulsions group in INTEVEP was addressing the transport of heavy crudes through pipelines, which required lowering their viscosity by preparing emulsions of hydrocarbons in water. A third group was working on the direct combustion of crudes:

> I was hired as part of the Combustion Group to develop efficient ways to use heavy crudes directly as fuels for steam and power generation. Since there was no prior experience firing these hydrocarbons, INTEVEP had installed a fully instrumented 1 million BTU pilot plant, with a combustion chamber, a fuel handling system, steam atomization, etc. Firing emulsions that contained a major quantity of water had not even crossed our mind until all of a sudden, so to speak, we were part of a team with the Emulsion Group and being supported by LAGOVEN. Our immediate goals were reformulated to test an emulsion of natural bitumen with 30% water. Expecting the test to fizzle out as soon as it started, we heated up the chamber, firing it first with gas. When we opened the valves and the flow of this emulsion entered the burner tip, we could not believe we were watching such a steady and luminous flame. We stood around in silence, mouths agape with incredulity and excitement, until someone ran to fetch INTEVEP's top authorities. Some of them left their offices to stand side by side with us, awed witnesses to the birth of a new fuel. I forget the exact date, but it was July 1985.
>
> Euler Jiménez, Researcher, Combustion Group, ex-INTEVEP[33]

All these parallel research efforts had converged on a singularity, a stable emulsion of evenly sized tiny droplets of natural bitumen dispersed in water, capable of being used in standard commercial boilers installed in power plants around the world.[34] However, this new technical invention had to be converted into a viable commercial product.

> I still remember a venerable Spanish manager affronted by our proposal that he allow water in a fuel into his power plant. The challenges we faced were both internal and external. Our small office space in LAGOVEN was disparaged as 'Fantasy Island', and we knew we all risked our future in PDVSA if we failed in a project very few believed in. You ask me what kept us going, in spite of such scepticism, bordering on ridicule at times. Support was scarce, but vital. Brigido Natera, President of PDVSA at the time, believed in the vision. Horacio Quintero, Director, told us to persevere based on facts and results. Without Manuel de Oliveira, who headed all our effort, and used his seniority, prior experience in the power generation sector and respect within LAGOVEN to bulldoze our way forward, the project would have foundered. He fought for, and obtained, market acceptance even before production was up to par. We persevered because we knew we had something new and exciting in our hands, even if few within PDVSA could perceive it yet. The fact that major companies such as BP, ENEL and others showed a keen interest in our results showed that we were not being delusional.
>
> Eugenia Vásquez[35]

Three years after the first successful test combustion of the emulsion, PDVSA created a new subsidiary, Bitumenes Orinoco S.A. (BITOR), to manufacture and market INTEVEP's new fuel worldwide.[36] No other NOC at the time had set up a commercial corporation to market globally, on a scale of millions of tons, one of its R&D innovations. The registered trade name of the commercial product was Orimulsion®, the emulsion of 30 per cent water with 70 per cent natural bitumen from the Orinoco Oil Belt.

The history of Orimulsion® is a cautionary tale of the factors that during a brief window of time came into perfect conjunction, thus paving the way for the successful entry of a uniquely Venezuelan product into the global fuel market for power generation (Figure 3). First, the market had to believe the supply was long-term and sustainable. Venezuela possessed the Orinoco Oil Belt, a vast source of natural bitumen.[37] Second, the market required a respected research institute that would provide support to clients and constantly improve the product. INTEVEP had such technical credibility. Third, market share had to be wrenched from very strong traditional suppliers of coal and gas. The marketing challenge was a novel one for PDVSA: the fuel could only be sourced from Venezuela, and it contained just under 3 per cent sulphur, immediately flagging up its potential environmental impact. Though commercial supply contracts obliged the client to install Flue Gas Desulphurization (FGD) facilities, which made its environmental compliance second only to gas-powered plants, the criticism on environmental grounds in countries such as the United Kingdom and the USA, with strong domestic coal lobbies, became very fierce. The threat posed to coal was very real: Orimulsion® offered a 20 per cent reduction in $CO_2$ emissions, its pricing allowed for the installation of FGDs, and plant operators prefer to handle a liquid fuel, instead of a solid one. The fact that Denmark, a country with one of the strictest environmental legislations in the world, came to generate up to 15 per cent of its electricity at one point using Orimulsion® was overlooked in these attacks.[38]

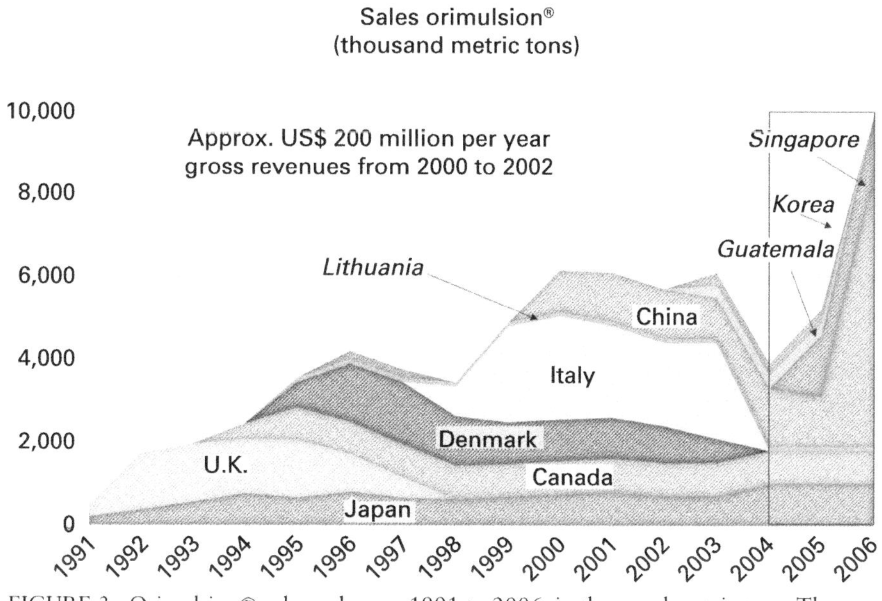


FIGURE 3: Orimulsion® sales volumes, 1991 to 2006, in thousand metric tons. The business was wound down 2004 to 2006, with pending commitments as indicated.

The fourth and crucial factor was support from the Venezuelan State and PDVSA. The power generation sector in any country is of strategic national importance, and to allow the use of a fuel that could only be sourced from a single country subject to political instability was a major policy decision for any government. BITOR had to rely on commercial diplomacy to support its message that thanks to Venezuela's reserves and the track record of PDVSA, it would honour its supply contracts. At the time the corporate solidity of PDVSA outweighed any perceived weakness in Venezuela's political history. By the late 1990s PDVSA figured among the top five major oil corporations of the world and had won a reputation for strictly honouring its supply contracts, its payments to third parties for services rendered and other international obligations. PDVSA backed its R&D child not only with its reputation but also with its financial resources. It provided the funds for the construction of the first Orimulsion® production module, at a cost of over US$350 million, and also the crucial guarantee that if BITOR failed to supply Orimulsion® to its customers, PDVSA would step in and supply fuel-oil at the equivalent price. But there was one more condition that PDVSA had to accept. If Orimulsion® was not priced at a level that would allow the client's plant to dispatch at base load into the grid, then it had no future in the market.

## *The price of entry to the electricity market*

The economics of power generation was shifting at the end of the twentieth century, from the State-owned power plants who passed on fuel costs to their customers, or received State subsidies to control their retail pricing, to competition between power plants who had to compete by bidding a suitable price for the supply of electricity to the grid. The pricing options for Orimulsion® were just two: to be offered in the range of coal prices, which would allow it to be considered for existing or new generation plants, or to promote its use in plants that used fuel-oil, where it could also compete at a higher price against new-built Combined Cycle Gas Turbine (CCGT) plants using Liquid Natural Gas (LNG) as fuel.[39]

The initial marketing plan focused first on establishing an arm's-length showcase that would dispel any doubts on its technical qualities and its environmental compliance. In parallel it aimed to incorporate Orimulsion® as the fuel of choice in the wave of new Independent Power Projects (IPP) projects.[40] On both fronts it captured the attention of the market with the incentive of coal-based pricing. Plants in Canada, Japan and the UK tested it from 1988 onwards, with commercial supply contracts as of 1991. The supply to SK Power in Denmark (1995) would be a very important milestone in establishing the environmental credentials of the fuel. Then the focus on IPPs led to an unexpected hiatus in the growth of the Orimulsion® market. Some were postponed repeatedly (Japan), others generated too much negative publicity where domestic coal mustered a strong opposition (UK and USA), and failed.

## *Breakthrough*

The marketing breakthrough in the early years of the twenty-first century came via three new fronts. First, China had approached BITOR directly, which highlighted the reticence of many countries in Asia to allow Japan to be an intermediary in its sourcing.[41] Chinese companies would negotiate a deal with BITOR to build a new production module in Venezuela at a price still tied to coal. It would be the last contract negotiated at the lower price band, due to two new marketing opportunities: BITOR now also targeted existing fuel-oil plants that were still in operation (Italy, Singapore), and Wärtsilä of Finland successfully tested Orimulsion® as a fuel for small power generators (<250 MW) based on their internal combustion engine.[42] By allowing the client to revert to fuel-oil, and limiting investment to the cost of new FGDs (if at all), the price bands of Orimulsion® were raised substantially, and there was no further need by PDVSA to guarantee an alternate fuel supply.

PDVSA's 20-F filing to the Securities and Exchange Commission (SEC) of the USA for the year ending 31 December 2002, envisaged a capital investment plan of $1.36 billion required to substantially expand production of Orimulsion® by 2006. INTEVEP had entered the select club of oil R&D institutes with a global commercial success in their portfolio. Meanwhile, in 1999 Venezuela had elected Lieutenant Colonel Hugo Chavez as its president. The radically new political context would now define the future of INTEVEP and Orimulsion®.

## 1999 TO 2003: THE WATERSHED YEARS

The irony of Orimulsion® is that President Chavez quickly became its most ardent supporter.[43] One of the unresolved issues of this history is how the most powerful and charismatic political figure within Venezuela, allowed an eventually stronger personality within his energy advisers to force the closure of the business, precisely at the moment of its market breakthrough. By September 2003 Orimulsion® was being denigrated in public by the highest oil authorities of Venezuela.[44] Two arguments, based exclusively on the ideas of Bernard Mommer, a senior energy figure in the government, stood out: a) the opportunity cost to PDVSA of allowing Orimulsion® to be sold in the pricing band of coal was unacceptable, as well as the preferential royalties allowed to BITOR in the past. It was argued that the State would receive higher revenues if PDVSA simply mixed its heavy crudes with a lighter fraction, a process that required very little capital investment, and the resulting blend was sold as a *Merey 16*; b) since natural bitumen was now decreed not to exist in the Orinoco Oil Belt, all Orimulsion® production had to be included within the OPEC quota, which meant it would displace higher priced oil products from the Venezuelan basket.[45]

To judge the soundness of the arguments being used against Orimulsion®, it is necessary to review briefly PDVSA's pre-1999 strategy with regards to the reserves of the Orinoco Oil Belt. On the one hand it targeted the transport sector market by building four delayed cokers, whose role was to produce various grades of synthetic crude, which would then be used as feedstock in traditional refineries. This route would consume in its first stage around 600 thousand barrels per day (tb/d) of reserves of heavy crudes. In parallel, the development of the Orimulsion® business targeted the power generation market, while not affecting the OPEC quota since natural bitumen was exempt, and was set by 2006 to consume at least 210 tb/d of reserves of natural bitumen. From a market point of view, the transport and power generation sectors consumed roughly similar amounts of primary hydrocarbon energy sources, though oil has traditionally dominated the former, and coal the latter, with gas the newcomer to both. Orimulsion® introduced natural bitumen to the hydrocarbon basket for power generation. The new energy authorities now stated that *Merey 16* blends had always been a better option than Orimulsion®, but the logical conclusion to this line of argument was that the Cerro Negro JV should have never been implemented either, since it had cost PDVSA and its partners US$2.8 billion to produce just 92 tb/d of a syncrude which was no better than a very sour (high sulphur) *Merey 16*, plus around 0.8 million tons per year of unwanted coke and annual payments of technical royalties to foreign holders of patents and know-how. As to the second objection, since natural bitumen has always been present in the Orinoco Oil Belt, the whole issue was a moot point, related to OPEC politics and not based on technical facts.[46]

The passage of years has shown that the main argument raised by Mommer against Orimulsion® was false. Blends like *Merey 16* were never a growth market segment for Venezuela's oil reserves. The reason why the Cerro Negro project went ahead was simply that Venezuela never possessed enough domestic light crude to manufacture sustained high volumes of blends with its HC or XHC, even less to substitute its Orimulsion® sales volumes.[47] The cost of opportunity never existed. As to the elimination of natural bitumen by decree, it only served to catapult Venezuela on paper to the top of the OPEC countries with the largest oil reserves, to no practical advantage. By May 2018, Venezuela had regressed to producing the same level of oil as in 1950 (<1.5 million barrels per day). When the planners of LAGOVEN foresaw the rise of electric cars in the twenty-first century, they correctly identified the importance of finding a way to use Venezuela's hydrocarbon reserves as fuel to generate electricity. The Orimulsion® project was the key to this market. However, under the onslaught by the political owner of the business, it unravelled in 2005–2006, even though President Chavez in December 2004 was still offering resistance, promising China his support to accelerate the construction of the second Orimulsion® production module.[48] The unilateral decision by PDVSA to stop contractual shipments of Orimulsion® generated monetary penalties whose amount have not been made public.[49] Orimulsion® became just one of many casualties of the seismic shifts in the politics and economy of Venezuela after 1999. During the first quarter of 2003, more than 23,000 oil workers would be fired, out of a total PDVSA workforce of 56,000, as a consequence of civil opposition to the Chavez government. In the case of INTEVEP, 881 professionals from its research staff were fired, an unprecedented action in the history of R&D institutes in democratic countries, a harbinger of an intellectual diaspora that has gained momentum as conditions within the scientific sector in Venezuela have deteriorated with time.[50]

## A SWAN SONG OF WATER AND FIRE

Fifteen years have passed since the beginning of the end of the Orimulsion® project. In a counterfactual scenario, the rise in oil and coal prices in the twenty-first century would have lifted all Orimulsion® contracts to even higher levels of profit, and spearheaded PDVSA's entry as energy supplier (Orimulsion®, coal) to the global power generation market, in parallel to its strong presence in the transport sector.

The business acumen gained would have provided confidence in the ability of PDVSA/INTEVEP to market successfully other viable technological solutions for the global energy sector. In the parlance of the analysts of research and innovation, Orimulsion® had met the criteria of a 'radical product innovation'.[51] The timescale of its development had been well within the sixteen-year average reported for the sector.[52] All the above became irrelevant to influence the course of political events. The deep divisions and mistrust within Venezuelan society were already in place when the decision to end the Orimulsion® project was taken. In the end it was the enemy within, and not the market or any technical drawback, that decided its fate.

Underlying these events was a fundamental contradiction between the highest political authority of Venezuela extolling the benefits of Orimulsion®, and the executive order that led to the intellectual beheading of INTEVEP in early 2003. Orimulsion®, AQUACONVERSION™ and HDH Plus™ were the outcome of a specific structure of human relations within PDVSA, and of the social relations of this NOC within its public sphere. INTEVEP's R&D products were not simply technical examples of the commodity fetishism criticized by Marx, innovations defined solely by their capacity to monetize the reserves of the Orinoco Oil Belt, thus divorcing them from the work ethic and social relations of the human labour that begot them. To extol Orimulsion® was in essence to extol the corporate values of the pre-1999 PDVSA that had made it possible, since the former could not have existed without the specific idiosyncrasy of the latter. Orimulsion®, whose global market success signalled INTEVEP's coming of age as an oil R&D institute, became unexpectedly its swan song. It expired as a viable long-term business the moment PDVSA fired the majority of INTEVEP's researchers. The year 1999 marked a point of discontinuity in the mentalité of the political and social will that had driven the sector of science and technology in Venezuela, since at least the mid-twentieth century. In 2008 PDVSA created a new subsidiary, not to promote R&D breakthroughs, but to distribute and market food (PDVAL), which in a way was another first for a major NOC. As to INTEVEP, the trickle of patents filed in the USPTO after 2013 finally dried up altogether as of 2015. Between a forced retirement from the market with its technical reputation intact, or a steady deterioration of its R&D support and market credibility, Orimulsion® may have been ultimately spared the ignominy of the latter.

## EPITAPH, *POR AHORA*

Forty years after the OPEC seminar in Vienna, some NOCs in the Middle East are being advised on how best to develop their in-house R&D capacity, Mexico now argues that foreign private investment is the key to technical innovation for its nationalized oil sector, Petrobras (Brazil) became a leader in deep-water drilling technology, investing over US$1 billion in R&D just in 2014, whilst PDVSA is claimed to owe OFS companies around US$2 billion in unpaid debt for recent services rendered up to 2016.[53] Faced with such a wide spectrum of responses to the conundrum of in-house R&D institutes for NOCs, why did PDVSA in 1979 include one within its corporate structure, when there was no major precedent among its NOC peers to do so? The answer lies both in the spirit of that age and the pragmatism of the decision. The historical pioneering role of Venezuela's oil authorities should not come as a surprise, since they had been at the forefront of many other difficult decisions in the power struggle that took place in the mid-twentieth century, between multinational oil companies and the developing countries that owned the oil resources. They were the first to implement the fifty–fifty sharing formula for oil profits by 1948, which was increased to over 60 per cent in 1959. They were the intellect and driving force that led to the creation of OPEC in 1960, convincing the Arab producers of the need to create a common front to face the power of the transnational oil conglomerates.[54] The highly trained human resource that PDVSA was able to build on, generated the excitement of achieving breakthroughs in uncharted territories, that still pulses strongly in the memories I have probed in the previous sections.[55] A deep sense of loss is also present in these and other voices, a remembrance of a technical Camelot that transcends the regret of careers and lives derailed. Whether a perception of hubris, real or imagined, may have played a role in the events of 1999–2003 can be debated, but without a strong sense of technical and scientific empowerment within its staff, INTEVEP would never have innovated.

Together with this intangible and anecdotal factor, there was also a more quantitative reason why PDVSA could take such a decision. Though the investment in INTEVEP was very high compared to the budgets of other Venezuelan R&D centres, PDVSA recognized in its annual reports that: 'the amount that we have spent on our research and development activities has not been material'.[56] In other words, PDVSA took a calculated risk with INTEVEP that did not dent its budget. The modest (by PDVSA standards) investment in INTEVEP had borne fruit, since it provided important geological support, tested additives and certified catalysts, stored and gave access to technical information, gave on-demand technical support to the PDVSA group and acted as gatekeeper to international cooperation agreements. Since 1984, it had focused more than 50 per cent of its effort into finding the technologies to unlock the potential of the reserves of the Orinoco Oil Belt. By 1996, INTEVEP reported spending 1.5 million man-hours in its activities supporting PDVSA. Though the PDVSA Annual Reports are stronger on the qualitative than on the quantitative milestones of INTEVEP's activity, they do offer one tantalizing glimpse of the financial effect of its technical support for the year 1997: savings of US$16 million in exploration and US$75 million in production, in addition to generating revenues of nearly US$9 million in technical assistance charged to third parties, plus a further US$1 million in royalties.[57] The creation of INTEVEP coincides with a substantial decrease of expenses on third-party technical support from the heights of the early period after statization (Figure 1).

What in the end mattered more than the actual R&D intensity of PDVSA, was the culture within INTEVEP that allowed small groups of very competent researchers, highly qualified in their field, to harness their intellect to the novel ideas generated in the operating companies such as LAGOVEN, whose funding allowed them to develop ambitious and ground-breaking projects, side by side with the bread-and-butter technical support required by the operating companies. Orimulsion® was in a way the ugly duckling, the natural bitumen fuel born within the clutch of PDVSA's traditional oil products, that against all odds became a swan.

Over a span of some fifteen years, corporate PDVSA had taken the decision to back its most successful R&D creation, through eight successive cohorts of the senior corporate hierarchy, and in the process showed the discipline and perseverance required to create a global commercial product. The very willing, even when led at times by the not totally convinced, had been allowed to do the unbelievable.

## ACKNOWLEDGEMENTS

The author wishes to thank all who shared their recollections and documents on INTEVEP, and who suggested amendments to the text, and for the corrections pointed out by the anonymous reviewers. All errors that remain are the author's alone. I regret that for reasons of length I have not been able to include the names of all who provided important contributions to the projects I have mentioned. The author worked for fifteen years within the PDVSA group (INTEVEP, PEQUIVEN and BITOR) until early 2003.

## NOTES AND REFERENCES

1. Ali Mohammed Jaidah, 'Introduction Speech', Proceedings of the OPEC Seminar 'The Present and Future Role of the NOC', OPEC Headquarters, Vienna, 10–12 October 1977.
2. Robert K. Perrons, 'How Innovation and R&D Happen in the Upstream Oil & Gas Industry: Insights from a Global Survey', *Journal of Petroleum Science and Engineering* 124 (2014): 301–312.
3. 'Oil Field Services: The Unsung Masters of the Oil Industry', *The Economist*, 21 July 2012, http://www.economist.com/node/21559358. See also Javier Martínez-Romero, *Innovation as an Imperative for the Mexican Oil Industry post Energy Reform*, Mexico Center, James A. Baker III Institute for Public Policy, Rice University, 2017: 5–6, and Renato Lima de Oliveira and Timothy Sturgeon, 'From Resource Extraction to Knowledge Creation: Oil-Rich States, Oil Companies and the Promotion of Local R&D', MIT-IPC Working Paper 17-003, 2017: 3–14, accessed on http://ipc.mit.edu/sites/default/files/documents/Internationalization%20of%20R%26D%20in%20Oil%20%26%20Gas.pdf
4. Data on revenues and capitalization from https://www.slb.com/news/press_releases/2018/ 2018_0119_q4_earnings.aspx
5. At the time only Iran's NOC coordinated its oil-related R&D activity. Algeria's and Indonesia's reported to the Ministry of Petroleum, Saudi Arabia's to the Ministry of Higher Education, Iraq's to the National Science Foundation and Kuwait's to the Prime Minister, see Jean-Claude Balaceanu and Jean Favre, 'The Prospects for Transfer of Technology to, and the Establishment of Advanced Research Centres in OPEC Member Countries', paper presented at the OPEC Seminar, 'The Present and Future Role of the NOC'.
6. Rómulo Betancourt, *El Petróleo de Venezuela* (Barcelona: Editorial Seix Barral, 1978): 29. All translations by the author.

7. The support for the creation of INTEVEP not only came from the political parties of the status quo, but also from voices located at their ideological antipode. In a book otherwise strongly critical of PDVSA's actions after its creation, the author stated in 1977: 'this initiative [the creation of INTEVEP] we deem to be of great importance and constitutes a positive act: it is the correct policy'. In Gastón Parra Luzardo, *De la nacionalización a la apertura petrolera* (Caracas: Banco Central de Venezuela, 2012): 97. The designation of Parra Luzardo as President of PDVSA in 2002 triggered the events that led to the temporary ouster from power of President Chavez that year. For further details on the creation of INTEVEP see Emma Brossard, *Petroleum Research and Venezuela's INTEVEP, The Clash of the Giants* (Houston: PennWell Books/ INTEVEP, 1993).
8. Magdalena Ramírez, 'Nuestra tecnología petrolera: catalizadores y adsorbentes', *Interciencia* 29 (2004): 4–6.
9. The terms under which such technical assistance continued to be supplied by the major oil corporations were the subject of strong criticism on legal and ideological grounds. See Parra Luzardo, *De la nacionalización a la apertura*: 91–108.
10. The Instituto Mexicano del Petróleo (IMP) was founded in 1965, well after the Mexican oil industry was nationalized, by the Federal government, as an institution independent of Pemex, and obeys directives from CONACYT, the Mexican government regulatory body for science and technology, which has also provided partial funding at times. Since inception it differs substantially from INTEVEP. For a summary of IMP's evolution, see Martínez-Romero, *Mexican Oil Industry*, 9–14.
11. INTEVEP budget figures in current bolivares, courtesy Jaime Requena. Other raw data, including yearly exchange rates, from *Annual Reports*, Petróleos de Venezuela, S.A., 1979 to 2000, and *Petróleo y otros datos estadísticos (PODE)*, Ministerio de Energía y Minas, 1985 to 2003. The accounting of total R&D costs by PDVSA are not easy to track. Costs incurred by INTEVEP were included within the corporation's total operating costs or as capital costs of new projects. From 1979 to 1993, PDVSA itemizes 'Technical Assistance' from third parties separately from operational expenses. In 1994 this itemized expense disappears, just after it had shown a major increase in value for 1993 and 1994, and from 1995 in PODE. However, from 1994 PDVSA reports within its accounting notes an amount titled 'R&D costs', which I assume corresponds also to third-party costs. Since in 1994 the PODE figure for 'Technical Assistance' and the PDVSA amount for 'R&D costs' do not match, they are two different accounting items, thus their separate treatment in Figure 1.
12. OFS companies registered an R&D intensity of 2.7 per cent in 2015. George Sarraf, Anil Pandey and Yahya Anouti, 'From Technology Adopters to Innovators. How R&D Can Catalyze Innovation in Middle East National Oil Companies', https://www.strategyand.pwc.com/me/home/thought_leadership_strategy/reports_and_white_papersme/display/technology-adopters
13. Robert Perrons, 'Innovation': 302.
14. The international oil companies by 1975 had reduced the presence of foreign professionals in Venezuela to less than 1 per cent to 5 per cent of the total workforce.

Eddie A. Ramírez and Rafael Gallegos, *Petróleo y Gas: el caso Venezuela* (Caracas: Editorial Lector Cómplice, 2015): 66, 68; Daniel Yergin, *The Prize* (New York: Simon & Schuster, 1991): 650.

15. Jaime Requena, private communication.
16. Jaime Requena, '¿Cuánto cuesta hacer ciencia en Venezuela?', *Interciencia* 28 (2003): 21–28.
17. Even before INTEVEP had been integrated within the PDVSA corporate structure, FGMA had already provided 18,755 scholarships at the technical, university and postgraduate levels, around 40 per cent in Venezuela and 60 per cent overseas, at a cost of just under US$290 million (2000 $) by the end of 1978. For 1979 it was aiming at 40 per cent technical and 25 per cent postgraduate degrees. FGMA would over the years change substantially its mission and vision, but the sheer ambition of scale of the original idea is evident from the fact its initial goal of creating 10,000 scholarships from scratch, was equivalent to the total number of university graduates in 1974–1975 in Venezuela. FGMA was created with the express intention of providing the professional manpower that would be required after the statization of Venezuela's mining and oil industry. See *Fundación 'Gran Mariscal de Ayacucho'. Memoria 1978. Exposición de Motivos* (Caracas: Fundación Gran Mariscal de Ayacucho, 1979): 45, 47, and end tables, and Ministerio de Educación, *Memoria y Cuenta 1980* (Caracas: Ministerio de Educación, 1981): 616. To this effort must be added the sudden influx of expertise from Chile, Argentina and Uruguay, a 'brain gain' of exiles of the right-wing military dictatorships of the 1970s and 1980s.
18. See for example Ignacio Avalos and Marcel Antonorsi, 'Análisis de la Ley Orgánica de Ciencia y Tecnología (LOCTI)', 2009, http//www.varen.org/quepasa/tecnopod.pdf
19. Brossard, *Clash of Giants*: 180.
20. Jaime Requena, private communication.
21. Jaime Requena, 'Ciencia en Venezuela': Table I.
22. Alfonso Ravard, interviewed by Sofia Imber and Carlos Rangel, in Buenos Dias, Venevisión, 12 January 1978. Transcript available at http://saber.ucab.edu.ve/jspui/bitstream/123456789/38694/2/sicr123719780112.pdf. For a description of the current installations, see www.pdvsa.com
23. Because of the nature of its work, the number of publications in peer-reviewed journals would not be a relevant yardstick. In effect, INTEVEP in the period lagged far behind IVIC and the universities in this regard, 3 per cent to 27 per cent and over 54 per cent, respectively. Jaime Requena, 'Decay of Technological Research and Development in Venezuela', *Interciencia* 36 (2011): 341–347.
24. Raw data from USPTO website, graph based on Jaime Requena, 'Decay of Research': 341–347. See also Martínez-Romero, *Mexican Oil Industry* and *Annual Report 2000*, PDVSA. By the second decade of the twenty-first century, PDVSA/INTEVEP disappears from academic studies on oil and gas R&D by NOCs, while Petrobras becomes the point of reference in Latin America. See for example Lima de Oliveira and Sturgeon, 'From Resource Extraction to Knowledge Creation'.

25. According to the *OPEC Annual Statistical Bulletin 2017*, in 2016 OPEC had 81.5 per cent of the total oil reserves of the world, of which Venezuela led with 24.8 per cent, 302,250 million barrels.

26. 'I felt my work at INTEVEP allowed me to pay back my country for all the opportunities it had given me. I was born and raised in the *23 de Enero* [one of the rough neighborhoods in Caracas]. From the fourth grade, all the way to my doctorate, I had to rely on *becas* (State scholarships) to fund my studies. Without this educational lifeline I would never have fulfilled my professional potential'. Magdalena Ramírez, personal communication. For more technical information on refining processes, see María Magdalena Ramírez-Corredores, *The Science and Technology of Unconventional Oils* (London: Academic Press, 2017).

27. Pedro Pereira, Cauri Flores, Hugo Zbinden, José Guitián, Rodolfo Bruno Solari, Howard Feintuch and Dan Gillis, 'Aquaconversion Technology Offers Added Value to E. Venezuela Synthetic Crude Oil Production', *Oil & Gas Journal* 99, 20 (2001): 79–85. Pereira was a project leader both for HDH Plus™ and AQUACONVERSION™. The API (American Petroleum Institute) scale, expressed in degrees, indicates the specific gravity of a crude, single figure values indicating very high specific gravities.

28. Joint Ventures as initially structured (name, JV partners, target production, specific gravity syncrude): PetroZuata, ConocoPhillips, 104tb/d (thousand barrels per day), 19–25° API; Cerro Negro, ExxonMobil, BP, 105 tb/d, 16° API; Sincor, TotalFinaElf, Statoil, 180 tb/d, 32° API; Ameriven, ConocoPhillips, ChevronTexaco, 180 tb/d, 26° API.

29. Manuel Pulido, 'A propósito de la Orimulsión', *Interciencia* 29 (2004): 179.

30. Magdalena Ramírez, 'Catalizadores y adsorbentes': 10. 'A single catalytic technology can have three or four inventions associated with it: the method to prepare the catalyst, the specific use for catalyst, and the processes employed with the catalyst'. Ramírez, personal communication.

31. Personal communication. Vásquez would ultimately head the marketing office of BITOR in Europe, having been one of the founding members of the Orimulsion® project in LAGOVEN, one of the main operating companies of the PDVSA group, active in the Orinoco Oil Belt.

32. Ignacio Layrisse of INTEVEP, and Prof. Jean-Louis Salager (Universidad de los Andes, Venezuela) played critical roles in this stage of the project.

33. Private communication. Jiménez figures in seven of the INTEVEP patents on Orimulsion®, including US Patent 04795478: 'Viscous hydrocarbon-in-water emulsions', 1989.

34. For another view of the development of Orimulsion®, and the people involved, see Hebe Vessuri and Maria Victoria Canino, 'Restricciones y oportunidades en la conformación de la tecnología: el caso Orimulsión', in Arnoldo Pirela (ed.), *Venezuela: el Desafío de Innovar* (Caracas: Fundación POLAR-CENDES, 2003): 181–201.

35. Personal communication.

36. By creating a separate commercial entity, distinct from LAGOVEN, it underlined the fact that Orimulsion®, based on natural bitumen, lay outside the traditional Venezuelan oil basket, and also of OPEC quotas.
37. According to Anibal Martínez, the foremost Venezuelan expert in the classification of oil reserves, a third of these reserves consist of natural bitumen. As quoted in Diego González, 'Orimulsión y algo mas', *Petroleo YV* 25 (2006): 46.
38. The definitive technical vindication of the environmental compliance of Orimulsion® when used in conjunction with FGD equipment was the report titled *Environmental Impacts of the Use of Orimulsion* (Washington, DC: Environmental Protection Agency, July 2001).
39. For information on the comparative economics of power generation in new-built and existing power plants using different fuels, contact the author: guerrero.saul@gmail.com
40. By the late 1990s independent power project developers were funding 30 per cent of new power projects, as reported in Ulf Hansen, 'Technological Options for Power Generation', *The Energy Journal* (1998): 63–87.
41. BITOR had established a JV with Mitsubishi Corporation (MC-BITOR Ltd) for marketing in Asia. Orimulsion®'s ties to Japan led to its cameo appearance in a best-selling spy novel, albeit incognito and with the wrong paternity: 'The Japanese . . . were drilling their heavy oil long before they discovered how to use the stuff . . . their top scientists hunted day and night for a f____ formula to break it down . . . the Nips have *found* their magic emulsion . . . you pour it in, you shake it all about, and bingo, you've got oil like all the other boys . . . able to flood the f____ world with it', in John Le Carré, *The Tailor of Panama* (New York: Alfred A. Knopf, 1996): 208–209.
42. Thomas Hägglund (Wärtsilä Finland OY), 'Comparative Advantages of Orimulsion®, LNG and Petcoke', presentation given at 'Private Power in Central America', Miami, 13–14 June 2002.
43. I was present when President Chavez met with the highest authorities of the energy sector of Singapore during his trip to Asia in 1999. The strong personal commitment of President Chavez to the success of Orimulsion® paved the way for Singapore to accept in principle the purchase and use of a single-source fuel, even with no alternate fuel guarantee from PDVSA. He also lobbied for Orimulsion® in his keynote luncheon speech titled 'Energy: Challenges and Opportunities in the 21st Century', at the meeting of the Pacific Economic Cooperation Council (PECC), held in Manila, Philippines, 22 October 1999.
44. The attacks on Orimulsion® in interviews on the State channel Venezolana de Television, and in newspapers and websites (*El Nacional*, *El Universal*, *Ultimas Noticias*, www.soberania.org), are too many to detail. Conflicting reports on the curtailing of the business begin to appear from March 2004 onwards. In early 2005, the Minister of Energy and Oil, and President of PDVSA, Rafael Ramirez, was still declaring that PDVSA would honour all its Orimulsion® contracts (*El Universal*, 7 February 2005). Two months later the same Minister would announce the closure of the business, *Reporte Diario de la Economia*, 4 April 2005. The new authorities in charge of the oil policy of Venezuela believed that all of PDVSA's actions between 1976

and 1999 were tainted by the original sin behind Article 5 in the statization decree of the Venezuelan oil industry of 1976, which they denounced as a legal back-door left open on purpose to allow transnational oil companies a way back into a future privatized Venezuelan oil industry. As a consequence, nearly all of PDVSA's projects prior to 1999 were grist to this particular political mill. This explains in part why the Orimulsion® project was dragged into an ideological maelstrom, even though it was conceived and entered the market well before the *Apertura* (Opening) drive under Luis Giusti, President of PDVSA between 1994 and 1999.Whether Orimulsion® was implicated out of conviction or convenience remains an open question. The ideological arguments and criticism previously directed to pre-1999 PDVSA by those who came into power in 1999 (see for example Parra Luzardo, *De la nacionalización a la apertura*, and the introduction by Carlos Mendoza Potellá) have not aged well. The greatest contradiction and ideological volte-face has been the eagerness of post-1999 PDVSA to apply the spirit and letter of Article 5 to attract foreign investment in Venezuela's oil industry (Francisco Monaldi, 'The Impact of the Decline in Oil Prices on the Economics, Politics and Oil Industry of Venezuela', Center on Global Energy Policy, Columbia University (2015); Luis Pacheco, 'Venezuela's Oil Mythologies Have Hindered Its Development', Issue brief no. 02.05.18. Rice University's Baker Institute for Public Policy, Houston, Texas (2018)). There is a dearth of positive academic articles on the current state (2018) of Venezuela's oil industry, after a period of very tight ideological control after 1999 that will soon equal in length the span of time that saw the creation and consolidation of PDVSA between 1976 and 1999.

45. Bernard Mommer, 'The Value of Extra-Heavy Crude Oil from the Orinoco Oil Belt', *Middle East Economic Survey* 47, 11 (2004): D1–D11; 'La Orimulsión: verdades cientificas y mentiras políticas', *Interciencia* 29 (2004): 6–7; *El mito de la Orimulsión. La valorización del crudo extrapesado de la Faja Petrolífera del Orinoco*, (Caracas: Fondo Editorial Darío Ramírez, Ministerio de Energía y Minas, 2004).

46. For more details see Rafael Quiroz, interviewed by Marianna Parraga, 'Barril de bitumen rinde mas que crudo mejorado de la Faja', *El Universal*, 30 October 2003; Saul Guerrero, Luis Pacheco and Ignacio Layrisse, 'The Optimal Use of Venezuela's Hydrocarbon. Reserves: A Critical Analysis', *Middle East Economic Survey* 47, 24 (2004): D1–D4; Saul Guerrero, Leslie Jones-Parra, Emilio Abreu, Manuel Urbano, Luis Montefusco and Luis Gil, 'Orimulsión', *Interciencia* 29 (2004): 2–3; Saul Guerrero, 'Emulsion Fuel Options Still Viable for Heavy Oil', *Oil & Gas Journal* 22 (2008): 24–30. For the data on the Cerro Negro JV see PDVSA 20-F SEC filing for the year ending 31 December 2003.

47. Diego González, 'Orimulsión y algo mas', *Petroleo YV* 25 (2006–2007): 46; Saul Guerrero, 'Emulsion Fuel Options': 26.

48. http://www.abc.com.py/edicion-impresa/internacionales/chavez-ofrece-ayuda-a-china-para-que-cree-su-reserva-estrategica-de-petroleo-804111.html; www.chinatoday.com.cn/hoy/2005n/5hn2/21n.htm; BBC News, 'Venezuela and China Sign Oil Deal', 24 December 2004.

49. The Statement of Claim between New Brunswick Power Corporation (Canada) and Petróleos de Venezuela S.A. and Bitúmenes Orinoco S.A., filed on April 2004, is a

useful guide to the potential damages faced by PDVSA and also of the contradictory messages issued by the authorities of PDVSA on the future of Orimulsion® during this period.

50. Jaime Requena, 'Decay of Research', 341–347; 'Science Meltdown in Venezuela', *Interciencia* 35 (2010): 437–444.

51. Martínez-Romero, *Mexican Oil Industry*: 313.

52. Robert Perrons, 'Innovation in Oil': 302.

53. George Sarraf et al., 'Middle East'; on Mexico's recent policy, see Martínez-Romero, *Mexican Oil Industry*; for Petrobras expenditure on R&D, see https://ycharts.com/companies/PBR/r_and_d_expense; on the PDVSA debt, see https://www.bloomberg.com/news/articles/2017-08-23/schlumberger-halliburton-ious-preserve-venezuela-link-for-now

54. Yergin, *The Prize*: 431–437, 445, 510–518.

55. 'One of the key things was excitement. It was really contagious.' Quote attributed to John Tully, Professor of Chemistry at Yale, as an explanation for the success of Bell Labs (where he spent twenty-five years) as one of the major industrial R&D centres of its time, as cited in Daniel Yergin, *The Quest* (New York: The Penguin Press, 2011): 550.

56. SEC 20-F filing for the year ending 31 December 2000: 56.

57. Petróleos de Venezuela, S.A., *Annual Reports*, 1979 to 2000. No details are provided on the methodology to arrive at the estimate on savings in 1997.

# LIST OF CONTRIBUTORS

**Christiane Berth** is a historian at the University of Bern, Switzerland. Most recently, she investigated the social and cultural history of the telephone in Mexico. In general, her research focuses on global economic and cultural entanglements among the fields of technology, development cooperation, food, consumption, migration and trade, for which she has conducted research in Central America and Mexico since 2005. Prior to attending the University of Bern, Christiane Berth has worked at the Institute of European Global Studies in Basel, the University of St. Gallen and the Institute of Contemporary History in Hamburg.

**Eve Buckley** is Associate Professor of History and director of the Latin American & Iberian Studies program at the University of Delaware. She is a graduate of the University of Chicago and received her PhD in History and Sociology of Science from the University of Pennsylvania. Her book *Technocrats and the Politics of Drought and Development in Twentieth-Century Brazil* was published by University of North Carolina Press in 2017 and received recognition from the Brazil Section of the Latin American Studies Association and the Conference on Latin American History. Research for that book was supported by the National Endowment for the Humanities, the National Science Foundation, the University of Delaware, the Rockefeller Archive Center and the Josephine De Kármán foundation. Dr. Buckley's current research focuses on overpopulation debates during the early Cold War and the role of intellectuals from the Global South in contesting that paradigm, with a focus on the career of Brazilian physician Josué de Castro. She serves as co-convener of the Earth & Environmental Sciences working group organized by the Philadelphia Consortium for the History of Science, Technology, and Medicine, which hosts a monthly videoconference; interested readers are welcome to participate! www.chstm.org/groups

**Debora Gerstenberger** is assistant professor (junior professor) in Latin American History at the Institute for Latin American Studies at Freie Universität Berlin. Her work spans time periods and geographies but is often engaged with the central question of what techniques and technologies states use to secure and increase their power. Her PhD dissertation *Gouvernementalität im Zeichen der globalen Krise. Der Transfer des portugiesischen Hofes nach Brasilien* (Governmentality in times of global crisis. The transfer of the Portuguese Court to Brazil, 2013) analysed the relocation of the Portuguese royal court to the Brazilian city of Rio de Janeiro during the early part of the nineteenth century. In her current research project she examines the introduction of computer technology into the systems of state security (military, police, intelligence) in the Southern Cone in the 1960s to 1980s, during the time of military dictatorships. The aim is to broaden the geography of how we understand computing while highlighting the importance of computing in how we understand political history. Her research is based on perspectives taken from global history and actor-network-theory. Her contributions to the theory of (global) history include the edited volume *Techniken der Globalisierung. Globalgeschichte meets Akteur-Netzwerk-Theorie* (Techniques of Globalization. Global History meets Actor-Network-Theory, 2016).

**Saul Guerrero** started his early professional training in chemistry (Universidad Simon Bolivar, Caracas) with postgraduate degrees (Strasbourg, Bristol) in the field of plastics. His first professional stage was in academia, research and teaching six cohorts of Materials Science engineers. He then worked for 15 years within a major transnational energy and petrochemical corporation (PDVSA), at times negotiating transfers of technology, the commercial and legal embodiment of know-how for sale. He came late into the Orimulsion® project, becoming involved in its marketing in Asia. This taught him not only the barriers of mistrust and unfair competition that can hinder entry into a global market of an innovation engendered outside the traditional centres of knowledge, but also how to overcome them. He is now a historian (Warwick, McGill) who has just published his first monograph at the age of 66. His area of research involves applying chemical principles to interpret certain historical events, such as the history of silver and gold refining prior to the 20c, and its environmental consequences. He has also published previously on Venezuelan energy policy, thus his interest in contributing with an article on the Venezuelan experience in R&D in the oil and energy sector.

**Ian Inkster** is a professorial research associate at the Center of Taiwan Studies, SOAS, University of London; a senior fellow at the Taiwan Studies Program, China Policy Institute, University of Nottingham; and the editor of the international journal *History of Technology*. He is author of over 100 journal and chapter publications, 13 books and around 300 book reviews and notices. His many publications include the books *Science and Technology in History: An Approach to Industrial Development* (Macmillan, 1991) and *Technology and Industrialization: Historical Case Studies and International Perspectives* (Ashgate, 1998). Other publications include the book chapter 'Indigenous Resistance and the Technological Imperative: From Chemistry in Birmingham to Camphor Wars in Formosa, 1860s–1914' (Palgrave-Studies in Economic History, 2018) and the article 'Potentially Global: "Useful and Reliable Knowledge" and Material Progress in Europe, 1474–1914' (*The International History Review*, 2006).

**Diana J. Montaño** is Assistant Professor in History at Washington University in St. Louis. Her teaching and research interests broadly include the construction of modern Latin American societies with a focus on technology and its relationship to nationalism, everyday life and domesticity. She has received the Center for the Humanities' First Book Fellowship for Fall 2019 to complete the manuscript entitled *Electrifying Mexico: Cultural Responses to a New Technology, 1880s–1960s*. It looks at how ordinary citizens (businessmen, salespersons, inventors, doctors, housewives, maids and domestic advisers, etc.) used electricity, both symbolically and physically, in the construction of a modern nation. It follows the 'electrifying agents' who first crafted a discourse for an electrified future and secondly, how they shaped its consumption. It shows how these agents of modernity promoted and created both imaginary and tangible notions of this technology. Taking a user-based perspective, this study reconstructs how electricity was lived, consumed, rejected and shaped in everyday life.

**David Pretel** is Visiting Professor at the Centre for Historical Studies at El Colegio de México (Colmex). He is a historian and an economist specializing in international political economy, science and technology studies, and the history of capitalism, with an ever-increasing interest in Latin American international history. His first book *Institutionalising Patents in Nineteenth-Century Spain* (Palgrave Macmillan) examined the development of the Spanish patent system in the years 1826 to 1902, providing a fundamental reassessment of its evolution in an international context. He is also a co-editor of, and contributor to, the volumes *The Caribbean and the Atlantic World Economy* (Palgrave Macmillan, 2015) and *Technology and Globalisation* (Palgrave Macmillan, 2018). His recent publications include articles in the journals *Business History*, *America Latina en la Historia Económica* and *Economic History Research*.

**Daniel B. Rood** is associate professor of history at the University of Georgia, where he co-organizes 'Dirty History: An Interdisciplinary Workshop in Agriculture, Environment, and Capitalism'. He is also author of *The Reinvention of Atlantic Slavery: Technology, Labor, Race, and Capitalism in the Greater Caribbean* (Oxford University Press, 2016).

**Bárbara Silva** is Postdoctoral Fellow in History of Science and Lecturer of Contemporary Latin American History at Universidad Católica de Chile. Her research combines the analysis of nation building and collective identities, and the role of science and technology in contemporary history. She also studies history of science and its relationship with global history, integrating a cultural and environmental perspective. Her current research focuses in the international history of astronomy in Chile during the global Cold War. Some of her recent books are *Una identidad terremoteada. Comunidad y territorio en el Chile de 1960* (2018), the Public History best seller *Chile. 100 días en la historia del siglo XX* (2015), and *Astronomy at the Turn of the Twentieth Century in Chile and the United States. Chasing Southern Stars* (forthcoming).

**Josep Simon** is Ramon y Cajal Research Fellow at the Institut interuniversitari López Piñero, Universitat de València. He has taught and researched the history of science, technology and medicine in Colombia, Mexico, Brazil and the US. His work integrates insights from the history of science, technology and medicine, history of education, science and technology studies, book history, museum studies and science education. His research is comparative, cross-national and transnational. He is the author of the award-winning *Communicating Physics: The Production, Circulation* *and Appropriation of Ganot's Textbooks in France and England (1851–1887)* (Pickering & Chatto, 2011; University of Pittsburgh Press, 2016) and co-editor of *Beyond Borders: Fresh Perspectives in History of Science* (Cambridge Scholars Publishing, 2008) and special issues for *History of Science, Science & Education* and *Endeavour*. Recent contributions include chapters 'Textbooks' in the *Companion to the History of Science* (Wiley-Blackwell, 2016) and, 'The Transnational Physical Science Study Committee: The Evolving Nation in the World of Science and Education (1945–1975)' in *How Knowledge Moves: Writing the Transnational History of Science and Technology* (University of Chicago Press, 2019).

**Helge Wendt** is a Research Fellow at the Max Planck Institute for the History of Science in Berlin, where he is associated with the project 'Knowledge in and of the Anthropocene'. He is also member of the CRC 'Episteme in Motion'. He studied Latin American History and Anthropology at the Freie Universität Berlin and received his PhD from the University of Mannheim in 2009, where he taught early modern history. He is frequently teaching at Humboldt Universität and Technische Universität at Berlin. He currently works on the global history of knowledge of black coal (eighteenth and nineteenth centuries). Wendt published a book on the global history of colonial missions *Die missionarische Gesellschaft. Mikrostrukturen einer kolonialen Globalisierung* (Franz Steiner, 2011) and on a revision of European history in a global context *Geschichte des mestizischen Europas* (Springer, 2019). He published several papers on different aspects of history of colonial mission, global history of coal and history of science. He is co-editor of *The History of Physics in Cuba* (Springer, 2014) and editor of *The Globalization of Knowledge in the Iberian Colonial World (1500–1900)* (Edition Open Access, 2016).

**Helen Yaffe** is a Lecturer in Economic and Social History at the University of Glasgow, specializing in Cuba and Latin America, the history of economics, and the political economy of late development. Her first book *Che Guevara: The Economics of Revolution* (Palgrave Macmillan) examined the least known aspect of one of the twentieth century's best-known icons; analysing Guevara's economic work as a member of Cuba's revolutionary government, including his promotion of science and technology, and his contribution to socialist political economy debates. Subsequently, Helen has continued to analyse Cuban political economy and projects of regional integration within Latin America. Her pending publication with Yale University Press examines developments in diverse fields in Cuba, about which little is known outside the island, but which have contributed towards Cuba's survival as a socialist state to the post-Soviet era. This includes chapters on Cuban medical science and medical internationalism, renewable energies and sustainable development, the restructuring of education and the battle of ideas, the economic reforms and relations with the United States. Helen is also the co-author, with Dr Gavin Brown from the Geography Department at the University of Leicester, of *Youth Activism and Solidarity: The Non-Stop Picket against Apartheid* about the most radical section of the British anti-apartheid movement.